Encyclopedia of Water

Encyclopedia of Water

David E. Newton

An Oryx Book

Greenwood Press
Westport, Connecticut • London

Library of Congress Cataloging-in-Publication Data

Newton, David E.
   Encyclopedia of water / David E. Newton.
        p. cm.
   Includes bibliographical references and index.
   ISBN 1-57356-304-8 (alk. paper)
   1. Water—Encyclopedias. I. Title.

GB655.N38 2003
553.7′03—dc21   2002070031

British Library Cataloguing in Publication Data is available.

Library of Congress Catalog Card Number: 2002070031
ISBN: 1-57356-304-8

First published in 2003

Greenwood Press, 88 Post Road West, Westport, CT 06881
An imprint of Greenwood Publishing Group, Inc.
www.greenwood.com

Printed in the United States of America

∞™

The paper used in this book complies with the
Permanent Paper Standard issued by the National
Information Standards Organization (Z39.48-1984).

10   9   8   7   6   5   4   3   2   1

To the Hart women—
Kathy, Heidi, Gemini, Poppy, and Courtney—
for many years of help and friendship.

Contents

Guide to Selected Topics

Biographical Sketches

Biological Sciences

Earth and Space Sciences (including Geography)

History

Mythology and Religion

Physical Sciences

Preface

The purpose of the *Encyclopedia of Water* is to provide an overview of the many ways in which water has played a role in human society and the natural world over the centuries. Many of the topics in this Encyclopedia deal with scientific and technical issues, such as the physical and chemical properties of water, steam, and ice; the uses to which water and steam are put; and the purification, transfer, and use of water by communities. The book also describes the role of water in the natural world, including the hydrologic cycle, the properties and distribution of ice masses, and the properties of oceans and rivers.

Attention is also paid to the philosophical, mystical, metaphorical, and symbolic role that water has played for humans in literature, mythology, religion, the visual arts, and music. Finally, a number of individuals and organizations that either have made or are currently making important contributions to our understanding of water and its role in human life and in the natural world are also included. Because no book of this size can possibly include mention of every such individual or organization, a select few are presented here to represent the much larger number who have been essential in water-related areas.

Finally, I strongly recommend that the *Encyclopedia of Water* be used in conjunction with another essential resource on the subject of water, *The Water Encyclopedia*, by Frits van der Leeden, Fred L. Troise, and David Keith Todd (Lewis Publishers, 1990). This reference contains tables, charts, graphs, data, and statistics on virtually any aspect of water that one might imagine. It should prove a superb companion volume to the present work.

One matter of relatively minor significance should be mentioned. A variety of measurement systems are used in the discussion of water-related issues, including not only the metric and English systems of measurement, but also special nautical systems, such as knots, for the measurement of speed. Each measurement has been converted to at least one other system. These conversions have been made consistently with the principle of significant figures—that is, a measurement given as 500 meters in the metric system has a precision of only one digit, expressed by the 5 in

500. To convert that measurement to feet in the English system, one would multiply 500 by 3.2808, the number of feet in a meter. The numerical answer obtained is 1,640.4 feet. But the precision of the answer in the English system cannot be greater than the original metric value of 500 meters—that is, it can have only one significant figure. The correct conversion, therefore, would be 2,000 feet. Since measurements this imprecise are usually given as estimates, their value in another system must also be expected to be imprecise.

Introduction

The twentieth century has sometimes been called the golden age of petroleum because of the critical importance of that liquid to human civilization. Some observers now believe that the twenty-first century will become known as the golden age of water because humans will once more come to appreciate the extent to which our survival depends on this precious liquid.

The importance of water as an avenue of transportation, a source of food, an element in the growth of plants, and an essential medium for human life itself has long been appreciated by humans. When the earliest men and women created the spirits and gods to whom they gave praise and offered worship, they almost without exception included gods and goddesses of water.

Some of the earliest technological feats involved the design and construction of equipment and devices to move water from one place to another for irrigation, public water supplies, transportation routes, or other purposes. With the invention of the steam engine, a whole new range of applications for water became possible.

Over the centuries, human societies have expanded into areas where water is scarce, and new and more sophisticated systems have been developed to deliver water to homes, businesses, and industries. With the growth of such communities, controversies have arisen as to who should have priority in the use of water resources, especially as those resources become increasingly depleted and inadequate. Today's newspapers often carry stories about disagreements—and sometimes battles—over precious water resources among farmers, generators of hydroelectric power, communities, industries, and those concerned with the protection of wildlife. As human population increases, there seems every reason to believe that such debates will continue and intensify over the next decades.

A

Acid Rain

Acid rain is a term used to describe rain that has a lower pH than that of normal rain, pH being its measure of acidity. The more acidic water is, the *lower* its pH; the less acidic, the *higher* its pH. A term more accurate than acid rain is *acid deposition*, of which there are two kinds: wet deposition and dry deposition. *Wet deposition* includes all forms of natural deposition that are more acidic than normal: rain, snow, and fog, for example. *Dry deposition* includes gases and particles that also occur in wet deposition—such as sulfur dioxide and oxides of nitrogen—but that fall to earth in a dry form.

Formation

Acid deposition is formed by both natural and anthropogenic processes. When rain, snow, or other forms of precipitation fall to the earth, they tend to absorb certain gases found naturally in the atmosphere, such as carbon dioxide. Those gases react with water molecules in precipitation to form acids. For example, carbon dioxide reacts with water to form carbonic acid (H_2CO_3):

$$H_2O + CO_2 \Rightarrow H_2CO_3$$

Because of such reactions, normal precipitation has an acidity somewhat greater than that of pure water. Its pH is about 5.5, about that of black coffee.

Acid deposition is also caused by a variety of human activities, the most important of which is the combustion of fossil fuels (coal, oil, and natural gas). Such fuels normally contain small amounts of impurities, the most important of which (from the standpoint of acid-deposition production) are compounds of sulfur and nitrogen. When these compounds burn, they form, respectively, sulfur dioxide (SO_2) and a mixture of nitrogen oxides, usually represented by the formula NO_x.

Researchers have now determined that the major sources of sulfur and nitrogen oxides in the atmosphere are fossil-fueled electric power generating plants. In the United States, these plants are located primarily in the midwestern states. When sulfur and nitrogen oxides escape from smokestacks of such plants, they rise into the atmosphere and are carried eastward by prevailing winds. As they travel through the air, these oxides are absorbed by and react with water molecules, forming sulfuric acid (H_2SO_4) and nitric acid (HNO_3). It is these two acids that are primarily responsible for the increase in the acidity of rain, snow, and other forms of natural precipitation over their natural levels.

Effects

Scientists have been observing and measuring the effects of acid deposition for over 150 years. The term *acid rain* was first used by the English scientist Robert Angus Smith (1817–1884), who described his studies of the phenomenon in his book *Air and Rain: The Beginnings of a Chemical Climatology*. The subject did not receive widespread public attention, however, until the early 1960s, when the Swedish scientist Svante Oden began writing about acid rain in popular magazines and newspapers.

We now know that acid deposition can have harmful effect on lakes, rivers, and other bodies of water; on trees and other kinds of plant life; on building materials; and on atmospheric visibility.

Lakes and Rivers

Acid deposition that falls into rivers and lakes affects the acidity of those bodies of water. In the northeastern United States and eastern Canada, for example, the acidity of about half of all the lakes that have been studied and a quarter of all the rivers has increased significantly in the last fifty years, apparently because of acid deposition. Some lakes in the region now have a pH of less than 5.0.

The effect of acid deposition on a body of water is dependent on a number of factors, one of the most important of which is the nature of the rocks and soil that surround the water. Some earthy materials react with acids in acid deposition, reducing their effect on water. Limestone, for example, is made of calcium carbonate ($CaCO_3$), which reacts with acids to produce less acidic products. Granite, by contrast, does not react with such acids. A lake with a granite basin, therefore, will be more acidic than one only a few miles away whose basin is made of limestone.

Acidic water has a variety of harmful effects on organisms that live in lakes and rivers. Each organism has a particular level of acidity that it can tolerate without becoming ill or dying. Frogs, for example, are particularly hardy, able to withstand pHs of less than 4.0 in many instances. Most fish, on the other hand, can survive a pH of no more than about 5.0, while some begin to die off at a pH level of 5.5. Shellfish tend to be even more sensitive to acidity. Clams and snails are generally unable to live in water with a pH of less than 6.0.

The effects of acid water are actually more complex than the above summary suggests. For example, young fish are more sensitive to high acidity than are adults. Fish eggs are even more sensitive and may not hatch at pH levels at which adults can remain healthy. Also, the interaction among species in an aquatic environment is complex, and increased pH levels may harm species in indirect ways. For example, although frogs can withstand relatively high levels of acidity, the organisms on which they live may not be able to do so. As their food supplies disappear, more resistant organisms (like frogs) may also begin to suffer and die off.

Plant Life

The effects of acid deposition on plant life are, like those in aquatic environments, complex and not fully understood. The pH of acid deposition is usually not low enough to kill plants outright. But even relatively low levels of acidity can cause damage to plants in indirect ways. For example, plants depend for their growth and development on nutrients stored in the soil. As the acidity of soil increases (for example, as the result of acid deposition), those nutrients may become more soluble and more likely to wash away before they are absorbed by plants. Scientists also believe that increased acidity may convert less soluble toxic compounds in the soil into more soluble forms, which are then absorbed by plants, resulting in cell damage.

Building Materials

Acid deposition can damage certain types of building materials in much the same way it reacts with certain kinds of naturally occurring rocks and minerals. For example, many buildings and other structures are made of limestone or marble, both of which consist primarily of calcium carbonate ($CaCO_3$). When acid deposition comes into contact with such materials, a chemical reaction occurs in which the building material is slowly changed into a more soluble form and washed away. For example:

$$HNO_3 \text{ (from acid deposition)} + CaCO_3 \rightarrow Ca(NO_3)_2 \text{ (soluble)} + H_2O + CO_2$$

Many buildings, statues, works of art, and other structures made of limestone and marble that have survived for hundreds or even thousands of years have, over the past few decades, suddenly begun to show dramatic erosion and damage, largely as the result of acid deposition.

Atmospheric Visibility

Tiny droplets of acid deposition suspended in the atmosphere have a tendency to scatter light and, therefore, reduce visibility. The Environmental Protection Agency (EPA) estimates that particles containing sulfate compounds are responsible for 50–70 percent of the reduction in visibility that has been observed in the eastern United States over the past half century. Reduced visibility can have effects on safety, as, for example, in restricted visibility for automobile drivers and airline pilots. But the more common complaint about reduced visibility is the limitation it has placed on the viewing of natural wonders, which has been noted at the Great Smoky Mountain, Grand Canyon, and Bryce Canyon national parks.

Remedies

Efforts to reduce acid deposition usually take one of two forms: the use of "cleaner" fuels, or fuels that contain lower concentrations of nitrogen and sulfur compounds; and emission-control devices. The first alternative has some serious limitations since clean fuels tend to be significantly more expensive than those that contain impurities. Most pollution-control programs, therefore, tend to focus on the use of some type of emission-control system.

Catalytic converters, for example, have not become standard equipment on all motor vehicles sold in the United States and other developed countries. A catalytic converter is a device that increases the efficiency with which automotive fuels are burned, thereby reducing the amount of pollutants—including oxides of nitrogen—released into the atmosphere.

One of the most successful systems for the removal of sulfur dioxide from waste gases in industrial facilities involves the use of scrubbers. Waste gases flow through a *scrubber*, which contains some chemical or combination of chemicals that will react with sulfur dioxide (and other pollutants) in the waste gases. A common type of scrubber, for example, contains powdered lime (calcium hydroxide, $Ca(OH)_2$), which reacts with sulfur dioxide to form calcium sulfate, $CaSO_4$. The calcium sulfate can be collected from the scrubber and sold or used for other purposes in the plant.

Acid Rain Program

The primary program through which acid deposition in the United States is monitored and controlled is the EPA's Acid Rain Program. The Acid Rain Program was mandated by Title IV of the Clean Air Act of 1990. That legislation required the reduction of sulfur dioxide emissions by 10 million tons and of nitrogen oxide emissions by 2 million tons below 1980 levels. The sulfur dioxide reductions were to occur in two stages, the first of which was completed in 2000. Phase I covered 445 industrial facilities, the majority of which were coal-burning power plants, and achieved 40 percent of the total reduction required by the act. Phase II, which began in 2000, will affect more than 2,000 facilities in an effort to achieve the remaining 60 percent of the act's goals. The target date for nitrogen oxide reduction was 2000, at which time a "significant portion" of the required reduction (according to the EPA) had been achieved.

Further Reading

"Acid Rain," http://www.epa.gov/airmarkets/acidrain/index.html.

Chetdav, Rain S. *Acid Rain*. Falls Church, VA: Writers Club, 2002.

Ellerman, A. Denny, et al. *Markets for Clean Air: The U.S. Acid Rain Program*. Cambridge: Cambridge University Press, 2000.

Satake, Kenichi, ed. *Acid Rain 2000: Proceedings from the 6th International Conference on Acid Deposition: Looking Back to the Past and Thinking of the Future, Tsukuba, Japan, 10–16 December 2000*. Dordrecht: Kluwer, 2001.

Aegir

Aegir was a character from Norse mythology. He was a Vanir, a race of giants whose importance diminished with the rise of the Aesir, the clan that included Odin, Balder, Loki, Thor, and Njord. Aegir was a god of the oceans who was notorious for causing storms and stealing sailors off boats. Before beginning a voyage, captains often sacrificed one or more of their crew to Aegir to placate the god. In other cases, sailors were given coins with which to buy their freedom from the god should they be captured.

Because of its association with the giant, the ocean was sometimes referred to as Aegir's wide jaws.

Aegir is often depicted as an older man with white hair and clawlike fingers. He lived at the bottom of the sea with his wife and sister, Ran. They had nine daughters who were said to be responsible for causing the white-capped waves.

One of the most famous legends associated with Aegir's name describes the enormous kettle given to him by Thor. Thor had obtained the kettle from the sea god Hymir, who was reputed to have the largest kettle in the world, more than a mile deep. Thor gives the kettle to Aegir so that he can brew enough beer to serve all the gods who visit him in his underwater castle. In connection with this story, the ocean is also referred to on occasion as Aegir's kettle.

An especially interesting feature of Aegir's castle is the method of lighting it. Instead of using candles or lamps, a large piece of gold is placed on the floor of the castle. The light it gives off is sufficient to illuminate the whole structure. As a remembrance of this tale, gold is still sometimes referred to in Scandinavia as Aegir's fire.

Other names by which Aegir was known are Eagor, Hler, and Gymir.

Further Reading

Lindow, John. *Handbook of Norse Mythology*. Santa Barbara, CA: ABC-CLIO, 2001.

Picard, Barbara Leonie. *Tales of the Norse Gods*. New York: Oxford University Press, 2001.

Agriculture

See Irrigation

Alvariño, Angeles (1916–   )

Angeles Alvariño is a marine biologist with special interest in the ecology and distribution of zooplankton. One of her greatest contributions was the recognition of the complex interaction among zooplankton and other marine organisms and the ways in which zooplankton could be followed as "indicator species"—organisms that could be used to follow changes in ecosystems caused by human activity.

Personal History

Alvariño was born in El Ferrol, Spain, on 3 October 1916. From an early age, she was interested in becoming a doctor, like her father. After being discouraged from this occupation by her father, she decided to study sciences and literature at the University of Santiago de Compostela. After graduation in 1933, she entered the University of Madrid to continue her studies in natural sciences. The Spanish Civil War interrupted her academic career from 1936 to 1939, but in 1941, she was awarded a master's degree in natural science from Madrid. Alvariño then accepted a post of instructor of biology, botany, zoology, and geology at various academic institutions in El Ferrol.

Alvariño's progress in the academic world was hindered to some extent by the sexism that permeated much of society. Nonetheless, she was accepted as a research biologist at the Spanish Department of Sea Fisheries in Madrid in 1948. Two years later, she was admitted as a graduate student at the University of Madrid, from which she received her Ph.D. in 1951. A year later, she earned an appointment as marine biologist and oceanographer at the Spanish Institute of Oceanography.

Major Accomplishments

Over time, Alvariño became especially interested in three types of marine organisms: chaetognaths (arrowworms), hydromedusae (jellyfish), and siphonophores (a form of zooplankton). She developed equipment with which to collect these organisms and encour-

aged the aid of fishing vessels and research ships to obtain specimens for her. She eventually became one of the world's experts on these organisms.

In 1956, Alvariño was awarded a Fulbright Fellowship at the Woods Hole Oceanographic Institute. Her work there prompted Roger Revelle, then director of the Scripps Institute of Oceanography (SIO) in La Jolla, California, to offer her a position as marine biologist. Alvariño accepted and remained at SIO until her retirement in 1987. Between 1970 and 1987, she also worked at the Southwest Fisheries Science Center at La Jolla.

American Institute of Hydrology

The American Institute of Hydrology (AIH) was established in 1981 as a nonprofit scientific and educational organization. The primary function of AIH is to offer certification to professionals in all fields of hydrology. The four major activities of the organization are (1) to establish standards and procedures for the certification of individuals in the fields of hydrology and hydrogeology; (2) to establish and maintain ethical standards that will protect the general public from irresponsible work by hydrologists; (3) to provide education and training in the field of hydrology; and (4) to provide advice and guidance to government agencies and the general public in activities involving hydrologic issues. The association publishes a quarterly bulletin and maintains the AIH Registry of Professionals.

Further Information
American Institute of Hydrology
2499 Rice St., Suite 135
St. Paul, MN 55113-3724
Telephone: (651) 484-8169
Fax: (651) 484-8357
URL: http://www.aihydro.org
e-mail: aihydro@aol.com

American Water Resources Association

The goal of the American Water Resources Association (AWRA) is to attain recognition as the leading multidisciplinary association for information exchange, professional development, and education about water resources and related issues. The association was founded in 1964 as a nonprofit scientific organization for the advancement of knowledge and communication about water resources. AWRA currently has about 2,800 members in the United States and elsewhere, and over 700 subscribers to its professional publication, *Journal of the American Water Resources Association*. In 1999, AWRA began publication of another journal dealing with practical issues in water resource management, *IMPACT*. *IMPACT* carries less technical articles than does the *Journal*. The association also sponsors regular conferences and symposia on topics such as water quality monitoring and modeling, decision support systems for water resources management, and globalization and water management.

Further Information

American Water Resources Association
4 West Federal Street
P.O. Box 1626
Middleburg, VA 20118-1626
Telephone: (540) 687-8390
Fax: (540) 687-8395
URL: http://www.awra.org
e-mail: info@awra.org

Angling

See Aquatic Sports, Angling

Aphorisms and Sayings

An aphorism is a short phrase supposed to contain some fundamental truth or senti-
ment. The origin of many aphorisms goes back hundreds of years and is largely unknown to etymologists—scholars who study the origin of words and phrases. Listed below are some aphorisms that mention some aspect of water about which there is some evidence or, at least, some speculation as to its origin.

Many of these aphorisms grew out of some particular occupation, field of interest, or other unique source. Others are related to the topic of water in more random ways.

The Sea and Ships

Given the constant association that sailors have with the ocean, it should hardly be surprising to find a number of aphorisms that grew out of naval situations. Some examples include the following:

All at sea. Etymologist Robert Hendrickson explains that this phrase arose at a time when navigational skills were not well developed and most ships tried to sail within sight of the shore. If they traveled too far away from the coast, they stood the risk of losing track of where they were. Thus, the modern phrase "all at sea" refers to someone who doesn't quite know what's going on around them.

Between the Devil and the deep blue sea. Etymologists have suggested two possible explanations for this aphorism. The first dates to the story in chapter 8 of the book of Matthew in the New Testament. That story tells of Jesus casting the Devil out of two men whom he encountered. The Devil then entered a nearby herd of swine, who proceeded to fling themselves into the ocean. The swine were in a very difficult position, risking being possessed by the Devil or destroying themselves in the ocean.

A second explanation associates the term with the care of a sailing ship. At one time, ships had two seams located in places that were very difficult to get to. One was the seam between planks nearest the edge of the ship and the other was the seam between planks on the side of the ship just above the waterline. Both seams were said to be "the Devil to get to" or, simply, "the Devil." Any sailor who had the task of caulking these seams was in a very risky position and was said to be "between the Devil and the deep blue sea."

There is no way of knowing which of these explanations is accurate, but the phrase today does refer to someone who finds him- or herself in the difficult position of having to choose between two unpleasant and/or risky decisions.

Cut of his jib, I like (or don't like) the. The jib is a sail usually found at the front of a ship. It is, therefore, one of the first objects seen by other ships approaching it. Sailors often made quick judgments about the nature of the oncoming ship by the appearance of its jib. They either approved of it ("I like the cut of his jib") or did not approve of it ("I don't like the cut of his jib") based on what they saw. Today, the phrase suggests that someone has made a quick judgment about someone to whom they have just been introduced, based on one or more obvious physical features.

Davy Jones's locker. This phrase first appeared in a book by English author Tobias Smollett, *The Adventures of Peregrine Pickle*, in 1751. Two possible origins for the phrase have evolved. The first claims that there was actually a person named Davy Jones who owned an English pub in the sixteenth century. Jones is said to have kept his rum stored in a tightly guarded locker in his back room. Sailors who visited the pub are said to have been frightened of making any effort to visit Jones's locker.

A more widely accepted explanation for the origin of the phrase connects the name Jones with the biblical character Jonah, who was swallowed by a whale—hardly a fate to tempt any sailor. The name Davy is sometimes associated with the West Indian term *duppy*, meaning an evil ghost or devil, or

with St. David, the patron saint of Wales. How the combination of St. David and the biblical Jonah evolved to mean a sailor's worst possible fate is, however, unclear.

Oil on troubled waters. This phrase was apparently first mentioned by the Venerable Bede in his *Ecclesiastical History of England*, written circa 73. Bede repeats a phrase that seems to have been in use long before his time, suggesting that one can soothe or calm a person with gentle words and actions. Some authors have suggested that scientific research confirms the literal accuracy of the phrase. One Commodore Wilkes, according to Brewer's *Phrase and Fable*, observed that oil leaking from a whaling ship calmed the seas that had been produced by a violent storm near the Cape of Good Hope. The scientific basis for this claim, however, is unclear.

Rats deserting a sinking ship. This phrase has a long history and is related to the belief that rats leave a ship that is not seaworthy while they still can—for example, before the ship leaves port. Some scholars say the phrase evolved from a sixteenth century proverb claiming that rats desert a falling house, which conveys much the same sentiment as the "sinking ship" parallel. Today, the phrase refers to individuals who give up on a cause before the cause has actually failed, usually to save themselves from ruin or humiliation.

Sailing close to the wind. This nautical phrase refers to the situation in which a sailing ship is pointed directly or almost directly into the wind (with a wind angle of less than 30 degrees), with all sails and booms pulled in tightly. Under this condition, the boat is said to be *close-hauled*. This challenging sailing maneuver carries with it a certain degree of risk, which may explain the aphorism's common meaning. To sail close to the wind today means that one is acting just within the letter of the law or pursuing risky business practices. By contrast, sailing before the wind means that the bow of the boat is pointed in the same direction as the wind with sails unfurled, moving smoothly across the water. Therefore, a person who is sailing before the wind is experiencing easy progress to one's goals.

Sea change. The term *sea change* refers to some major transformation. It first appeared in 1610 in William Shakespeare's play *The Tempest*. In this play, Ferdinand's father has drowned, and his body has sunk to the bottom of the ocean. There, Ariel tells, his bones have been turned to coral and his eyes to pearls as the result of the sea's actions. "Nothing of him that doth fade," Ariel says, "But doth suffer a sea-change."

Three sheets to the wind. At one time, sailors had an informal system of indicating how inebriated they had become, ranging from one sheet to the wind to four sheets to the wind, the latter meaning that the person had lost consciousness. The "sheets" mentioned in this saying refer not to sails themselves but to the ropes that hold the sails. In nautical terms, a *sheet* is a rope that holds a sail or some other piece of equipment in position. The need for securing three sheets on a voyage meant that a ship was facing severe weather, reeling from side to side like a drunken sailor, as one might say. The connection between a rough ocean voyage and a rough bout of drinking thus becomes clear.

Fishing

As one of the most popular water sports, fishing has been the source of many aphorisms about water, such as the following:

Fish or cut bait. According to the Word Detective, this phrase first appeared in the *Congressional Record* in 1876. It refers to the fact that anglers usually have two tasks to complete: first, to cut up the bait that is to be placed on the hook; and second, to cast the line and actually begin fishing. The admonition to "fish or cut bait" suggests that a person has to make a decision to go forward with something or not—either to take action ("fish") or to give up the idea of action and stand aside ("cut bait") (Word Detective).

Like a fish out of water. If one can imagine a fish trying to survive on land, the meaning of this phrase becomes obvious: a person who is trying to operate in surroundings about which he or she knows little or nothing can be compared to a fish out of water. According to Robert Hendrickson, this phrase may first have been used in print by English theologian John Wycliffe in his *English Works*, published circa 1380, when he said, "And how thei weren out of their clositre as fish is withouten water" (Hendrickson, 254).

Weather

Some aphorisms contain references to rain or other meteorological phenomena, such as the following:

Raining cats and dogs. Up to a half dozen different explanations have been offered with respect to the origin of this phrase. The most popular explanation appears to be one that connects the phrase with Norse mythology. According to Norse legend, cats were thought to have a particularly strong influence on the weather. Also, when witches decided to travel in the midst of a storm, they did so in the form of a cat. The dog, on the other hand, was regarded by the Norse as a symbol of the wind and an attendant of Odin, the storm god. According to this explanation, then, the reference to "cats and dogs" suggests a rainstorm in which severe rain (cats) and wind (dogs) are present.

Another explanation for the phrase, suggested by some scholars, is that it refers to the horrible sewer conditions in England of the seventeenth century. During terrible rainstorms, dead cats and dogs were seen washing away through streets and rivers.

In any case, the phrase first appeared in English in Jonathan Swift's *A Complete Col-*

lection of *Polite and Ingenious Conversation*, published in 1738. It also appeared in a somewhat different form about a century earlier in the play *City Wit* (1653), by English writer Richard Brome. In this work, the phrase used is, "It shall raine . . . Dogs and Polecats."

Right as rain. In an essay on this phrase, writer Michael Quinion points out that phrases beginning with "right as . . . " have been popular since at least 1400. In all such cases, the phrases are meant to suggest that everything is entirely satisfactory and secure. The earliest of these phrases was "right as an adamant," an adamant being a form of magnet. Later variations took the form of "right as a line," "right as a gun," and "right as my leg."

Quinion says that the first appearance of "right as rain" in print occurred in Max Beerbohm's book *Yet Again*, published in 1909. It is unclear how rain sets a standard for satisfaction and security except that the phrase does have an alliterative appeal. Perhaps, Quinion surmises, the "rightness" of rain comes from the very straight vertical lines in which it sometimes falls (Quinion).

The Bible

A number of aphorisms still in use today can be traced to the Bible, such as the following:

I'm washing my hands of . . . . A person uses this phrase to indicate that he or she is finished with some activity and no longer wants to be associated with it in any way. The phrase probably had its origin in Pontius Pilate's act of washing his hands after refusing to have anything further to do with the fate of Jesus Christ. Matthew reports (chapter 27, verse 24) that "When Pilate saw that he could prevail nothing, but that rather a tumult was made, he took water, and washed his hands before the multitude, saying, I am innocent of the blood of this just person."

A drop in the bucket. This term is used to describe some small amount compared to a larger whole. It was first used in the Book of Isaiah, chapter 40, verse 15: "Behold, the nations are as a drop in a bucket, and are counted as the small dust on the scales" (New King James Version).

It rains on the just and the unjust. This phrase, describing the equality of God's actions on all people, first appears in the book of Matthew, chapter 5, verse 45, which says that God "sendeth the rain on the just and on the unjust."

Miscellaneous

Many aphorisms refer to some aspect of water not mentioned above, including the following:

All washed up. The final act for most workers at the end of the day is to wash their hands, indicating that their job is over. The phrase has since been extended to refer to any activity that has come to an end, such as a politician's career. Also see *I'm washing my hands of . . .* under "The Bible."

Don't throw the baby out with the bathwater. This phrase first appeared in print in a book by the German author Thomas Murner, *Narrenbeschwörung*, written in 1512. In the book, Murner tells of the folly of people who try so hard to rid themselves of bad things in their life that they also eliminate everything that is good. Critic Wolfgang Mieder suggests that the phrase was probably in use in Germany well before the appearance of Murner's book.

Mad as a wet hen. The origin of this phrase is not clear although it may arise from the presumption that some land animals (such as chickens) might be less at home in aquatic environments than others (such as ducks and geese). In any case, the expression is intended to describe someone with a high level of anger.

Of the first water. This phrase refers to a person, object, or event that is of the very highest caliber. The word *water* here has nothing to do with liquid water but arises from a phrase long used by jewelers to note a gem of the very highest quality. Gems of lesser quality were described as "of the second water," "of the third water," and so forth. The Word Detective suggests that this usage may have arisen from the Arabic word for *water*, which has a secondary meaning of "splendor" or "brilliance" (Word Detective).

Wet behind the ears. As with many aphorisms, this phrase first appeared in print quite recently, early in the twentieth century. However, it had probably been in popular use in language and music for a much longer period of time. It refers to a person who is new and inexperienced in some regard, and apparently arises out of the fact that the last part of a newborn animal to dry off is the area behind its ears.

Wet blanket. Just as a wet blanket is sometimes used to put out fires, a person who manages to suppress a conversation or event can be thought of as a wet blanket. The phrase first appeared in print in 1810.

Further Reading

Brewer, E. Cobham. *Dictionary of Phrase and Fable*. Philadelphia: Henry Altemus, 1898. New edition—revised, corrected, and enlarged—at http://www.bartleby.com.

Hendrickson, Robert. *The Facts on File Encyclopedia of Word and Phrase Origins*. Revised and expanded ed. New York: Facts on File, 1997.

Lurie, Charles N. *Everyday Sayings: Their Meanings Explained, Their Origins Given*. New York: G. P. Putnam's Sons, 1928.

Mieder, Wolfgang. "Don't Throw the Baby Out with the Bathwater," *De Proverbio: An Electronic Journal of International Proverb Studies*, 1, no. 1 (1995).

Quinion, Michael. "Right as Rain," http://www.worldwidewords.org/qa/qa-rg1.htm.

The Word Detective, http://www.word-detective.com.

Aquaculture and Mariculture

The terms *aquaculture* and *mariculture* refer to the raising and harvesting of fish and shellfish in land-based lakes and ponds and in fenced-off areas in lagoons, estuaries, and other parts of the ocean. Scientists believe that aquaculture was practiced as long ago as 1000 B.C. in China. In late 2000, researchers from the University of Pennsylvania found evidence of aquaculture farms in Bolivia dating back to at least 1700. In the past half century, aquaculture has become increasingly important because of the decline of natural supplies of fish and shellfish.

The Problem

Fish and shellfish make up a significant portion of the diet of people in many parts of the world, especially in those nations with large coastal areas. In recent years, fish and shellfish have become an increasingly important staple in the United States and other developed nations, as people attempt to reduce their intake of red meat and increase their intake of seafood. Between the mid-1980s and 1999, for example, the average consumption of seafood in the United States increased from 13.0 lb. to about 15 lb. This increase translates into an increase of about a half million metric tons of seafood in the United States alone ("Course Page," http://ag.ansc.purdue.edu/courses/aq448/content.htm). In 1994, production of fish and shellfish from aquaculture and mariculture in the United States reached 324 million kilograms (715 million pounds), accounting for about 13 percent of all seafood consumed ("Aquaculture Provides Answer to Nation's Diminishing Seafood Supply," http://www.aquanet.com/features/catfish_institute/catfish.htm).

At the same time, methods for taking fish and shellfish from the ocean have become increasingly sophisticated and efficient. The fishing industry has become so productive that natural supplies of fish and shellfish are declining in many parts of the world. The United Nations' Food and Agriculture Organization has estimated that 70 percent of the world's fish stocks are now completely exhausted, overfished, depleted, or rebuilding from previous overfishing ("Aquaculture," http://www.guardianunlimited.co.uk/Print/0,3858,4050639,00.html). A similar situation exists in U.S. coastal waters, where, according to the National Marine Fisheries Service, about 40 percent of all stock has been overfished or seriously depleted ("Aquaculture Provides Answer").

Experts believe that aquaculture and mariculture can significantly increase the world's supply of fish and shellfish. In fact, in 1996, about one-fifth of all the seafood consumed around the world came from such facilities. According to some authorities, greater attention to aquaculture and mariculture could double that output in the next two decades ("Aquaculture Provides Answer").

Methods

A wide variety of methods are used in fish and shellfish farming, some adapted for family and cooperative farms, and others adopted for large commercial facilities. In *intensive farms*, the habitat and feeding of fish and shellfish are completely controlled from the time the animals are born until they are harvested. In *semi-intensive farms*, natural habitats and food supplies are supplemented to provide maximum conditions for growth of fish and shellfish.

Fish farms today take advantage of the latest scientific information about the health and growth of fish and shellfish. Farms are designed to provide habitats with a good water source and containment systems with proper filtration, aeration, water flow, water temperature, and water quality. Feed provided to the stock contains all necessary nutrients, and antibiotics are supplied to prevent disease.

Many species of fish and shellfish have been raised in farms. The most common of these are catfish, tilapia, trout, crawfish, oysters, shrimp, and salmon. In the United States, 94 percent of the nation's catfish are raised in farms located in Alabama, Arkansas, Louisiana, and Mississippi. Tilapia are also popular farm fish. In 2000, the United States imported 15.0 million kilograms (33.2 million pounds) of the fish, primarily from China, Taiwan, Costa Rica, and Ecuador (American Tilapia Association).

Environmental Concerns

Environmentalists have voiced some concerns about expanding the use of aquaculture and mariculture. One problem is that fish raised in such systems are often genetically altered or hybridized, as is the practice with many agricultural crops. If these fish escape into the wild, they may compete with and displace natural populations. Since the gene pool in managed fish may be less diverse, that change may threaten the survival of some populations. There is also some concern about the release into the environment of uneaten food, antibiotics, and waste products from large farms.

Further Reading

Ackefors, Hans. *Introduction to the General Principles of Aquaculture*. New York: Food Products Press, 1994.

Fingerman, Milton, and Rachakonda Nagabhushanam, eds. *Aquaculture*. Enfield, NH: Science Publishers, 2000.

Pillay, T. V. R. *Aquaculture and the Environment*. New York: Halsted Press, 1992.

———. *Aquaculture Development: Progress and Prospects*. New York: Halsted Press, 1994.

Stickney, Robert R. *Principles of Aquaculture*. New York: Wiley, 1994.

Further Information

Aquaculture Magazine
P.O. Box 2329
Asheville, NC 28802
Telephone: (828) 254-7334
Fax: (828) 253-0677
URL: http://www.aquaculturemag.com
e-mail: comments@aquaculturemag.com

Aquatic Plants and Animals

Plants and animals that live in water can be subdivided into two major categories: those that live in freshwater and those that live in seawater. The latter are often referred to as marine plants and animals. Although organisms in the two categories have many characteristics in common, they differ from each other in various ways. Aquatic plants and animals are also significantly different from terrestrial plants and animals in many important ways.

Differences between Terrestrial and Aquatic Organisms

The physical and chemical environment in which aquatic and terrestrial organisms live accounts for a number of anatomical, physiological, biochemical, life history, and other differences. For example, water is about 800 times as dense as air and about 60 times as viscous. As a consequence, organisms that live in water are buoyed up to a much greater extent by the surrounding medium than are organisms that live in air. The skeletal system of bones and muscles that many terrestrial organisms have evolved is not required by aquatic organisms.

Such organisms have evolved other types of adaptations for surviving in water. For example, most fish have streamlined bodies that allow them to maneuver in moving water more efficiently than is needed by terrestrial animals (such as birds) who have to move through air.

Aquatic organisms have also developed biochemical systems substantially different from those found in terrestrial organisms. In the latter, carbohydrates are the primary biochemical family used for structural and energy-storage purposes, while aquatic organisms depend to a greater extent on proteins for these functions. The tendency of carbohydrate-based biochemical systems to survive for longer periods of time means that land-based plants and animals tend to live longer than plants and animals that live in fresh- and salt water.

Physical factors such as light, sound, and electrical currents play an important role in the differences between terrestrial and aquatic plants and animals. For example, light is essentially not absorbed by air, so that sunlight passes through very thick layers of the earth's atmosphere, reaching almost every part of the land-based environment. Water absorbs light much more efficiently than air, however, so that sunlight penetrates to depths of no more than a few hundred meters in oceans and lakes. It is within this upper layer, therefore, that the vast majority of aquatic organisms are able to survive.

Sound and electricity both travel through water much more efficiently than they do through air. The speed of sound at 20°C in air, for example, is 346 meters per second (about 1,100 feet per second), but 1,518 meters per second (about 4,800 feet per second) in water at the same temperature. In conditions where light intensity is low and the speed of sound high, aquatic organisms are more likely to depend on the latter rather than the former for communication and observation.

The same pattern is true for electrical conductivity. The electrical resistivity of air is about 10^{16} (ten quadrillion times) that of water. This fact means that aquatic organisms are likely to be much more sensitive to electrical currents than are their terrestrial cousins. Currents that can be detected on land only with relatively sophisticated measuring instruments are easily detected by marine organisms.

Biological Zones

Aquatic organisms are often classified into three major categories, according to their mode of transport. The *plankton* group consists of organisms that are unable to swim on their own, or that can swim only very weakly. The members of this group depend largely on ocean currents, wave motion, and other forms of water movement for their transportation. The plankton consist of two major groups: *zooplankton* (animals) and *phytoplankton* (plants). Most plankton are very small in size, ranging from a few millimeters to a few centimeters in diameter, with the vast majority at the lower end of this range. In spite of their small individual sizes, the plankton make up the largest group of organisms, both in numbers and in total mass, found in aquatic environments. The plankton include not only mature organisms but also immature forms of organisms that become members of another group (benthos or nekton) later in life.

A second group of aquatic organisms, the *benthos*, includes those plants and animals that live on or beneath the ocean bottom and on the bottom of lakes, ponds, rivers, and streams. Examples of such organisms include the barnacle, oyster, mussel, starfish, tunicate, periwinkle, limpet, coral, and sponge. Some of these organisms move about on the ocean or lake floor, while others, such as various types of worms and clams, actually burrow into the floor. Plants can survive in the bethnic environment to a depth where sunlight is still sufficient to allow photosynthesis, but not below.

The third group of organisms, the *nekton*, consists of animals that have evolved the ability to swim freely and, thus, do not rely on wave motion or ocean currents for mobility. Common examples of the nekton include all types of fish and aquatic mammals such as whales and dolphins. The ability to move about on their own means that, unlike the plankton and benthos, the nekton can actively pursue their prey; avoid their predators; and migrate over limited or extensive areas of the oceans, lakes, and rivers.

Aquatic Plants

Botanists classify plants into about ten major subgroups, or *divisions*. (Systems of classification differ, so the designation of divisions may differ from authority to authority.) Plants in six of those divisions make up the vast majority of plants found in fresh- and seawater. These divisions include Euglenophyta (euglenoids), Chlorophyta (green algae), Chrysophyta (yellow-green and golden-brown algae and diatoms), Pyrrophyta (dinoflagellates), Phaeophyta (brown algae), and Rhodophyta (red algae). The members of these divisions vary in size from less than a millimeter in diameter to more than 40 meters (130 ft) in length. Red, brown, and green algae make up the familiar group of plants commonly known as *seaweed*.

The six divisions listed above are not equally distributed in fresh- and saltwater environments. For example, more than 90 percent of all red and brown algae and dinoflagellates are found in marine environments. By contrast, green, golden-brown, and yellow-green algae and euglenoids are more commonly found in fresh water environments.

A number of common terrestrial plants either do not occur in aquatic environments or occur only to a modest degree. For example, there are essentially no members of the division Bryophyta (liverworts, hornworts, and mosses) and few members of the division Tracheophyta (vascular plants) found in aquatic environments. Among the subdivision Spermosida (seed plants), there are virtually no representatives of the large class of Coniferae (conifers; only about 700 species) and only a small number of Angiospermae (flowering plants; about 235,000 species). Aquatic specimens from the latter group make up no more than about 0.1 percent of all existing members of the class on earth.

Although relatively few in number, angiosperms—partly because of their size—are of considerable significance in some aquatic environments. They make up a large part of the group of plants known as Macrophytes (large plants), which includes such examples as *Pistia* (water lettuce), *Eichhornia* (water hyacinth), and *Lemna* (duckweed), all free-floating plants, and *Rorippa* (watercress), *Myriophyllum* (milfoil), *Zostera* (eel grass), and *Potamogeton* (pond weed), all rooted plants. Shallow-water ecosystems in which such plants predominate provide food and shelter for a large array of organisms, making them a highly productive source of biomass.

Aquatic Animals

All animal phyla are to be found in aquatic environments, some to a considerably greater extent than in terrestrial environments. Some phyla consist of aquatic representatives exclusively, or almost exclusively. For example, the phylum Porifera consists of a great variety of sponges that occur only in ocean and saltwater environments. Sponges are attached to the ocean bottom at all depths. The Cnidaria are also marine and freshwater animals, consisting of hydras, sea anemones, jellyfishes, and corals. The phylum Echinodermata, including sea stars, brittle stars, sea urchins, sea lilies, and sea cucumbers, are found only in marine environments.

Three phyla of unsegmented worms show an interesting transition from aquatic to terrestrial environments. These phyla—the Platyhelminthes, Nemathelminthes, and Trochelminthes—are found primarily in marine and freshwater environments. They

occur on land in the body fluids of host animals.

The phylum Mollusca consists of three classes, two of which live exclusively in aquatic environments. These are the Bivalvia (mussels, clams, oysters, and scallops) and Cephalopoda (squids, octopus, nautilus, and cuttlefishes). The third class, Gastropoda, includes animals such as snails, slugs, whelks, conchs, abalones, nudibranches, and periwinkles, some of which live on land, others of which live on the ocean floor and on the floor of lakes, ponds, and other bodies of freshwater.

The largest phylum of aquatic animals, in terms of both numbers and total biomass, is Arthropoda. The two most abundant classes of this phylum on dry land, Insecta and Arachnida (spiders, ticks, mites, scorpions) are largely absent from aquatic environments. Among the insects, only the water strider (*Halobates*) spends all of its life on water. Horseshoe crabs (*Limulus polyphemus*) and the relatively rare sea spiders (*Nymphon* and *Ascorhynchus*) have arthropod-like characteristics, but belong to separate and distinct classes (*Xiphosura* and *Pycnogonida*, respectively).

The predominance of arthropods in aquatic environments is due almost entirely to the presence of the crustaceans, a class that includes water fleas, barnacles, shrimps, crabs, lobsters, and the common zooplankton Euphausiacea. These organisms are especially important in aquatic food webs because they feed on phytoplankton and, in turn, are fed on by larger marine and freshwater animals.

The two phyla consisting of the most highly developed animals are the Hemichordata and Chordata, indicating the presence of a complete or partially developed notochord (an axial structure that supports a skeleton). The Hemichordata include the acorn worms (*Balanoglossus, Glossobalanus*, and *Saccoglossus*), that burrow under

Male walruses at Round Island, Alaska. *Robert Holmes/Corbis.*

the ocean floor. The Chordata consists of three subphyla, the Urochordata (or Tunicata), Cephalochordata, and Vertebrata. The first two subphyla are relatively unimportant in the aquatic environment, consisting of the tunicates and lancelets, both of which are filter feeders that swim on the ocean or lake bottom or submerge themselves with only their mouth parts exposed.

The subphylum Vertebrata is a large and complex group of animals that includes fish, amphibians, reptiles, birds, and mammals. Fish, of course, are exclusively aquatic animals, while many members of the remaining classes of vertebrates spend part or all of their time in or around the water. Amphibians, for example, share many characteristics of both fishes and reptiles, spending their early life cycle in water and their later life cycle on land. While birds do not strictly live on or in the water, many depend on the ocean or freshwater pods for their food and return to land only to breed.

The class Mammalia includes the most highly developed animals found in aquatic environments. These animals belong to two classes, Carnivora and Cetacea. Aquatic members of the Carnivora class are placed in their own suborder, Pinnipedia, which includes the Otariidae (sea lion, eared seal, and fur seal), the Odobenidae (walrus), and the Phocidae (true, or earless, seal and elephant seal). The Cetacea are subdivided into two suborders, the Odontoceti (toothed whale and porpoise) and Mysticeti (whalebone whale).

In some respects, the animals most frequently associated with aquatic environments are the fishes. The fishes are members of the class Pisces, consisting of at least three subclasses. (Experts differ widely on the classification of the fishes. The description that follows is only one of many possible taxa.) The three subclasses are Agnatha (jawless fishes), Chondrichthyes (cartilaginous fishes), and Osteichthyes (bony fishes).

The Agnatha include lampreys and hagfish. They have a circular jawless sucking disk for a mouth, lined with rasping teeth. They are primarily scavengers or parasites, often attaching themselves to another fish, cutting a hole in its body, and sucking out its bodily fluids.

The Chondrichthyes include sharks, rays, and chimaeras. Their bodies are covered with scales, although those scales do not overlap as they do in bony fish. Sharks are among the largest of all marine animals, some reaching a length of more than 15 meters (50 feet).

The Osteichthyes have a bony skeleton and a body covered by overlapping scales. The subclass includes all of the important commercial and recreational fish found in lakes, oceans, rivers, and streams, including sturgeon, dogfish, herring, salmon, pike, eel, carp, trout, perch, bass, tuna, and flounder. The members of this subclass are by far the most important aquatic animal in terms of human consumption.

Significant numbers of humans depend on fish as their main source of food, while many others enjoy fishing as a recreational sport. The United Nations Food and Agricultural Organization (FAO) has estimated that 16 percent of the animal protein consumed by humans around the world each year comes from the oceans. One billion people in Asia alone rely on fish as their main source of protein. The commercial value of all fish taken annually, according to the FAO, is about $80 billion, with a comparable estimate of about $3.5 billion for the United States alone ("Ensuring the Sustainability of Ocean Living Resources," http://222.yoto98 .noaa.gov/yoto/meeting/liv_res_316.html). *See also* Limnology; Oceanography; Plankton; Whales and Whaling

Further Reading

Dawson, E. Yale. *Marine Botany*. 2nd ed. New York: John Wiley, 1998.

Kalff, Jacob. *Limnology*. Upper Saddle River, NJ: Prentice Hall, 2001.

Nybakken, James Willard. *Marine Biology: An Ecological Approach*. 5th ed. San Francisco: Benjamin Cummings, 2001.

Wetzel, Robert G. *Limnology: Lake and River Ecosystems*. 3rd ed. New York: Academic Press, 2001.

Aquatic Sports

In this entry, the term *water sport* refers to any athletic event played on water or ice. Some popular water sports are boating, bobsledding, canoeing and kayaking, curling, diving, downhill and cross-country snow skiing, figure skating, fishing, hockey, ice skating, luge, powerboat racing, rowing, sailing, scuba diving, sled-dog racing, snowboarding, snowmobiling, speed skating, surfing, swimming, synchronized diving and swimming, water polo, waterskiing, windsurfing, and yachting.

The origin of many of these sports almost certainly extends back to antiquity. Humans have probably been swimming, boating, and fishing since the dawn of civilization, if not before. This entry, however, focuses, on the development of these activities as sports—that is, competitions between individuals or forms of recreation.

Space does not permit an extended treatment of any of these sports. However, the "Further Reading" section below lists two excellent references on this topic. The focus of this entry is to provide a brief overview of the history and development of some popular water sports and to summarize the types of activities included in each sport.

Further Reading

Hickok, Ralph. *New Encyclopedia of Sports*. New York: McGraw-Hill, 1977. Much of this work is available online at http://www. hickoksports.com.

Menke, Frank G. *The Encyclopedia of Sports*. 6th rev. ed. Revisions by Suzanne Treat. South Brunswick, NJ: A.S. Barnes, 1958.

Angling

See Aquatic Sports, Fly Casting

Bobsledding, Luge, and Skeleton

Bobsledding, luge, and skeleton are three closely related sports that involve one or more persons sliding down a hill on some type of sled. Bobsledding is often called *bobsleighing*. Sled-type devices have long been used in regions where snow and ice make transportation by wheeled vehicles, animals, or other means inconvenient or difficult. Toboggans have also been popular for recreation.

The bobsled was apparently first developed in the late 1880s in Switzerland when someone decided to add runners to the bottom of a toboggan to increase its speed. The sport got its name from early athletes' belief that they could increase their downhill speed by bobbing their bodies back and forth. That supposition turned out not to be true, but the sport and the sled retained their names.

The first organized competition involving bobsleds occurred in 1897, on the Cresta Run at St. Moritz. According to rules of the time, each sled carried five riders, two of whom had to be women. The sleds were steered with four runners attached to the sleds two axles. The international governing board for the sport, the Fédération Internationale de Bobsleigh et Tobagganing, was formed in 1923, and the sport was introduced in the Winter Olympic Games the following year in Chamonix, France.

The bobsled consists of two sections joined together by a flexible connection, with both a steering device and a hand brake. It is designed to hold either two- or four-person crews. The 2002 Winter Olympics held in Salt Lake City included competition for two- and four-men and two-women sleds.

Luge and skeleton are simpler versions of bobsledding. In both sports, riders lie flat on the sled, feet forward in luge and head forward in skeleton. Luge and skeleton are both forms of age-old sledding, familiar to the young in any nation that has snowy winters, adapted for the purpose of competition. The 2002 Olympics offered a men's singles and women's singles competition in both the luge and the skeleton and a doubles competition in the luge.

Further Information

Fédération Internationale de Bobsleigh et Tobagganing
Ermanno Gardella
Via Piranesi, 44/B
120137 Milano
Italy
Telephone: (+39.02) 757-3319
Fax: (+39.02) 757-3384
URL: http://www.bobsleigh.com
e-mail: egarde@tin.it

US Bobsled and Skeleton Federation
421 Old Military Road
P.O. Box 828
Lake Placid, NY 12946
Telephone: (518) 523-1842
Fax: (518) 523-9491
URL: http://www.usbsf.org
e-mail: info@usbsf.com

Canoeing and Kayaking

A canoe is a light boat, considerably longer than wide, usually propelled by paddling. Kayaks differ from canoes in that they are covered with an opening only large enough to permit operation by a single person with a paddle that has a blade on each end. Canoes may be operated by one or more (usually two) people with paddles that have a blade on only one end. The word *canoe* comes from a term used by the Arawak Indi-

Woman kayaking off Wye Island, Delaware, in the Chesapeake, August 1999. *Mary Hollinger, NODC biologist, NOAA.*

ans of the West Indies, while the term *kayak* originated with the Inuit Indians, who used the word *qayaq* for the vessel.

The origin of canoes dates at least to 6000 B.C. Early humans made canoes ("dugouts") by hollowing logs with stone tools and setting fires within prescribed parts of a tree trunk. These canoes were useful for transporting goods and people in regions where land travel was difficult, but they were at first somewhat clumsy in design and not easily transported between waterways.

Canoes became popular in areas where water travel was either essential or the most convenient mode of transportation, as in the South Pacific Islands, parts of the British Isles, and North America. The highest level of development was probably achieved in North America, where raw materials such as animal skins and tree bark were available. Early Indian tribes developed methods for

building canoes from wooden frames covered with hides or bark stitched together to cover the frame.

Early kayaks were often made of wooden frames tied together with sinew and covered with seal skin. They were especially popular in Arctic regions, such as Greenland Island, Labrador, the Aleutian Islands, and the Bering Straits. Each region developed slightly different kayak models that could be distinguished from those of other regions.

A Scottish lawyer, John MacGregor, is generally acknowledged as the founder of canoeing as a sport. In 1845, he designed a canoe fitted with a mast and a sail with which he traveled around Europe and to the Mideast. In 1866, he was one of the founders of the Royal Canoe Club (originally, the Canoe Club) in England, an organization that sponsored the world's first canoe competition a year later.

Canoeing and kayaking appeared as demonstration sports in the 1924 Olympics and became a regular part of the Summer Olympics in 1936. Competitive events are now designated with a letter/number combination that indicates the type of craft being used and the number of paddlers. For example, a C-1 race is one for canoes (now called *Canadian canoes*) operated by a single person. A K-2 race in one for kayaks operated by two persons. Modern Olympic games include C-1, C-2, K-1, and K-2 races at 500 and 1,000 meters, a K-4 race for men at 1,000 meters, and K-1, K-2, and K-4 races for women at 500 meters. White-water and long-distance canoe and kayak races are also available.

Further Information

American Canoe Association
7432 Alban Station Blvd., Suite B0232
Springfield, VA 22150
Telephone: (703) 451-0141
Fax: (703) 451-2245
URL: http://www.acanet.org
e-mail: aca@acanet.org

International Canoe Federation
Calle de Antarctica 7,4 pta
28045 Madrid
Spain
Telephone: (34) 91 506 11 50
Fax: (34) 91 506 11 55
e-mail: canoedg@ctv.es

Diving

Humans have probably been diving for recreational purposes off cliffs and other promontories, over oceans and lakes, for hundreds or thousands of years. But diving did not become a sport until the seventeenth century, primarily in Germany and Sweden. Gymnasts in those countries often moved their athletic equipment to the beach in good weather and performed their exercises over the water. In some regards, then, diving is as much a gymnastic event as it is an aquatic sport.

Competitive and recreational diving consists of two categories: platform and springboard. In the first category, divers jump from a fixed platform set at 5 to 10 meters above the surface of the water. The required dimensions of the platform vary depending on the distance of the platform above the water. A 5-meter (above the water) platform, for example, must be 1.5 meters wide and 6.0 meters long, while a 10-meter platform must be 3.0 meters wide and 6.0 meters long. Springboard diving is conducted from a long plank attached at one end, with dimensions of at least 4.8 meters in length and 0.5 meters in width.

Platform diving was introduced into the Olympic Games in 1904. Competition was limited to men using a 10-meter platform. Springboard diving was added in the 1908 Olympic Games, and women's diving was added in 1920.

Divers are judged on four aspects of their dive: approach, takeoff, execution, and entry into the water. A score of 10 is regarded as a perfect dive in every respect. The score awarded a dive is multiplied by a factor known as the *degree of difficulty*, a factor that indicates the relative difficulty of the dive. The degree of difficulty of a dive is decided by the sport's ruling body, the Federation Internationale de Natation.

Early diving competitions consisted of relatively simple dives, with a moderate amount of twisting and tumbling added to the diver's fall into the water. Today, the complexity of diving has advanced to the point that there are now sixty-three different dives that can be performed from a 1-meter springboard, sixty-seven dives from a 3-meter springboard, and eighty-five dives from a platform.

Further Information

U.S. Diving, Inc.
201 South Capitol Avenue, Suite 430

Indianapolis, IN 46225
Telephone: (317) 237-5252
Fax: (317) 237-5257
URL: http://www.usdiving.org
e-mail: usdiving@usdiving.org

Federation Internationale de Natation
Av. de l'Avant-Poste 4
1005 Lausanne
Switzerland
Telephone: (+41) 21-310-4710
Fax: (+41) 21-312-6610
URL: http://www.fina.org

Fishing

See Aquatic Sports, Fly Casting

Fly Casting

Fly casting is a competitive sport that has evolved out of the more general sport known as *fishing* or *angling*. The term *fly casting* refers to a form of fishing in which an artificial lure (the "fly") is attached to the end of a fishing line for the purpose of either catching a fish or coming as close as possible to some target.

The practice of catching fish for food dates to the earliest days of human existence in areas where people lived near rivers, streams, lakes, or oceans. A number of specialized techniques were developed to catch fish, including the use of hands, spears, nets and lines, and fishing rods of various levels of sophistication.

It seems likely that some anglers throughout history must have compared the results of their activities with their friends and companions, debating as to who had caught the heaviest, longest, or greatest number of fish. However, records of formal fishing competitions are rare prior to the mid-eighteenth century. At that time, fly-casting competitions became a popular way of determining the champion angler in various districts in England. By the beginning of the nineteenth century, angling clubs had been formed nearly everywhere in England, with fly casting by far

the most popular form of fishing—both for food and for competition.

The first angling club in the United States, the Schuylkill Fishing Company, was formed in 1732 in Philadelphia, and the first casting tournaments were organized during annual meetings of the New York Sportsmen's Club in the early 1860s. Probably the most important impetus to the growth of competitive fly-casting events in the United States was the first national tournament held at the Chicago World's Fair in 1893. The competition was sponsored by the Chicago Fly-Casting Club and included contests in accuracy, distance, and delicacy fly casting. The contests were held on a lawn because at that time, there was no method available for measuring casts made on water.

The second, third, and fourth national tournaments were held in Chicago in 1897, 1903, and 1905, and a fifth contest, in Kalamazoo in 1906. At the sixth national competition held in Racine, Wisconsin, in 1907, the National Association of Scientific Angling Clubs was established. In 1960, the association changed its name to the American Casting Association, which remains the parent body for the sport in the United States.

The international governing body of the sport is the International Casting Federation, which sponsors the European Championships once every other year. Fly casting is not included in the Olympic Games schedule, but is a part of the quadrennial World Games, most recently held in Akita, Japan, in August 2001. Events in the 2001 games included fly accuracy, fly distance single handed, and multiplier accuracy skish for men and women.

Formal competitions to determine the largest (heaviest) fish caught by anglers have also become popular in many parts of the world. The competitions, known as *matchfishing*, originated in Europe in the mid-twentieth century and have slowly spread to the United States. In matchfishing, anglers fish from a shoreline within a section of the waterway specifically designated for their use. The angler catching the heaviest fish is declared the winner of the competition. Rules for American Casting Association tournaments can be found at http://www.longbeachcastingclub.org/rules.html.

Further Information

American Casting Association
1773 Lance End Lane
Fenton, MO 63026
Telephone: (636) 225-9443
Fax: (636) 225-7238
URL: http://www.americancastingassoc.org
e-mail: ddddlanswer@aol.com

Ice Hockey

Humans have played games in which balls or other objects are maneuvered across a field with a stick for centuries. One of the precursors of ice hockey is hurling, a sport played in Ireland at least a thousand years before the birth of Christ. In hurling, a 9–10-inch ball is moved across a field about 80 yards wide and 140 yards long with a curved stick (the *hurley*), with the goal of hurling it under a crossbar 8 feet high and 21 feet wide. The rules of hurling were eventually applied to ice hockey.

Games similar to modern ice hockey have probably been played in frigid regions of the world, such as northern Europe, for hundreds of years. A Dutch game known as *kolven*, for example, was popular as far back as the early seventeenth century. A similar game known as *bandy* had spread to England by the 1820s.

Most authorities believe that modern-day ice hockey evolved from a game played by North American Indians. French observers of the game called it *hoquet*, a term used for a shepherd's crook. Debate continues as to the birthplace of modern hockey, with at least three Canadian cities—Halifax, Kingston, and Montreal—contending for that honor.

The first formal game appears to have been played in Kingston in 1855, pitting teams from the Royal Canadian Rifles against each other. In 1865, the first recorded game of ice hockey was organized by a student at Montreal's McGill University, W. F. Robertson. Robertson later wrote out the first set of rules for the game, basing those rules on the game of field hockey. Robertson's rules called for two teams of nine players each, and the use of a square puck.

The first amateur hockey league was organized in Kingston in 1885 and consisted of teams from the Royal Military College, Queen's University, the Kingston Athletics, and the Kingston Hockey Club. The first professional league was organized two decades later in 1904. The league was named the International Professional Hockey League (IPHL) because it included three teams from Northern Michigan and one from Canada.

The IHA lasted only three years and was soon replaced by another league, the National Hockey Association (NHA), founded in 1909, and the Pacific Coast Hockey Association (PCHA), founded in 1911. The NHA discontinued operations in 1916, but the best teams in the league went on to form another league, the National Hockey League (NHL), which continues to operate today. Between 1912 and 1925, winners of the PCHA and NHA competed with each other for dominance in the sport and for the Stanley Cup.

The Stanley Cup was established in 1893 by Lord Stanley of Preston, then governor-general of Canada. Stanley donated the cup to the team that won the professional championship of the world, which first meant the winner of the PCHA/NHL playoffs, and since has meant the winner of the NHL playoffs. Teams who are not members of the NHL are not eligible to compete for the "world championship" in hockey, as represented by the Stanley Cup.

The modern game of ice hockey is played by teams of six players—three forwards (left wing, right wing, and center), two defensemen, and one goalie. The playing area is known as a *rink*, usually about 61 meters (200 ft) long and 26–30 meters (85–98 ft) wide. The rink is divided in half by a "red line," and two "blue lines" about 18 meters (60 ft) from each goal mark the team's defensive zones. The red and blue lines indicate areas where pucks and players may or may not be passed or carried without penalty. Two goals at each end of the rink consist of metal frames about 1.2 meters (4 ft) high and 1.8 meters (6 ft) wide, supporting a rope net. The object of the game is for a player to shoot the puck into the opposite team's net.

The NHL now consists of 30 teams organized into two conferences (the eastern and western), each of which includes three divisions. Members represent not only traditional centers of hockey competitions, such as Montreal and Toronto, but also cities in areas where snow and ice are seldom if ever seen, such as Phoenix, San Jose, and Tampa.

Further Reading

Diamond, Dan, et al., eds. *Total Hockey: The Official Encyclopedia of the National Hockey League*. San Francisco: Total Sports, 1998.

Further Information

National Hockey League
1251 Avenue of the Americas
New York, NY 10020
URL: http://www.nhl.com

Rowing

Rowing has presumably been used as a mode of transportation for peoples living on or near bodies of water from time immemorial. The *Grolier Encyclopedia* says that the earliest record of rowing as an aquatic sport, is to be found in Virgil's *Aeneid*, where a race was held at the funeral games held in honor of Aeneas' father (*1998 Grolier Multimedia Encyclopedia*, "rowing").

By the early 1700s, competitive rowing had become a popular sport in England, especially on the Thames River. As an example, the actor Thomas Doggett (died 1715) provided a stipend in his will for a prize to be given annually to the winning six-man crew in a race from London Bridge to Chelsea, a distance of 4 ½ miles. That race is still held today and is known as the Doggett's Coat and Badge Race. The race between Oxford and Cambridge universities—arguably the most famous rowing contest in the world—was first held in 1829, and continues as an annual event today.

The first recorded rowing competition in the United States was held in New York city in 1811, and the first intercollegiate competition in 1852, between Harvard and Yale universities. The first formal rowing club, the Philadelphia Schuylkill Navy, was formed in 1858. The club was also the first amateur sports organization established in the United States.

Rowing was introduced to the Olympic Games in 1896, although no competition was held that year due to bad weather. Women were largely excluded from early rowing clubs and competition, and the first women's rowing organization in the United States, the National Women's Rowing Association, was not established until the early 1960s. The first Women's World Rowing Championship was held in 1974, and two years later, the sport was introduced into the Olympic Games.

Modern rowing competitions are divided into two major groups: sweeps and sculling. Sweep events consist of crews of 2, 4, or 8 rowers, each of whom holds a single oar. Sculling events consist of 1, 2, or 4 rowers, each of whom holds two oars. Crews of two or more rowers may also have a coxswain who directs the crew and steers the boat.

Rowing events are classified according to the type of boat (sweep or scull), the number of rowers, and the presence or absence of a coxswain. For example, an event for sculls rowed by four men without a coxswain is called an *M4-event*; one involving eight women with a coxswain is known as an *W8+event*.

Further Information

United States Rowing Association (US Rowing)
201 S. Capitol Avenue, Suite 400
Indianapolis, IN 46225-1068
Telephone: (317) 237-5656 or (800) 314-4ROW
Fax: (317) 237-5646
URL: http://www.usrowing.org
e-mail: members@usrowing.org

Swimming

For peoples who have lived on or near bodies of water, swimming has historically been as important a mode of transportation as has been walking. Mention of swimming is made in the Bible in Ezekiel 47:5, Acts 27:42, and Isaiah 25:11. As early as 2500 B.C., the Egyptians had developed a hieroglyph for the act of swimming. Among the ancient Greeks and Romans, swimming was an important part of military training programs, and the philosopher Plato taught that any person who didn't know how to swim was uneducated. A number of mosaics and paintings from the pre-Christian era suggest that the most common swimming style used by early humans was the so-called *dog paddle*, or dog crawl.

The first published work on the art of swimming was written in 1538 by a German professor of languages, Nicolas Wynman. In 1696, a French author by the name of Thevenot described in his *Art of Swimming* a style that was widely popular among peoples of the time, the *breaststroke*. According to Thevenot, the breaststroke is done by sweeping both arms away from the chest

while propelling oneself with a "frog kick." The swimmer is advised to keep his or her head out of the water as much as possible, at least partly because of a general fear of the (sometimes very real) health risks posed by public waterways. Thevenot's work is credited with helping to make the breaststroke the most popular swimming stroke for at least two centuries.

An evolution in swimming styles in Great Britain began in the 1840s. At the time, the breaststroke was still the most popular swimming style in England. In 1844, however, the British public was exposed to a new and different swimming stroke, now called the *crawl*, by a pair of American Indians named Flying Gull and Tobacco. Observers were appalled by the apparent recklessness of the swimming style, described by one writer as "totally un-European," in which the swimmers "thrashed the water violently with their arms, like sails of a windmill, and beat downward with their feet, blowing with force and forming grotesque antics" ("History of the Sport of Swimming," http://www.usswim.org/media_services/template.pl?opt=new&pubid=210).

In fact, it was not until 1873 that the English public began to adopt the crawl for their own swimming style. One J. Arthur Trudgen returned to England from a trip to South America with reports of a type of crawl used by South American Indians. Before long, many British swimmers had adopted the overhand stroke introduced by Trudgen, although they continued to use the frog kick for propulsion.

The final stage in the revolution in English swimming came three decades later when an Englishman-turned-Australian, Frederick Cavill, demonstrated the efficiency of combining the overhand stroke with an up-and-down "flutter kick." The final step needed to convince the English came in 1902 when one of Cavill's sons, Richard, set a new record in London using the new stroke—58.8 minutes for the 100-yard swim. The stroke soon became called the *Australian crawl*. Today, the Australian crawl is best known as the stroke most commonly used in *freestyle swimming*.

The fourth swimming stroke that is now a part of the Olympic Games competition, the *butterfly stroke*, was invented in 1934 by David Armbruster, coach at the University of Iowa. In the butterfly, the swimmer raises both arms out of the water at the same time, while being propelled forward with a vigorous kicking action.

The origin of competitive swimming is not known, although there is evidence of swimming races being held in Japan in 36 B.C. Most authorities credit the English with developing the modern sport of competitive swimming in the 1830s. At the time, a number of indoor pools had been built in London, and the National Swimming Society had been organized to regulate competition.

Swimming was included in the first modern Olympic Games held in Athens in 1896. The Hungarian Alfred Hajos won both the 100-meter and the 1500-meter freestyle events, and the Austrian Paul Neumann won the 400-meter freestyle, the only swimming style then included in the games. The backstroke first became an Olympic event in 1900, and the breaststroke and butterfly stroke were added in 1968.

Today's competitive events involve some combination of the four primary strokes with a short distance, a long distance, or a team of swimmers for both men and women. Some of the 35 swimming events in the 2000 Olympic Games were the 50-, 100-, 200-, and 400-meter freestyle for men and women, the 100- and 200-meter butterfly for men and women, and the 4 × 200-meter freestyle relay for men and for women.

Further Reading

Water Sports. New York: Arno Press, 1979.

Further Information

USA Swimming
One Olympic Plaza
Colorado Springs, CO 80909
Telephone: (719) 578-4578
URL: http://www.usa-swimming.org

Synchronized Swimming

Synchronized swimming is an aquatic sport in which athletes perform choreographed exercises to music. The synchronization suggested by the name refers not only to the way in which two or more swimmers perform with each other but also to their ability to swim along with the music. Synchronized swimming involves single performers, duets, and teams of eight swimmers.

Swimmers are judged in three categories: figures, technical routine, and freestyle. The figures portion of a competition consists of twenty combinations of movements, of which swimmers are required to perform four; the technical competition requires swimmers to demonstrate proficiency in certain basic elements of the sport; and the freestyle portion of the competition allows swimmers to create and carry out their own choreographic composition. Each portion of the competition is judged on a scale of 1 to 10, with points based on technical merit and artistic expression.

Synchronized swimming grew out of an earlier swimming style known as *water ballet*, first performed in the United States in the early twentieth century. In water ballet, swimmers (usually women) performed creative underwater "dances" for spectators. The first water-ballet club was formed by Katherine Curtis in 1923 at the University of Chicago. Curtis's swimmers performed as the Modern Mermaids at the 1934 World's Fair in Chicago.

The first synchronized swimming competition was held in 1939 between Wright Junior College of Iowa and the Chicago Teacher's College, and the first competition involving more than two teams was held a year later in Wilmette, Illinois. In 1941, the U.S. Amateur Athletic Union accepted synchronized swimming as a part of its approved program of sports.

The first formal international competition in the sport was held at the 1955 Pan-American Games, and it became part of the Summer Olympic Games in 1984. At those games, solo and duet events were held, with Americans winning both competitions. The last games, held in Sydney, Australia, offered competition in duet and team events, but not in the solo event.

Further Information

Federation Internationale de Natation
Av. de l'Avant-Poste 4
1005 Lausanne
Switzerland
Telephone: (+41) 21-310-4710
Fax: (+41) 21-312-6610
URL: http://www.fina.org

Synchro Swimming USA
201 S. Capitol Ave., Suite 901
Indianapolis, IN 46225
Telephone: (317) 237-5700
URL: http://www.usasynchro.org
e-mail: webmaster@usasynchro.org

Water Polo

Water polo arose in England during the second half of the nineteenth century when a number of land-based games were being adapted to aquatic conditions. Some of those games became known as water rugby, water handball, water soccer, and water polo. In some instances, water polo was played on water much like it was played on land, with players riding floating barrels painted to look like horses and striking a floating ball with wooden sticks. Over time, however, the game we now know of as water polo evolved more as a form of rugby than of land-based polo.

From its beginning, water polo was one of the most aggressive and roughest sports ever designed by humans. At first, the objective of rugby-style water polo was for a player to place a small Indian rubber ball at one end of the pool in which the game was being played. The goalie was allowed to stand on the edge of the pool and prevent a score by whatever means possible.

In 1880, the Scots introduced a ten-by-three-foot cage for the goal into which the ball had to be thrown, and the small ball was replaced with a soccer ball. Also, soccer-style rules, rather than those used in rugby, were adopted, allowing for the use of more sophisticated ball handling, passing, and teamwork. Late in the same decade, the Scottish rules were adopted throughout Great Britain.

Water polo was introduced into the United States in 1888 when an English swimming teacher, John Robinson, organized a team at the Boston Athletic Association. Shortly thereafter, teams were also formed in Providence, Rhode Island, and New York City. The first competition was held in this country on 28 January 1890, when the Providence team defeated the Boston team 2 to 1.

The game also spread to Europe in the late nineteenth century, with the first teams being formed in Hungary in 1889, Belgium in 1890, Austria and Germany in 1894, and France in 1895. The first Olympic Games competition in water polo was held as an exhibition sport at the Paris Games in 1900, with Great Britain taking home the gold medal. The United States was the only competitor in the 1904 game and was awarded the gold medal.

Water polo became popular among women in the early twentieth century, but was dropped by most swimming clubs and colleges in the 1920s because it was thought to be too rough for women. The game was revived in the late 1950s by Rose Mary Dawson, women's coach at the Ann Arbor, Michigan, Swim Club. The first world's championship competition for women was held in 1979 in Merced, California, with the U.S. team winning first place.

The modern game of water polo consists of four quarters, each lasting between five and eight minutes, depending on the level of competition. The goal is to advance the ball by passing it from one swimmer to another or by carrying it on a wave between a swimmer's arms and propelling it into the goal. As in soccer, only the goalie is allowed to touch the ball with both hands.

Further Information

Federation Internationale de Natation
Av. de l'Avant-Poste 4
1005 Lausanne
Switzerland
Telephone: (+41) 21-310-4710
Fax: (+41) 21-312-6610
URL: http://www.fina.org

United States Water Polo, Inc.
1685 West Uintah
Colorado Springs, CO 80904-2921
Telephone: (719) 634-0699
Fax: (719) 634-0866
URL: http://www.usawaterpolo.com
e-mail: bwigo@uswp.org

Waterskiing

Waterskiing is a sport in which a person planes across the water's surface on a specially designed ski while being towed by a powerboat, a large sail, or some other device. Waterskiing was invented in 1922 by Ralph Samuelson, of Minnesota. Samuelson attached 8-foot-long barrel staves to his feet with leather straps and was towed by an outboard-powered boat across Lake Pepin, Minnesota. Samuelson later gave demonstrations of the new sport in Michigan and Florida, including the first jump over a 5-foot-high float. New Yorker Fred Waller received the first patent for water skis in 1925. The first waterskiing competition was

held in Massapequa, New York, in 1936, and was won by Jack Andresen. Three years later, the American Water Ski Association was organized and held its first national championship at Jones Beach on Long Island, New York.

Waterskiing consists of three basic events: slalom, jumping, and trick skiing. The sport has also expanded and evolved to produce a number of related events. These include:

Barefoot skiing, in which a person skis barefoot on the water without use of any type of skis

Hydrofoiling (also known as *air chairing* and *sky skiing*), in which the rider sits on a chair mounted on an extra-wide board and travels just above the surface of the water

Kite boarding (also known as *kite surfing*), in which the skier is pulled by a kite or sail rather than a boat

Kneeboarding (also known as *hydrosliding*), in which the rider, in a kneeling position, is pulled on an extra-wide board

Wakeboarding (also known as *skurfing*), in which the rider uses a single board, whose design is between that of water skis and a kneeboard.

Further Information

USA Water Ski
1251 Holy Cow Road
Polk City, FL 33868
Telephone: (863) 324-4351
Fax: (863) 325-8529
URL: http://www.usawaterski.org
e-mail: usawaterski@usawaterski.org

Windsurfing

Windsurfing is an aquatic sport in which an individual sails by standing on a flat board to which is attached a moveable mast and sail. Windsurfing is also known as boardsailing, sailboarding, windsailing, and windgliding. The principle of windsurfing was developed independently in the 1960s by Newman Darby and the team of Jim Drake and Hoyle Schweitzer. In both cases, the inventors were trying to adapt existing equipment for a new application.

Beginning in the 1940s, Darby explored the possibility of developing a new kind of sailboat that could be steered without a rudder. By the early 1960s, he had designed a surfboard-shaped craft to which was attached a mast and sail. The mast was connected to the board by means of a universal joint, allowing the rider to move the mast and sail in any direction. Darby and his wife Naomi began manufacturing and selling their so-called sailboards in 1964, but failed to attract a very large market for their product. By the end of the decade, the Darby's had discontinued production of their sailboards.

Shortly after the Darbys began building their sailboards, Drake and Schweitzer became interested in finding a way to combine the two sports in which they were interested, sailing (Drake) and surfing (Schweitzer). They started with a surfboard, to which they attached a mast and sail by means of a universal joint, similar to the sailboard built by the Darbys. They received a patent for their invention in 1970 and began construction of a craft which they named the Windsurfer. They experienced greater commercial success than had the Darbys, and within three years, windsurfing became widely popular throughout the United States and Europe. Indeed, according to one authority, one in every three European households owned at least one sailboard by the late 1970s ("History of Windsurfing," http://inventors.about.com/science/inventors/library/inventors/blwindsurfing.htm).

Today, windsurfing is classified into one of two general categories: light-wind and high-wind windsurfing. Light-wind windsurfing makes use of winds of 10 knots or less, and is an easier and more relaxing way to travel. High-wind windsurfing requires winds of at least 10 knots—usually between 15 and 25 knots. Among the special events that can take place at high winds are wave sailing, "bump and jump" sailing, and slalom sailing.

Windsurfing was introduced into the Olympic Games in 1984 at Los Angeles. The winner of that competition was Stephan Van Den Berg of the Netherlands. Competition for women began with the 1992 games in Barcelona. The winner of that competition was Barbara Kendall of New Zealand.

Further Information

American Windsurfing Industries Association
1099 Snowden Road
White Salmon, WA 98672
Telephone: (800) 963-7873
Fax: (509) 493-9464
URL: http://www.awia.org
e-mail: info@awia.org

Yachting

Merriam Webster's Collegiate Dictionary (tenth ed.) describes a yacht as "any of various recreational watercraft" or "a sailboat used for racing." The use of sailing boats for carrying people or cargo, for naval purposes, and for pleasure sailing and racing dates at least to Egyptian, Greek, and Roman civilizations. But the term *yacht* itself appears to have been derived from the Dutch term *jaght*, which first appeared in 1557. At first used as a fast naval ship, the jaght was later adapted for pleasure sailing and racing. The ship was then introduced to England by King Charles II in 1660 upon his return from exile in the Netherlands. The first yachting clubs in Great Britain were the Water Club of Cork Harbor (founded in 1720), the Cumberland Fleet Club (later, the Royal Thames Yacht Club; 1775), and the

Yacht Club (1815). The oldest yacht club in North America is the Royal Nova Scotia Yacht Squadron, founded in 1837.

The most famous yacht-racing contest in the world is the America's Cup. The America's Cup originated in 1851 when the Royal Yacht Club invited the New York Yacht Club (NYYC) to send a representative to its annual race around the Isle of Wight. The winner of that race was awarded the Hundred Guinea Cup. The NYYC boat, *America*, won the race in a time of 10 hours and 37 minutes, 18 minutes ahead of the next closest boat. The Hundred Guinea Cup was brought to the United States and awarded to the New York Yacht Club, to be retained as a challenge for future international competitions.

The next such competition did not take place for nearly two decades. In 1870, the English yacht *Cambria* competed against fourteen boats from the NYYC for the Hundred Guinea Cup, now renamed the America's Cup. The *Cambria* finished eighth in the contest.

Over the next 125 years, competitions for the America's Cup were held a total of twenty-nine times at irregular intervals. From 1870 through 1903, races were held about once every few years, but there was no competition between 1904 and 1919 and between 1921 and 1929. Between 1930 and 1957, only three races were held, and since 1958, there have been 13 more competitions. Each has taken place at an interval of about three years. During this period of time, a United States yacht won every contest except one against competitors from England, Canada, Scotland, Northern Ireland, Australia, New Zealand, and Italy. Finally, in 1995, the *Black Magic* from New Zealand wrested the cup from the United States.

The America's Cup races have involved some of the largest recreational sailing ships ever built, with lengths of up to about 50 meters (160 ft). Other types of sailing contests have and are conducted regularly for smaller ships. Such contests include the United States to Bermuda, California to Hawaii, Port Huron to Mackinac, Miami to Nassau, and similar races throughout the world. Ships in the 10-meter (35 ft) to 20-meter (70 ft) range generally take part in such races.

Yacht racing first appeared in the 1896 Olympic Games held in Athens. That competition involved sailing ships in the eight- and six-meter classes. Over the years, Olympic sailing contests have evolved to include a wide variety of boats ranging from one- and two-person dinghies to windsurfers and keelboats.

Further Information

American Sailing Association
13922 Marquesas Way
Marina del Rey, CA 90292
Telephone: (310) 822-7171
Fax: (310) 822-4741
URL: http://www.american-sailing.com
e-mail: info@american-sailing.com

New York Yacht Club
37 West 44th Street
New York, NY 10036
Telephone: (212) 382-1000
Fax: (212) 302-1295
URL: http://www.nyyc.org
e-mail: library;llnyyc@msn.com

Aqueducts

An aqueduct is an artificial channel built to carry water from some distant area to a populated area. The word *aqueduct* comes from two Latin terms meaning "to carry" (*-duct*) "water" (*aqua-*).

History of Aqueducts

In the most primitive human civilizations, people lived close to a surface water supply, such as a lake or river. Where such water sources were not available, they may have had to dig a well to bring groundwater to the surface for their use.

Eventually, some communities found that the water supplies close at hand were not sufficient to meet their needs. In such cases, they needed to find ways to bring water to the community from more remote sources. Aqueducts were the solution to this problem.

The first aqueduct was built in Assyria in 691 B.C. It consisted of a 55-kilometer-long channel made of stone and supported on an arch built across a valley. The aqueduct was the solution to moving water across the valley to a city. As urban areas grew in size, aqueducts became increasingly necessary components of public water supplies. The ancient Greeks, in particular, became very skilled at building aqueducts that made use of siphons, underground tunnels, pumps, and other hydraulic devices.

The most famous system of aqueducts in the ancient world was the one built by the Romans. Rome was a city that used huge amounts of water. In addition to personal and community applications, water was used in a number of large public fountains. To bring all this water to the city, a system of eleven aqueduct systems were built, delivering 38 million gallons of water to the city each day. The system made use of buried pipes, channels supported by huge arches, and surface canals and pipelines. Roman architects constructed aqueduct systems in every part of the empire. Remaining examples of their work can be found across Europe, from Great Britain to Spain to Turkey.

Modern Aqueducts

Aqueducts have continued to be an important part of municipal water systems in many parts of the world. One of the first modern systems was built in 1853 to bring water to Washington, D.C. The city had experienced two serious fires in the preced-

ing year and found that its water supplies were inadequate to fight them. Over the next decade, a huge system was built to bring water from Georgetown, 12 miles upstream on the Potomac. The aqueduct system built during that period is still in operation and brings 300 million gallons of water to Washington each day.

Probably the largest aqueduct system in use today is the one that brings water to the metropolitan Los Angeles area. The original system was completed in 1913 and consisted of 233 miles of pipes, tunnels, canals, and other water-carrying systems. It drew primarily on waters available from the Colorado River Basin. In 1940, the system was extended an additional 105 miles to Mono Lake. In 1970, a second aqueduct system was constructed to bring water from the eastern Sierra watershed to the city. It increased capacity of the system by 50 percent and now provides the metropolitan Los Angeles area with about 430 million gallons of water per day. *See also* Canals; Irrigation; Plumbing

Further Reading

Derry, T. K., and Trevor I. Williams. *A Short History of Technology from the Earliest Times to A.D. 1900.* New York: Dover Publications, 1993, passim.

Van Deman, Esther Boise. *The Building of the Roman Aqueducts.* Washington, D.C.: Carnegie Institution of Washington, 1934.

Aqueous Solution

An aqueous solution is a homogeneous mixture in which a solid, liquid, or gas is dissolved in water. A solution, in general, is a homogeneous mixture of any two or more substances in the gaseous, liquid, or solid phase. For example, an alloy is a solid solution in which two or more metals are mixed homogeneously with each other. Aqueous solutions, those containing water, are by far the most common type of solution known.

A solution consists of two parts: (1) the component present in the largest quantity and said to be responsible for the dissolving process, and (2) the component present in the smaller quantity and said to have been dissolved. The former is called the *solvent* and the latter, the *solute*. In a saline solution of salt dissolved in water, for example, water is the solvent and salt, the solute.

The Process of Dissolving

A common rule of thumb for the formation of solutions is that "like dissolves like." In chemical terms, this rule means that substances composed of particles carrying an electrical charge tend to form solutions with other substances composed of particles carrying an electrical charge, while those composed of electrically neutral particles tend to form solutions with others composed of electrically neutral particles.

This rule can be used to explain how two "like" substances dissolve in each other, as, for example, the dissolving of salt in water. Salt (sodium chloride, NaCl) consists of tiny charged particles known as *ions*. An ion is an atom with an electrical charge. Salt consists of sodium ions (Na^+) and chloride ions (Cl^-). Water molecules are *polar* molecules—that is, molecules that are positively charged at one end and negatively charged at the other end. A simplistic way of representing a water molecule is as $^{\delta+}H - O^{\delta-} - H^{\delta+}$, where $^{\delta+}$ and $^{\delta-}$ represent a partial positive and partial negative charge, respectively.

When water is added to salt, water molecules are attracted to the sodium ions and chloride ions of which salt is made. The partially positive end of a water molecule is attracted to the negatively charged chloride ion ($^{\delta+}H - O^{\delta-} - H^{\delta+} \ldots Cl^-$) and the partially negative end of a water molecule is attracted to the positively charged sodium ion.

$$^{\delta+}H - O^{\delta-} \cdots Na^+$$
$$|$$
$$H^{\delta+}$$

Because of these forces of attraction, each chloride ion is eventually surrounded by a number of water molecules and each sodium ion is surrounded by a number of other water molecules. The ions are said to be *hydrated*. One can express this condition by showing the number of water molecules surrounding any given hydrated ion, as, for example: $Cl^-(H_2O)_6$. When ions are hydrated by this process, they are isolated from each other and remain in suspension as they are carried by thermal motion throughout the solution.

The "like dissolves like" rule also explains why many substances made of nonpolar (uncharged) particles do not dissolve in water. When oil is added to water, for example, the two liquids separate into two layers rather than forming a solution. The reason is that the force of attraction between water molecules is much stronger than the force of attraction between a water molecule and an uncharged molecule of oil.

Factors Affecting Solubility

The *solubility* of a substance can be defined as the extent to which that substance will dissolve in a solvent or, in the case of aqueous solution, the extent to which it will dissolve in water. The solubility of sodium chloride in water, for example, is 36 g/100 mL (grams per 100 milliliter) at 20°C.

The solubility of a substance may depend on three factors. The first factor, as is obvious from the preceding discussion, is the nature of the solvent in which it is to be dissolved. Sodium chloride dissolves well in water because both consist of charged particles, but sodium chloride does not dissolve in ether because ether consists of nonpolar particles.

A second factor affecting the solubility of a substance is temperature. This fact explains the necessity of listing the solubility of a substance at some given temperature, as in the case for sodium chloride above. Generally speaking, an increase in the temperature increases the solubility of a solid in water. For example, the solubility of sodium chloride in water is about 40 g/100 mL at 100°C compared to only 36 g/100 mL at 20°C.

The solubility of some substances increases quite dramatically with a rise in temperature. For example, the solubility of sodium hydroxide (NaOH) goes from about 42 g/100 mL at 0°C to 347 g/100 mL at 100°C. In a few cases, an increase in temperature may result in a *decrease* in solubility. The solubility of cerium sulfate ($Ce_2(SO_4)_3$), for example, drops from about 10 g/100 mL at 0°C to 2.5 g/100 mL at 100°C.

The effect of temperature on solubility for gases is the reverse of that for solids. That is, gases tend to be less soluble at higher temperatures than at lower temperatures. Carbon dioxide (CO_2), for example, has a solubility of about 0.335 g/100 mL at 0°C and 0.169 g/100 mL at 20°C.

For gases, pressure is also a factor in determining solubility. In general, the greater the pressure on an aqueous solution, the greater the solubility of a gas in water. For example, the solubility of nitrogen gas (N_2) at normal air pressure (one atmosphere) is about 16 ml/l, while at six times that pressure, its solubility is about 80 ml/l.

Expressions of Concentrations

Expressing the relative amounts of solute and solvent in a solution is important to chemists and other scientists. A rather extensive terminology has developed to deal with this issue.

Qualitative Terminology

In the first place, solutions can be described as *dilute* or *concentrated*. A dilute solution is one that contains a relatively small amount of solute compared to solvent, while a concentrated solution is one that contains a relatively large amount of solute compared to solvent. These terms are satisfactory for general descriptions, but have little or no value when specific quantitative descriptions are needed.

Solutions may also be described as *saturated, unsaturated,* or *supersaturated* depending on the amount of solute they contain compared to the maximum amount they *could* contain. Consider the case of a saltwater solution made from 36 g of sodium chloride dissolved in 100 mL of water at 20°C. From information provided above, it is clear that the maximum amount of sodium chloride has been dissolved in this amount of water at this temperature. The solution, then, is said to be *saturated*.

But consider a solution made from any quantity of sodium chloride *less than* 36 g, such as a solution made from 5 g of sodium chloride in 100 mL of water at 20°C. A solution of this concentration is said to be *unsaturated* because additional sodium chloride could still be dissolved in it.

Finally, it is sometimes possible to make a solution that is supersaturated. In the case of a sodium chloride solution, this apparently oxymoronic condition can be produced by adding 40 g of sodium chloride to 100 mL of water and then raising the temperature of the mixture to 100°C. From information provided above, it is apparent that the 40 g of sodium chloride will dissolve completely at 100°C.

Now, assume that this saturated (at 100°C) solution can be cooled very slowly and very carefully so that no sodium chloride deposits out of solution. This step is possible with some solutes, but not with others. If the solution *could* be cooled to 20°C without deposition of any solute, however, one would have a solution containing 40 g of sodium chloride in 100 mL of water at 20°C, a *supersaturated* solution. Supersaturated solutions are very unstable. Simply agitating them or adding a tiny crystal more of solute will cause the excess solute to settle out, leaving behind a saturated solution.

Quantitative Terminology

Scientists usually do not find terms such as *dilute* or *concentrated, saturated* or *unsaturated*, to be very helpful in their work. They require other methods for designating the concentration of a solution that give the precise amount of solute per unit of solvent. A number of systems have been developed to meet this need.

A *percentage concentration* solution is one that tells the ratio of solute to solvent, expressed in mass (weight) or volume or some combination of the two. For example, suppose that a solution is made by dissolving 5 grams of sodium chloride in enough water to make a solution weighing 100 grams. The *weight-weight (w/w) percentage concentration* of that solution is then 5 g (NaCl) ÷ 100 g solution = 5%. Similarly, a solution could be made by dissolving 7.5 mL of ethyl alcohol in enough water to make a solution with a volume of 100 ml. The *volume-volume (v/v) concentration percentage* of this solution is 7.5 mL (alcohol) ÷ 100 mL solution = 7.5%.

Variations on these two kinds of percentage composition solutions have been devised to meet special needs in various areas of science. For example, biologists and those in the health sciences sometimes use *weight/volume (w/v)* solutions, which express concentrations in terms of a certain number of grams per 100 mL of solution.

Perhaps the most widely used system for designating solution concentrations in science is that of *molarity*. In chemistry, a *mole* is the basic unit for measuring the amount of an element, compound, or other substance. A *molar solution*, then, is one that contains one mole of any given substance per liter of solu-

tion. A closely related system is called *molality*, and is based on the number of moles per kilogram (rather than liter) of solution. Since the mass of one liter of water is one kilogram, molarity and molality are very nearly the same for all aqueous solutions.

Colligative Properties

The addition of a solute to pure water changes some of the physical properties of the water. For example, the boiling point of pure water is 100°C. But water to which sodium chloride has been added boils at a temperature higher than 100°C. The amount by which the boiling point is elevated depends on the kind and amount of substance added.

The reason for this effect is not difficult to understand. To boil water, enough heat must be added to make water molecules move fast enough to escape from the liquid. The presence of molecules and ions *other than* water molecules means that some of the heat that has been added goes to making these molecules and ions move faster also. To achieve the desired result (getting water molecules moving fast enough to escape), then, more heat must be added than is the case with pure water.

Changes that occur in the physical properties of a solvent, such as water, because of the addition of a solute are called *colligative properties*. Elevation of boiling point, depression of freezing point and vapor pressure, and changes in the osmotic pressure exerted by a liquid are typical colligative properties.

Practical Applications

The significance of aqueous solutions in the natural world, in medicine, in industry, and in other aspects of human life is immeasurable. In the natural world, for example, pure water is rarely if ever found. Water is such a good solvent that it dissolves at least a small amount of many different substances whenever it comes into contact with air, the ground, or any other substance. When one speaks about water in the natural world, then, one is really referring to some type of aqueous solution.

Aqueous solutions are important in research, industry, and other fields because they provide a way of bringing about a chemical reaction between two or more other substances. Most chemical reactions used to produce some product occur in solution, often aqueous solution. Aqueous solutions are also essential in medicine and the health sciences because they provide a way of delivering some substance, such as a drug, to a patient. *See also* Hydrate; Hydrogen Bonding; pH; Properties of Water

Aquifer

See Groundwater

Archimedes (287–212 B.C.)

Archimedes was a Greek mathematician, natural philosopher, and engineer. One of his most lasting discoveries was the law of floating bodies, now known as *Archimedes' principle*. In addition, he is reputed to have invented a device for raising water from a lower level to a higher level, a device we now call the *Archimedes' screw*.

Archimedes was born in Syracuse—a Greek city-state in Sicily—in 287 B.C., the son of Phidias, an astronomer. For his education, he traveled to Alexandria, where he studied with the Greek mathematician Conon, once a pupil of the great Euclid. No biography of Archimedes currently exists, although one was written during his lifetime by his friend Heracleides. Still, a great deal is known about Archimedes because of the frequent mention of his work in the writings of his contemporaries. In addition, a number of Archimedes' own works remain, including *The Sandreckoner, On Floating Bodies, On the Sphere and the Cylinder, Measurement of a Circle, On Spirals, On Concoids and Spheroids*, and *The Method*.

One of the many stories about Archimedes' life found in contemporary works tells of how he was killed. It is said that he was working on a mathematical problem at the time a Roman army landed in Syracuse. To aid in the solution of this problem, Archimedes had drawn some geometrical figures in the sand. When a Roman soldier seemed about to walk across those figures, Archimedes is said to have warned him off with a cry of "Don't disturb my circles!" Unwilling to be ordered about by an old man, the solider drew his sword and killed Archimedes on the spot.

Archimedes' Screw

Archimedes is often credited with the invention of a device for lifting water that bears his name. There is evidence, however, that the device had been used by the Egyptians long before Archimedes was born. The screw consisted of a long hollow cylinder containing a rod through its center. Attached to the rod were thin strips of wood arranged in a spiral pattern. The strips of wood were arranged so that they fit tightly against the inside walls of the cylinder. A crank was attached to the rod so that it could be turned within the cylinder.

To use the screw, one end was placed in a source of water, such as an irrigation ditch; the other end was placed in a trench, a container, or some other place to which water was to be transferred. The handle on the rod was then turned. As the screw turned, water was trapped by the wooden strips and lifted upward to the higher end of the cylinder. When it reached the top of the cylinder, it flowed out into the trench or container.

The screw is one of the simplest methods available for lifting water. It is still used

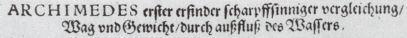

Eighteenth-century depiction of Archimedes experimenting with water. *Archivo Iconografico, S.A./Corbis.*

widely in many parts of the world where water pumps are too expensive or not available for use.

Archimedes' Principle

Many middle and high school students today have heard about the discovery of Archimedes' principle. The philosopher had been asked by the King of Syracuse, Hieron II, to find out whether a crown he had been given was actually made of pure gold, as the king had requested. Archimedes was faced with solving this problem, without taking the crown apart or otherwise damaging it.

For some time, the philosopher was unable to think of a way to meet Hieron's request. Then one day, the answer came to

him in a flash. He is reputed to have been sitting in his bathtub when the solution came to him. As he lowered himself into the tub, a certain amount of water overflowed and spilled out of the tub. The amount of water his body displaced, Archimedes realized, was a function of his own weight and volume.

If that were true, he could use the same process to test the purity of gold in the king's crown. The first step was to place the crown in a container of water filled to the brim and to measure the amount of water displaced. The second step was to place an equal amount of gold in the same container of water, refilled to the brim, and measure the amount of water displaced. If the two volumes of displaced water were the same, the crown was really made of pure gold. If not, the gold had been mixed with a less valuable metal such as silver, which turned out to be the case.

Archimedes was so excited about his discovery that he is said to have jumped from the bathtub and run through the streets of Syracuse naked, shouting "Eureka, eureka"—that is, "I have it, I have it." His solution to King Hieron's puzzle brought him everlasting fame, though it also brought death to the dishonest goldsmith who had made the crown.

Today, Archimedes' principle is stated in a more formal way, although its meaning is the same as in the original story. A modern statement of the principle is that an object immersed in a fluid is pushed up with a force that is equal to the weight of the displaced fluid.

Further Reading

Archimedes Home Page, http://www.mcs.drexel. edu/~corres/Archimedes.

Dijksterhuis, E. J. *Archimedes*. Translated by C. Dikshoorn. Princeton, NJ: Princeton University Press, 1987.

Heath, Sir Thomas Little. *Archimedes*. New York: Macmillan, 1920.

Stein, Sherman K. *Archimedes: What Did He Do Besides Cry Eureka*? Washington, D.C.: Mathematical Association of America, 1999.

Archimedes' Principle

See Archimedes

Archimedes' Screw

See Archimedes

Artesian Well

See Groundwater

Atlantis

Atlantis is a mythical island-continent originally described by Plato in two of his dialogues, *Timaeus* and *Critas*, in about 360 B.C. The island was supposed to have existed in the Atlantic Ocean, west of the Strait of Gibraltar.

Plato's Descriptions

Plato tells two somewhat different stories about Atlantis in the two dialogues. In *Timaeus*, he describes a rich nation that controlled not only the island of Atlantis but also regions of the European and African continents. He said that the rulers of Atlantis made a terrible mistake when they tried to conquer Greece as well, as they were defeated, and had to retreat to their island home. Shortly thereafter, the island was destroyed by a terrible earthquake in which all traces of the nation disappeared.

In *Critas*, Plato says that Atlantis was ruled by Poseidon, the Greek god of the oceans. He provides extensive details of a rich, powerful, and beautiful island with fertile plains, extensive mountains, villages, lakes and rivers, and the primary city of Atlantis itself. For many centuries, the people who occupied Atlantis were good and virtuous. Over time, however, they became immoral and sinful. Finally, Zeus became so angry with the state of affairs on the island that he sent a terrible earthquake that destroyed the whole nation.

Plato claims that he learned about Atlantis from earlier legends dating back at least two thousand years. Some authorities think that those legends may have been based on actual volcanic eruptions that tore apart islands in the Mediterranean or Aegean seas. For example, in 1740 B.C., the Cycladean island of Santorini was wracked by a volcanic eruption, which completely demolished the central core of the island. Today, all that remains of Santorini is the shell of the mountain in which the volcano existed.

Later Stories about Atlantis

Plato's stories about the legendary island have captured the imagination of individuals for centuries. Over the years, many stories have sprouted up that embellish on his original descriptions. Late in the nineteenth century, for example, Ignatius Donnelly, a U.S. congressman from Minnesota, wrote a book entitled *Atlantis: The Antediluvian World*, which became widely popular. People throughout the United States and other parts of the world developed a renewed interest in the possibility of this "lost continent."

Many other writers have added their own interpretations to Plato's original stories. For example, some people today believe that Atlantis was originally founded by extraterrestrial beings who knew about aircraft, explosives, advanced methods of farming, and other modern developments. Some writers report that they have had extrasensory contact with those beings and can prophesy based on information obtained from such contacts. Many of these individuals believe that other now-lost continents existed in the past. The three most often mentioned are Mu and Mar in the Pacific Ocean and Lemuria in the Indian Ocean.

Scientific expeditions have been organized for the purpose of obtaining evidence for the existence of Atlantis. One website in 1999, for example, reported on the findings of a research team from Pepperdine University that had located vast numbers of remains from the island-continent at the bottom of the Atlantic Ocean. The report described "roads, domes, rectangular buildings, unidentified metallic instruments, and a statue holding a 'mysterious' crystal containing miniature pyramids" ("Atlantis—An Introduction").

The vast majority of mainstream scientists have not been convinced by this research, however, and few if any would give credence to any factual basis for the story of Atlantis.

Further Reading

Asimov, Isaac, Martin H. Greenberg, and Charles G. Waugh, eds. *Atlantis*. New York: New American Library, 1988.

"Atlantis—An Introduction," htty://www. newage.com.au/library/atlantis2.html.

Donnelly, Ignatius. *Atlantis: The Antediluvian World*. New York: Dover Publications, 1985.

Ellis, Richard. *Imagining Atlantis*. New York: Knopf, 1998.

Autoclave

See Steam Pressure Sterilizer

B

Baptism

See Biblical References to Water

Barges

A barge is a large boat, usually flat-bottomed, once used primarily for transporting goods on inland waterways. The earliest barges of which we have concrete evidence are those used by the Egyptians in the third millennium B.C. They were employed to transport large pieces of stone from Upper Egypt to building sites in Lower Egypt, such as those near the ancient capital of Memphis. Barges were also used extensively by the Romans to deliver grain from seaports to cities and villages throughout the country.

By the late eighteenth century, barges were becoming a major means of transportation in many European countries. In England, for example, the inventor John Wilkinson (1728–1808) built a 70-foot barge made of bolted cast-iron plates to be used on the Severn River. Barges were used on both natural rivers and man-made canals that connected towns not otherwise reached by rivers.

In most cases, barges have no independent means of locomotion. In earlier times, they were pulled along a waterway by one or more horses, donkeys, or humans. Over time, some regions developed towboats, which pulled themselves along by means of a chain laid on the bottom of the river.

With the invention of steam power in the late eighteenth century, a new type of boat was invented for the sole purpose of pulling barges—the *tugboat*. Tugboats were small vessels with pointed bows designed to cut easily through the water. They were capable of pulling up to a half dozen barges at a time. Comparable vessels today are powered by diesel engines and capable of pulling hundreds of tons of cargo packed on barges along inland waterways and along coastal areas of oceans and large lakes. They are still used to move large shipments of bulk cargo, such as coal and iron ore.

One of the most important modern uses for barges is in supplying LASH (lighter aboard ship) cargo vessels. Barges designed for this purpose are known as *lighters* and are used to transport cargo down rivers for loading onto larger ocean-going cargo ships.

In many areas where barges were once the primary method of moving cargo—such as the inland waterways in Europe—they have been replaced by other means of transportation, such as trucks and railway cars. Many of those obsolete barges have since been converted to tourist facilities, offering slow trips through rivers and canal systems on which cargo no longer travels.

A 1964 photo of the Thai royal barge gliding up the Chao Phya River in Bangkok.
Bettmann/Corbis.

Throughout history, some types of barges have been provided with their own means of locomotion—that is, oars, sails, or engines. A number of barges traveling on the River Thames and other English waterways, for example, have long been fitted with sails. A number of these sail-powered barges still exist and are most frequently used as pleasure crafts by private owners.

The term *barge* is sometimes used to describe other types of vessels not designed primarily for commercial purposes. For example, large, elaborate, highly decorated ships powered by rowers have long been used for ceremonial purposes in Thailand (formerly Siam). The earliest of these royal barges dates to the Sukhothai period (1238–1438). During the Ayutthaya period (1351–1776), processions of more than thirty royal barges were held during the annual Royal Kathin Ceremony, during which the king and his retinue traveled to royal monasteries. A parade of royal barges is still held in Bangkok each year. *See also* Boats and Ships

Benthos

See Aquatic Plants and Animals

Bernoulli, Daniel (1700–1782)

Bernoulli is one of the founders of the science of hydrodynamics, the study of the motion of fluids and its effect on bodies in contact with those fluids.

Personal History

Bernoulli was born in Groningen, the Netherlands, on 8 February 1700. He came from a long line of eminent mathematicians and physicists. His father, an uncle, two brothers, a cousin, and two nephews were also respected professional mathematicians or physicists. He was educated at the University of Basel, from which he received his bachelor's degree at the age of fifteen and his master's degree at the age of sixteen. He then began to study medicine and was awarded a doctorate at the age of twenty-one for research on the action of lungs.

In 1725, Bernoulli was appointed to the Chair of Mathematics at St. Petersburg Academy. He left Russia in 1732 to take the post of Professor of Anatomy and Botany at the University of Basel. He was later named Professor of Natural Philosophy at the same institution, where he remained until his retirement in 1777. He died in Basel, Switzerland, on 17 March 1782.

Major Accomplishments

Bernoulli's most famous discovery is one that still carries his name, *Bernoulli's principle*. This principle says that as the velocity of a fluid increases over a surface, the pressure it exerts on that surface is reduced. Bernoulli's principle is used in a large variety of phenomena found in nature as well as in many different kinds of machines. For example, air that rushes over the curved wing on an airplane travels more rapidly over the upper surface than the lower surface. As a result, it exerts less pressure on top of the wing than on the bottom. The greater pressure on the bottom of the wing pushes the plane upward, helping it to fly.

Bernoulli first expressed this principle in his most important book on liquids, *Hydrodynamica*. In the book, Bernoulli also devised the first mathematical analysis of the behavior of fluids, based on the assumption that such fluids are based on tiny, discrete particles. This analysis was the first step in the development of the kinetic theory of matter, fully expressed a century later.

Further Reading

Quinney, D. A., "Daniel Bernoulli and the Making of the Fluid Equation," ⟨http://pass.maths.org.uk/issue1/bern/index.html⟩.

Biblical References to Water

Water is mentioned in the Bible many times, both with reference to specific physical bodies of water and, more importantly, with symbolic allusions. Among the metaphorical representations of water in the Bible are those that deal with jealousy, separation, troubles, purification, and salvation. As with many sections of the Bible, references to water are often subject to a wide variety of interpretations. References provided here are taken from the King James Version (KJV) or the New Living Translation (NLT).

Water of Jealousy

The water of jealousy appears in Numbers 5:11–31 as a portion of Jewish law describing the testing of a woman thought to have been unfaithful to her husband. The woman was to be brought by her husband to the temple and made to drink a "bitter water" potion prepared by the priest. If she was guilty of infidelity, she was expected to die at the hand of God, while, if innocent, she was to be restored to her husband.

Water of Separation

The water of separation is another ritual act prescribed for the Jews in Numbers 19:9–23. The water of separation is prepared by mixing the ashes of a burnt sacrifice with water, and is to be used for purifying a person of sins. Any person who touches a dead human body, for example, was thought to be unclean. That person could be purified by being sprinkled with the water of separation.

Water of Troubles

The Bible also refers to water in a number of places as representative of the troubles and afflictions to which humans may be exposed. In Isaiah 43:2, for example, the prophet explains God's promise to protect the Jews from "deep waters and great trouble." "I will be with you," he says. "When you go through rivers of difficulty, you will not be drowned." A similar assurance is found in 2 Samuel 22:17–18. "He reached down from heaven and rescued me," the writer says; "he drew me out of deep waters. He delivered me from my powerful enemies, from those who hated me and were too strong for me" (NLT).

References of this kind sometimes compare floods and storms to the threats and attacks of enemies of the Jews. In Psalms 18, for example, the reader is assured that God will protect him or her from enemies that overcome one like "channels of water." "He sent from above," the writer says, "he took me, he drew me out of many waters. He delivered me from my strong enemy, and from them which hated me; for they were too strong for me" (NLT).

The prophet Jeremiah uses similar imagery. "Thus saith the Lord," he writes; "Behold, waters rise up out of the north, and shall be an overflowing flood, and shall overflow the land, and all that is therein; the city, and them that dwell therein; then the men shall cry, and all the inhabitants of the land shall howl" (NLT).

Water of Purification

Water of separation was only one specific example of a broad array of rituals in which water was used to purify humans and their clothing. Long sections of the book of Leviticus describe the circumstances under which washing with water is to be performed and the precise procedures to be used for such acts of cleansing.

Any form of sexual behavior, for example, was to be followed by ritual cleansing with water. Chapter 15 provides precise rules for the washing of men and women after sexual activity or during a woman's menstrual period. In the latter case, the instructions are that "every thing that she lieth upon in her separation shall be unclean; every thing also that she sitteth upon shall be unclean. And whosoever toucheth her bed shall wash his cloths, and bathe himself in water, and be unclean until the even. And whosoever toucheth any thing that she sat upon shall wash his clothes, and bathe himself in water, and be unclean until the even" (Leviticus 15:20–22 KJV).

The most extensive and complete provisions for cleansing apply to lepers. The whole of chapter 14 in Leviticus is devoted to a description of the procedures by which such washings must be conducted for those who come into contact with lepers.

Over time, the Jews extended the ritual of washing of objects and circumstances not specifically mentioned in the Bible. For example, Mark describes the ritual washing of cups, pots, and hands before a meal, a practice that becomes a source of contention between the scribes and Pharisees and Jesus during his visit to the temple in Jerusalem (Mark 7).

Rites of ablution were hardly unique to the Jews and the early Christians. But, according to some authorities, these acts were imbued with a higher level of symbolism than was the case with those of their contemporaries, such as the Babylonians, Canaanites, Philistines, and others who practiced polytheistic religions. According to one standard reference, "The idea of uncleanliness was not peculiar to the Jew; but with all other nations simple ablution sufficed; no sacrifices were demanded. The Jew alone was taught by the use of expiatory offerings to discern to its fullest extent the connection between the outward sign and the inward fount of impurity" ("Purification," in *Smith's Biblical Dictionary*).

Water of Salvation

Possibly the most significant metaphorical use of water in the Bible is as a symbol of God's spirit, his help and guidance, and his promise of salvation and redemption. This theme is introduced in the Old Testament when the prophets promised that God would wipe away human sin and provide everlasting life. Isaiah quotes God as saying, "For I will gather you up from all the nations and bring you home again to your land. Then I will sprinkle clean water on you, and you will be clean. Your filth will be washed away, and you will no longer worship idols" (Isaiah 36:24–25 NLT).

That theme is developed and further explained in the teachings of Jesus, as reported in the New Testament. "On the last day, the climax of the festival, Jesus stood and shouted to the crowds," John writes, "'If you are thirsty, come to me! If you believe in me, come and drink! For the Scriptures declare that rivers of living water will flow out from within.' (When he said 'living water,' he was speaking of the Spirit, who would be given to everyone believing in him. But the Spirit had not yet been given, because Jesus had not yet entered in his glory)" (John 7:37–39 NLT).

The redemptive image of water is carried throughout the New Testament, even into the description of the final days of the world. "And the angel showed me a pure river with the water of life, clear as crystal, flowing from the throne of God and of the Lamb. . . . Let the thirsty ones come—anyone who wants to. Let them come and drink the water of life without charge" (Revelation 22:1, 17 NLT).

Baptism

Nowhere is the role of water in Christian ritualism more clearly seen than in the sacrament of baptism. Baptism is an act in which either water is sprinkled or poured on a person or the person is immersed in water. The act is an essential part of most (but not all) Christian sects and is common among other religious groups.

The significance of and procedures associated with baptism have been the subject of controversy within the Christian church for two millennia. Some theologians have taught that complete immersion is required for a proper baptism, while others say that sprinkling alone is sufficient for the act.

Perhaps more fundamental has been the controversy as to the precise meaning of the act. One common interpretation is that baptism can be seen as a way of "washing away" a person's sins and restoring him or her to a state of purity and oneness with God. In this act, the flow of water over a person may be seen as representative of the flow of God's spirit over the person's soul.

In addition, baptism is often regarded as a contract between God and a person. As such, it marks the precise moment at which a person joins the Christian community.

Other authorities disagree with either or both of these interpretations. The argument can be made, for example, that gaining forgiveness for one's sins and entering into a covenant with God can and should occur independently of and prior to baptism itself.

In August 1937, pastors of the Pentecostal Assemblies of God in Tynemouth, England, perform a full-immersion baptism before an audience of inquisitive bathers. *Hulton-Deutsch Collection/Corbis.*

Those who take this position may cite 1 Peter 3:21, where the apostle writes that "baptism is not the putting away of the filth of the flesh, but the answer of a good conscience with God" (KJV).

Whatever specific meaning is ascribed to the act of baptism, the ritual is an important ceremony in most Christian communities. It is often called *christening* and may be accompanied by other associated rituals, such as the assignment of a baptismal name and godparents. *See also* Sacred Waters

Further Reading

Butler, Trent C., ed. *Holman Bible Dictionary*. Nashville: Holman Bible Publishing, 1991.
Crosswalk.com Bible Study Tools, http://bible.crosswalk.com.
Smith, William. *Smith's Biblical Dictionary*. Nashville: Thomas Nelson Publishers, 1997.
Strong, James, ed. *The New Strong's Exhaustive Concordance of the Bible*. Nashville: Thomas Nelson Publishers, 1997.

Biological Functions of Water

Water plays a number of different functions in all forms of life. That fact should hardly be surprising; since life arose in the oceans, one might expect plants and animals to have evolved with some connection to seawater. Even today, some marine organisms are little more than packages of seawater surrounded by a membrane. Water makes up about 95 percent of the body weight of a jellyfish, for example. Organisms that have adapted to more arid environments have evolved body systems capable of functioning with lesser concentrations of water. The body of a kangaroo rat, for example, is about 65 percent water, while that of a pea weevil is only 48 percent water. Most organisms have body water content somewhere between these extremes: 74 percent for the chicken, 70 percent for the corn plant, and 65 percent for the human body, for example. On average, the water content of a living cell ranges between 75 and 85 percent.

Functions

Water has many essential functions in living organisms. These functions include:

Transport: For cells to function properly, they must be provided with nutrients. In addition, waste products produced by chemical reactions in the cell must be removed. Water plays an essential role in both processes because of its ability to dissolve such a large variety of other substances.

Chemical reactant: Water also plays a role in many chemical reactions that take place inside cells. For example, the process of digestion commonly involves the chemical reaction between a large molecule (such as a carbohydrate) and a molecule of water, known as *hydrolysis*. The product of such reactions is generally two or more smaller molecules.

Temperature regulation: Many different chemical reactions are needed in a living organism to produce new body parts and energy. Heat energy is always released as a by-product of these reactions. Water is an effective means of stabilizing and removing excess heat produced by such reactions because it has a high specific heat and an ability to transfer the heat from cells to the external environment (through perspiration, for example).

Turgidity: Water also helps maintain the shape and structure of cells and, hence, the shape and structure of plants and animals and their constituent parts. When an organism loses water (becomes *dehydrated*), its cells and body parts may become flaccid and, eventually, plasmolyzed—a condition in which the cytoplasm of the cell shrinks away from the cell membrane or cell wall. *Molecular structure*: Biochemical molecules, such as lipids, proteins, and carbohydrates, often contain water molecules as structural components that maintain the shape and size of organelles in the cell. When these molecules lose their water, they may no longer be able to perform the biological function for which they are normally responsible.

Cellular coating: The moist covering on the outside of cells facilitates the transfer of nutrients (minerals and gases) into and wastes out of cells.

Lubrication and cushioning: In animals, water may play the additional role of lubricating joints—allowing them to move with less friction—and soft tissues—preventing them from sticking to each other. In mammals, it also plays an essential role in protecting the fetus from damage during pregnancy by allowing it to "float" in a seawater-like environment.

Water in Plants

The movement of water in plants has long intrigued botanists because it appears to defy the law of gravity. Water enters the plant through its root systems, moves upward through stems into branches and leaves, and then is expelled into the atmosphere. This pattern holds as true for redwood trees a few hundred feet tall as it does for tomato plants only a foot tall. By what mechanism does this anti-gravity effect take place?

The first stages of this process are not mysterious. Water from the ground passes into a plant through the cell walls of root hairs. Water may also pass through the interstitial (between cells) space into the plant root.

Plants have adapted their root systems to reach water wherever it may be found in the ground. Shore plants may need roots no more than a few inches in length to reach the water table, while plants in more arid regions may require root systems that reach 30 feet or more into the earth. Other adaptations have made root systems more efficient. In some cases, such systems may consist of millions of tiny root hairs, each capable of absorbing water from the surrounding soil. In other cases, root hairs may be covered with fungi that live symbiotically with the plant. The fungi absorbs water from the soil, which it passes on to the root, and is "rewarded" by absorbing sugars, nitrogen-containing compounds, and other nutrients from the plant roots.

Once water has entered a root hair, it moves into the xylem layer of the plant. Xylem consists of long, tube-shaped cells that carry water upward into the plant. There is no problem in explaining the movement of water a distance of 30 feet or so. Atmospheric pressure is able to maintain a water column 34 feet in height at sea level. But how does water rise higher—often, much higher—in a tree?

One possible answer to this puzzle was suggested in 1894 by Irish plant physiologist Henry H. Dixon. Dixon's theory, now known as the *cohesion theory of water transport*, focuses on the role of transpiration in plants. Transpiration is the process by which water is lost from the stems and leaves of plants. Plants typically excrete very large amounts of water through their stomata (singular, *stoma*)—guard cells on the epidermis of leaves through which water and carbon dioxide are exchanged with the surrounding atmosphere.

According to Dixon, as plants lose water through transpiration, water is pulled from deeper within a leaf to replace the lost water. This "pulling" effect is possible because of the strong hydrogen bonding between water molecules. As one water molecule escapes through a stoma into the atmosphere, it pulls on an adjacent water molecule, which then takes it place. Similarly, this pulling effect (the *cohesion* that gives the theory its name) acts on a chain of water molecules running through the leaf vein, into the leaf xylem, and then into the stem and trunk xylem.

By whatever mechanism it is that water reaches the leaves of a plant, it is at these sites that it plays its most important role. In the process known as *photosynthesis*, it combines with carbon dioxide that enters the leaf through stomata, resulting in the formation of carbohydrates.

The amount of water lost by transpiration, compared to the amount used during photosynthesis, can be remarkably large. A single birch tree, for example, can lose up to 20 liters (80 gal) of water in a single day.

Still, plants have had to develop a variety of mechanisms to control the loss of water by transpiration to ensure an adequate supply for photosynthetic reactions. For example, plants that live in arid regions may have very small leaves with waxy coverings that reduce water loss from stomata. Deciduous trees have evolved the mechanism of losing their leaves during cold weather to virtually eliminate the loss of water through transpiration.

Water in Animals

All animals face essentially the same problems in obtaining and using the water they need to carry out the functions described above. They must have mechanisms for collecting or producing the water they need, circulating that water through their bodies and delivering it to individual cells, and excreting excess water along with wastes produced during body functions. The specific means used by various types of animals are, however, very different.

For example, marine animals like hydras, jellyfishes, and sea anemones have relatively simple water-exchange systems. Water, carrying with it most of the nutrients needed by the animal, passes back and forth across the animal's body surfaces by means of direct diffusion. Sponges, by contrast, have specialized cells that regulate the flow of water into and out of the animal. Platyhelminths (flatworms, flukes, and tapeworms) have evolved specialized excretory cells and excretory ducts to aid in the movement of water, while nematodes (roundworms) have bladderlike excretory organs. Arthropods (spiders and insects, for example) have well-developed digestive tracts and vascular systems that include secretory glands.

Chordates have the most complex systems for the transport of water of all animals, systems with complete digestive and vascular systems that include kidneys, bladders, and other organs for the transport of water in one form or another. Water in higher animals occurs in two forms: intracellular fluid (the watery mixture found inside cells) and extracellular fluid (the watery mixture that circulates around the outside of cells). Extracellular fluid occurs in the form of blood, which is circulated through the body by means of a vascular system consisting of the heart, veins, arteries, and capillaries, and as interstitial fluid, which occupies the spaces between cells and tissues.

The amount of water ingested by animals and the way that water is obtained depends on many factors, only one of which is the species itself. Environment is an important factor since some animals cannot depend on rain, lakes, rivers, or other natural sources for the water they need for their bodily functions. Humans, for example,

obtain about half the water they need through drinking, another 40 percent from food, and the remaining 10 percent through cellular respiration, in which water is a by-product. By contrast, the kangaroo rat, which lives in hot desert environments, gets about 90 percent of the water it needs through metabolism, about 10 percent from food, and essentially none through drinking.

Foods provide a significant amount of the water taken in by animals because they themselves consist of significant amounts of water. For example, most fruits and vegetables contain anywhere from 90 to 99 percent water. Poultry consists of about 60 to 65 percent water, while meats contain about 50 percent water by weight. Dairy products may contain anywhere from about 15 percent (for butter) to about 50 percent (for cheeses) to 90 percent (in the case of milk). Baked goods contain less water, ranging from less than 5 percent for crackers and hard biscuits to about 40 percent for bread.

Animals lose water from their bodies through a variety of mechanisms, primarily the excretion of urine and feces, and evaporation through perspiration. In humans, for example, about 60 percent of all water loss occurs in the form of urine, 4 percent as feces, and the remaining 36 percent as the result of evaporation. Kangaroo rats, by contrast, lose about 20 percent of their body water in the form of urine, about 4 percent as feces, and the remaining 76 percent through evaporation.

Controlling the gain and loss of water is obviously a critical body function. An inadequate amount of water in the body (*dehydration*) means that essential body functions will not occur, and an animal may become sick and die. Two mechanisms operate to control water balance in most animals: thirst and kidney function. In addition to controlling the proper amount of water in the body, the kidneys also perform another essential function: controlling the proper concentration of salts (*electrolytes*) dissolved in blood and interstitial fluid.

The primary mechanism for governing both thirst and kidney function in vertebrates is a region of the forebrain known as the *hypothalamus*. When the volume of water decreases or the concentration of electrolytes increases in the extracellular fluid (as, for example, would occur as a result of heavy sweating during exercise), a portion of the hypothalamus is stimulated. The hypothalamus then sends a neural (nerve) message to the pituitary gland, triggering the release of the hormone *vasopressin*, also known as *antidiuretic hormone*, or ADH. Vasopressin is carried from the pituitary gland through the bloodstream to the kidneys. In the kidneys, vasopressin brings about a change in the permeability of certain cells, causing the kidneys to release less water into the urine. As less water is released in the form of urine, the overall volume of extrastitial water increases and its concentration of electrolytes decreases. This change in water volume and electrolyte concentration is then detected by the hypothalamus, which reduces stimulation of the pituitary and, thereby, the output of vasopressin. *See also* Transpiration

Boats and Ships

The word *ship* comes from the Old English and Old Norse word *scip* and the German term *schiffe*, both meaning a vessel that travels on the water. The word was first used sometime before the twelfth century. The term *boat* is equally as old and is derived from the Old English term *bat* and the Old Norse *beit*, both meaning water-going vessels.

The distinction between these two terms is not clear. In general, ships are larger vessels—capable of carrying a boat, according to one traditional definition. Boats are often thought of as smaller vessels, frequently open and propelled by oars, paddles, sails, and/or small motors. The distinction between the two terms is not clear, however, and they are sometimes used interchangeably.

Ships and boats have been used for a variety of purposes for at least six thousand years. In addition to being a primary means of transporting people and goods, boats and ships have been used for exploration, military activities, and recreation.

Early History

The earliest boats were relatively simple in design, consisting of a hollowed-out log, bundles of grass or reeds tied to each other, or a wooden frame covered with animal skin or tree bark. Primitive boats of these designs are still used widely throughout the world. For example, natives on Lake Titicaca continue to make their boats by tying together bundles of totora reeds, just as their ancestors had done for many centuries.

The earliest boat still in existence today is one built for the funeral of Pharaoh Khufu (Cheops) (2590–2568 B.C.). The boat was discovered in 1954 by the Egyptian archaeologist Kamal al Mallakh near the south face of the great pyramid of Khufu. The boat—made of cedar wood brought from Lebanon and assembled without the use of nails—is thought to resemble boats commonly used on the Nile during the period. It is believed that the boat was built as a way for Khufu to travel to his eternal home.

Because of the central role of the Nile River in its culture, the Egyptians depended heavily on boats for the transport of goods and raw materials. Paintings on temple walls and ceramic bowls show that as early as the third millennium B.C., boats were being used to move the huge stones needed for constructing the pyramids and other important structures. The boats used for this purpose were often constructed of wooden planks joined to each other by rope and sealed with pitch, an invention developed by Egyptian naval architects. They were powered by a single square sail and steered by oarsmen seated at the stern of the boat. Over time, the square sails were replaced by rectangular sails, wider than tall in shape. The Egyptians later used ships of this design for transport on the Mediterranean, bringing cedar wood from Lebanon, for example.

The boat design that worked well on the Nile proved to be less effective on the open seas. It was the Minoans who occupied Crete in about 2500 B.C., instead, who developed the first truly effective ocean-going boats. No firsthand records of their work—either written or actual remains—exist. However, we know much about Minoan ships because of paintings and written records left by other cultures, especially the Egyptians. Minoan ships were apparently much larger than those of the Egyptians, long and narrow, and powered by oarsmen on both sides of the boat. They also used one large square sail at the center of the boat. With these vessels, the Minoans controlled the eastern Mediterranean for hundreds of years.

Much more is known about the shipbuilding methods of the Phoenicians and Greeks after about 1200 B.C. These two nations largely controlled the eastern Mediterranean from that date until the rise of the Roman Empire in about 100 B.C. The first development in shipbuilding in these two seafaring nations was the introduction of a second mast over the bow of the ship. By 500 B.C., two more masts had been introduced—a triangular sail over the main mast and a second over the stern.

Ships of this design measured about 30 meters (100 ft) in length, were capable of traveling at about 5 knots, and could carry up to about 200 metric tons of cargo. According to the Greek historian Herodotus (about 484–420 B.C.), sailing ships of this design were able to travel out of the Mediterranean, down the coast of Africa, around the

Cape of Good Hope, and into the Indian Ocean.

The Phoenicians and Greeks made two other modifications in the basic structure of their sailing ships. First, they added a sharp wooden point at the bow of the boat to use as a battering ram during naval battles. Second, they added additional rows of oarsmen on both sides of the boat. The Phoenician *bireme* consisted of two banks of oarsmen and first appeared in about 700 B.C.; about two hundred years later the Greeks added a third row of oarsmen to create their *trireme*.

By the end of the Greek era, the fundamental principles of shipbuilding had been well established. The framework of the boat consisted of a long backbone—the *keel*—running from bow to stern. Lying at right angles to the keel were ribs made of planks curved upward to make the sides of the boat. The basic skeleton was then covered with planks overlapping each other (*clinker* built) or flush with each other (*carvel* built). Boards were first fastened to each other with wooden pegs, which were replaced with iron nails when they became available, and made watertight with pitch.

Developments Elsewhere

Shipbuilding in other parts of the world took different directions. For example, the earliest Chinese sailing vessels were made by joining two dugout canoes to each other, catamaran style, and then constructing a boxlike structure on top. A watertight box fixed to the center of the boat held a sail, which, along with oars, provided the means for driving the boat. The Chinese were also using a stern rudder as early as 250 B.C., about 1,500 years before it was "invented" in Europe. This basic design is retained today in the vessel known as a *junk*, still widely popular in China and other Asian countries. During his travels to China in the 1260s, Marco Polo described an elaborate junk with fifty private staterooms.

Peruvian fishing boat, 1999. *Jose Cort, NOAA.*

In the Persian Gulf, boatbuilders early on invented the lateen sail for use on their vessels. The lateen sail has a triangular shape and is attached to a long beam that hangs from a short mast. The name *lateen* comes from the French word *latine*, meaning "Latin," since boats of this type were first seen by Europeans in the Mediterranean Sea, then controlled by the Romans.

The lateen sail is somewhat clumsy to work with since it has to be lowered each time the boat changes direction. It is used most commonly, therefore, in areas where the wind tends to blow consistently from the same direction. Both the Arabian dhow and the Egyptian felucca are powered by lateen sails, as are boats in some parts of the Pacific Ocean, such as the traditional lakatoi used in New Guinea.

Later Developments

From the eighth to the twelfth century, the most skilled boatbuilders in the Western world were the Vikings. Modern day knowledge of Viking shipbuilding practices are based to a large extent on a number of boats from this period found in burial sites where they were entombed along with Viking leaders. One of the most famous of these archaeological finds was uncovered in 1880 near Gokstad in southeastern Norway. The vessel was a type of ship known as a *longship*. It was built in about A.D. 900 and was about 25 meters (80 ft) long and 5 meters (17 ft) wide. It was propelled by a single square sail set amidships and by sixteen oarsmen on each side of the boat.

Another type of longship was the *knarr*, developed for the transport of heavy cargo in the punishing weather conditions faced by the Vikings at sea. One knarr uncovered in the late 1890s in Norway was about 16 meters (54 ft) long and about 5 meters (15 ft) wide. In vessels such as these, the Vikings traveled throughout the North Sea and the North Atlantic, reaching as far as Greenland and North America.

Although Viking dominance of the oceans had declined by about 1200, their construction techniques continued to influence boatbuilding practices in other nations. Around the British Isles and Northern Europe, for example, the successor to the knarr was the *cog*, a larger, more spacious vessel used for both commerce and warfare. The primary technological development incorporated into the cog was a steering rudder, invented to replace the steering oars traditionally used to guide a ship.

By the 1300s, shipbuilding in Southern Europe had become standardized, based largely on methods developed by the Greeks more than a thousand years earlier. The first step in the process was to lay down the keel and cross ribs, forming the skeleton of the ship. The planks forming the frame of the ship were then attached to the keel and ribs. Southern shipbuilders also adapted the lateen sails developed by the Arabs to provide better maneuvering ability for their ships. Ships were powered both by sail and by oars, depending on weather conditions and proximity to land.

Over the next few centuries, ships became larger and more complex. They were known by names such as the *hulk, caravel, nef, buss,* and *carrack.* The final product of this development was the *galleon,* which first appeared in the mid-sixteenth century. They had high structures at their stern (*sterncastles*) for housing travelers and some crew members and were powered by two or three masts, each of which carried one or two sails. Galleons were used for both commerce and warfare and may be best known to the general public because of their role in the defeat of the Spanish Armada in 1588 by the British navy.

Scheduled Ship Service

The Age of Exploration (about 1450–1750) opened up vast new parts of the world to European trade, conquest, and colonization. Ships were designed and built to take advantage of these opportunities. For example, in the early 1600s, a type of ship known as the *East Indiaman* was developed to carry on trade between Europe and the Far East. The East Indiaman vessels were usually three-masted ships with two or more sails per mast. They generally carried guns to protect themselves against attack by pirates. The East Indiaman design grew larger over time until the mid-nineteenth century, when a typical vessel might range up to 70 meters (220 ft) in length, with an overall weight of more than 2,000 tons.

Commerce between Europe and North America also created the need for regularly scheduled ship travel, a need met by the construction of the *packet ships.* Prior to the early 1800s, ships sailed on an irregular schedule, depending on the number of passengers who wanted to travel, the arrival time of cargo, and weather conditions. In 1818, however, there was sufficient demand to establish regularly scheduled service between Liverpool and New York. The first company to offer that service was the Black Ball Line, later known as the Old Line. The first ships used for this service were about 30 meters (100 ft) in length, carried about two hundred passengers, and took about three weeks to travel from New York to Liverpool. The trip in the opposite direction took longer because of the prevailing westerly winds on the Atlantic.

Passengers who could afford the best accommodations paid about £25 (a little over $100)—a price that included meals. Well over 95 percent of all passengers traveled "in steerage," however, where rates were variable, often depending on whatever the company could charge based on demand.

The culmination of sailing-ship design was reached in the mid-nineteenth century with the construction of the *clipper ships*. The clipper ships were the largest sailing ships built to the time, reaching a length of more than 60 meters (200 ft) and having a capacity of about 1,400 metric tons (1,500 short tons). They were powered by large arrays of sails attached to as many as six masts, each holding a maximum of six sails. The clipper ships could travel anywhere in the world at speeds of up to 20 knots.

By the end of the nineteenth century, the age of sailing ships as a means of transporting passengers and cargo had largely come to an end. The largest sailing ship ever built, the German vessel *Preussen*, was christened in 1902. It was 132 meters (433 ft) long and 16 meters (54 ft) wide, with a capacity of 7,300 metric tons (8,000 short tons). The *Preussen* was rammed by the channel steamer *Brighton* on 6 November 1910 and lost at sea.

Steam-Powered Ships

The invention of steam engines by James Watt (1736–1819) in 1765 revolutionized the construction of sailing ships, just as had so many other fields of technology. The first successful sailing vessel operated by steam was built by the French inventor Marquis Jouffroy d'Abbans in 1783. D'Abbans' 180-ton paddle wheeler, *Pyroscaphe*, traveled a short distance on the river Saône, near Lyons, but the inventor was unable to repeat his success. Four years later, the American inventor John Fitch (1743–1798) designed an early form of the side-wheeler and, a few years later, built one of the first stern-wheelers. In the latter ship, he began regular commercial service in 1790 between Philadelphia and Trenton.

Credit for the first ongoing commercially successful steam-operated ship is usually given to American inventor Robert Fulton (1765–1815), whose stern-wheeler *Clermont* began operation between New York and Albany in 1807. Two years later, Fulton added a second boat operating across Lake Champlain. At about the same time, American inventor John Stevens (1749–1838) built the first steamboat to operate on the open seas, the *Phoenix*. Stevens's vessel made the trip from New York City to Philadelphia in thirteen days, about six times the time required by the best sailing boats of the time. In 1819, the first steam-powered ship to cross the Atlantic, the *Savannah*, made the trip from New York to Liverpool in twenty-nine days.

Iron and Steel Ships

The change from sail to steam for the operation of ships was roughly paralleled by the change from wood to iron and steel for their construction. As early as 1746, the French mathematician and physicist Pierre Bouguer (1698–1758) calculated the types of stress to which a ship made of iron would be exposed on water. Bouguer was far ahead of his time, however, and the actual construction of iron ships did not begin in earnest until the end of the eighteenth century. One of the first of these ships was a 70-foot barge of bolted cast-iron plates built by English inventor John Wilkinson (1728–1808), first operated on the Severn River in 1787.

For a relatively brief period of time, shipbuilders used iron and steel in place of wood in the construction of sailing ships. The main impetus for this transition in Europe was the rapid decline in the availability of wood for construction. Therefore, as new steam ships were designed, they would most likely be made of iron at first, and steel during later periods.

The first commercial iron ship was the *Aaron Manby*, built by the British naval architect Isambard Kingdom Brunel (1806–1859) in 1821. Brunel also designed and built the first all-iron steamship intended for regular transatlantic travel, the *Great Western*, launched in 1837. The *Great Western* was 72 meters (236 ft) long and 25 meters (85 ft) wide. Before being scrapped it made a total of sixty-seven trips across the Atlantic in eight years.

Technological Advances

Three inventions have been of special significance in the development of ship design over the past century: the screw propeller, the turbine engine, and nuclear power.

The first ships to use steam power for propulsion operated by means of rotating paddle wheels attached to the sides or back of the ship. These paddle wheels worked satisfactorily in calm seas, but lost effectiveness when the ocean was rough enough to cause the ship to tip from side to side or front to back, lifting the wheel(s) out of the water.

One possible solution to this problem had been known for more than 2,000 years, the Archimedes' screw. The Archimedes' screw is a device for moving water and, hence, might reasonably be considered a mechanism for driving a ship through water. One of the first inventors to explore the use of a screw propeller was John Fitch in 1796, but nothing came of his experimentation. In fact, it was not until more than forty years later, in 1838, that two inventors, the Englishman Francis Peter Smith (1808–1874) and the Swede John Ericsson (1803–1889) independently invented a screw propeller for the propulsion of sea-going vessels. Smith used a screw propeller in the design of his 237-ton steamship *Archimedes*, while Ericsson incorporated essentially the same concept into one of the first iron-clad warships, the *Monitor*, famous for its Civil War battle against the warship *Merrimac* in 1862. The invention worked so well that screw propellers were adopted for use in nearly all new steamships.

The use of screw propellers was also an important impetus in the development of more improved engines that could drive the propellers more rapidly and, hence, increase the speed of the ship. Just when it appeared that no further improvements in the standard steam engine could be made, a new approach to power generation was proposed in 1894 by the English inventor Charles A. Parsons (1854–1931). Parsons designed a rotary turbine engine with many times the efficiency of the traditional steam engine. Parsons incorporated the turbine engine in the design of his new steamship, the *Turbinia*, launched in 1897. The *Turbinia* attained maximum speeds of more than 34 knots, greatly exceeding any previous record set by traditional steam engines. Before long, steam turbines, like screw propellers, had become part of the design of all new steamships.

The third great improvement in ship propulsion occurred in 1954 with the introduction of nuclear reactors as power sources. The first vessel to use nuclear power was the U.S. submarine *Nautilus*, which operated until 1979. Five years after the launch of the *Nautilus*, the first nuclear-powered surface ship, the *Savannah*, was launched. While nuclear power has become an important factor in the design and construction of military ships, it is still too expensive to be used in merchant vessels.

Modern Ships and Boats

During the first third of the twentieth century, the most impressive commercial ships were the great ocean liners. At first, it was Great Britain's ships that ruled the waves, with the *Queen Mary* and *Queen Elizabeth* leading the way. The larger of these, the *Queen Mary*, was 310.7 meters (1,019 ft) in length, with a capacity of 1,957 passengers served by 1,174 crew members.

Its maximum speed was 28.5 knots, and the ship made 1,001 transatlantic crossings between its launch in 1936 and its retirement in 1967.

By the end of World War II, the age of the great ocean liner had begun to decline. Air travel provided a much faster way to cover the same distances previously served by ocean liners. The last U.S. ocean liner, the *United States*, ended service in 1969, and the only remaining luxury liner still in service today is the *Queen Elizabeth 2*, which travels the Atlantic route as well as making round-the-world cruises off-season.

Cargo ships in the twentieth century became more specialized than the general purpose ships used prior to that time. One distinction that can be made among cargo ships is between regularly scheduled ships and *tramp steamers*. The former are usually owned by large companies and travel regular fixed routes, moving raw materials and finished products between specific ports. Tramp steamers, by comparison, are available for hire by a variety of agents and carry a variety of different products.

Cargo ships can also be classified according to the kind of cargo they carry and, hence, according to their design construction. One type of cargo ship is the *general cargo ship*, which carries virtually any type of packaged item, such as automobiles and trucks, furniture, machinery, steel and steel products, textiles, and any other product that can be loaded directly onto the ship. General cargo ships are often constructed for specific types of service. For example, *container ships* are essentially huge warehouses divided into separate compartments into which individual containers can be placed. Each container is of a standard size and is filled with some product before reaching the ship. The containers are often transported from ship to destination and vice versa on railroad cars specially designed to carry them.

A variation of the container ship is the *roll-on/roll-off* (or ro-ro) ship. In a ro-ro ship, containers are fitted with wheels, allowing them to be rolled directly from the dock, through a door in the side of the ship, into the cargo hold. Yet another variation of the container ship is the *LASH* (for lighter aboard ship), in which barges, known as *lighters*, are brought down a river to an ocean port and then loaded onto an ocean-going cargo ship by means of cranes.

A second type of cargo ship is the *tanker*—a ship built especially to carry liquid products, such as petroleum. At one time, such products were carried in barrels on board general purpose cargo ships, but in the late 1870s, Swedish industrialist Ludwig Nobel, brother of Alfred Nobel of Nobel Prize fame, built the first ship designed specifically to carry liquid cargo. Nobel's first tanker began operation on the Caspian Sea in 1879, transporting kerosene. In 1885, a second tanker began carrying petroleum products from Batum to London. Today, tankers are among the largest ships afloat. The largest among them, called *super-tankers*, can range up to 500 meters (1,600 ft) in length and 60 meters (200 ft) in width, with capacities of more than 500 kilograms (1 billion lb) of petroleum.

A third type of ship, similar to the tanker, is the *dry bulk carrier*. Like the tanker, the dry bulk carrier is essentially a large open space divided into compartments in which dry products—such as wheat, coal, or iron ore—can be carried. The first dry bulk carriers were built in the 1890s to carry iron ore across the Great Lakes. As of the late 1990s, there were more than 5,500 dry bulk carriers, each weighing more than 10,000 metric tons. The world's dry bulk carriers transport nearly 2 billion metric tons of cargo each year.

A fourth type of cargo ship is the *multipurpose cargo ship*. As its name implies, a ship of this design is capable of carrying a variety of different types of products. It may include separate areas with specific functions, such as compartments for dry or liquid storage, refrigerated areas, and flat decks for the transport of automobiles and trucks. Multipurpose cargo ships often combine the design of other types of ships, making possible both ro-ro and LASH operations, for example.

Military Ships

The story of the development of military ships parallels that of other types of ships until the seventeenth century. At that point, improvements in military ordnance design resulted in the manufacture of guns so heavy that special ships had to be built to carry them. From that point on, the construction of commercial and military vessels took significantly different paths.

The U.S. Navy now lists about a dozen major types of fighting and support ships, including aircraft carriers and escort aircraft carriers, destroyers, cruisers, battleships, frigates, torpedo boats, submarines, amphibious ships, mine-warfare vessels, patrol craft, coast-guard vessels, and yard craft.

Boats

The term *boat* today covers a vast array of water-going craft, ranging from the simplest rowboat to the most sophisticated yacht. Boats are used for a great variety of purposes, ranging from fishing to water patrol to recreation. The many different types of boats can be classified in a variety of ways. For example, some are powered by sails, others by motors of one kind or another, and still others by human energy (such as rowing or paddling). Boats can also be classified according to the type of hull they have. *Flat-bottom boats*, for example, are relatively easy and inexpensive to build, although they do not travel well in rough water. They also tend to be somewhat unstable. Some examples of flat-bottom boats

include johnboats, crawdads, scows, punts, skiffs, dories, some types of canoes, and some types of high-speed runabouts.

A second type of boat, the *vee-bottom boat*, has a V-shaped hull. They tend to offer a smoother ride in rough water than do flat-bottom boats, but since they are partly submerged within the water, they require more power to operate than a comparable flat-bottom boat. Many kinds of powerboats, some kayaks, and some canoes have vee-bottom hulls.

A third type of boat is the *round-bottom boat*, which, as the name suggests, has a curved bottom. Round-bottom boats maneuver easily in still water, especially at slower speeds, although they tend to roll if the waters become rough. Some types of canoes, many types of sailboats, and most trawlers have round bottoms.

A fourth type of boat, the *multi-hull boat*, has two or more separate and distinct parts to its hull. A *catamaran*, for example, has two hulls joined to each other by the overlying deck. A *trimaran* has a similar structure, except that it rides on three hulls rather than two. Pontoon boats and houseboats are usually built in catamaran style, with an open riding area in the former case and a house-like structure in the latter.

Other Types of Boats and Ships

For the most part, all boat and ship designs today are based on concepts developed hundreds of years ago, improved and expanded, no doubt, but fundamentally based on the concept of a solid vessel skimming through a body of water. Two approaches to shipbuilding that differ from this concept are the *hydrofoil* and the *hovercraft*. Both are based on the principle of having a vessel travel on a thin layer of air above the surface of water, without actually coming in contact with the water or only to a moderate degree.

One of the efforts to design such a watercraft was that of the Swedish philosopher

Emanuel Swedenborg (1688–1772). In 1716, Swedenborg suggested that a water-going vessel could be built in which humans scooped up air, rather than water, with a set of oars. The air would be pushed underneath the boat, Swedenborg said, providing a cushion on which the boat could travel. Although the concept was sound, humans could not scoop up air fast enough to make Swedenborg's concept work.

Scientists continued to pursue ideas like those of Swedenborg's for more than two centuries, but no practical watercraft was produced. The main problem was the lack of a motor powerful enough to produce the cushion of compressed air on which the craft could float.

Finally, in 1905, Enrico Forlanini (1848–1930), an Italian designer of helicopters and aircraft, built the first successful watercraft to operate on a cushion of air, a vessel he called the *hydrofoil*. There are two kinds of modern hydrofoils. In both types, powerful fans produce a cushion of compressed air that lifts the craft out of the water. It then travels across the surface on a pair of winglike foils that ride on the surface of the water (in one design) or just below the surface of the water (in the second design).

In spite of Forlanini's work, credit for the greatest advances in hydrofoil design usually goes to the German inventor Hanns von Schertal. One of von Schertal's early models, launched in 1940, traveled at a speed of 47 knots, a record that lasted until the late 1960s.

A closely related watercraft is the *hovercraft*, which travels completely out of the water on top of a cushion of compressed air created by powerful fans. The first successful hovercraft was designed and built by the British inventor Christopher Cockerell in 1959. Cockerell's hovercraft traveled well on calm seas, but could not be controlled when the waters became rough. He later developed an improved version of his invention that, in 1962, traveled at speeds of 50 knots

on calm seas and up to 40 knots in seas with waves of up to two meters (six ft).

Hydrofoils and hovercraft have become increasingly popular in the last few decades. They are currently used for short-range commuter service, fishery patrols, harbor and water police, firefighting teams, and submarine pursuit. For a complete listing of all active and historic U.S. naval ships, see Dictionary of American Naval Fighting Ships Online at http://www.hazegray .org/danfs. *See also* Aquatic Sports; Barges

Further Reading

Blackburn, Graham. *The Illustrated Encyclopedia of Ships, Boats, Vessels, and Other Water-Borne Craft*. Woodstock, VT: Overlook Press, 1999.

Paine, Lincoln P. *Ships of the World: An Historical Encyclopedia*. Boston: Houghton Mifflin, 1997.

Polmar, Norman. *The Naval Institute Guide to the Ships and Aircraft of the U.S. Fleet*. 17th ed. Annapolis, MD: United States Naval Institute, 2001.

Smith, A. G. *Traditional Boats from around the World*. New York: Dover, 2001.

Further Information

American Boating Association
P.O. Box 417
Harwich Port, MA 02646
Telephone: (508) 432-8846
Fax: (508) 430-2049
URL: http://www.americanboating.org
e-mail: admin@americanboating.org

Bobsledding

See Aquatic Sports, Bobsledding

Bramah, Joseph (1748–1814)

Bramah was a prolific inventor, whose most important creation was probably the hydraulic press. A hydraulic press is a device for magnifying the force exerted on a surface by means of a system of cylinders filled with water or some other fluid.

Personal History

Joseph Bramah was born in Stainborough, Yorkshire, England, on 13 April 1748 under the name of Joe Brammer. As a young man, he worked on his father's farm and then, at the age of sixteen, was apprenticed to a cabinetmaker. At the conclusion of that apprenticeship, he decided to seek his fortune in London, and began that adventure by walking 170 miles to the city. Bramah changed the spelling of his last name in 1778 when he took out a patent for his new water closet. He died in London on 9 December 1814.

Accomplishments

Early on in his life, Bramah understood the potential of using water to drive machines. When he was only thirty-six years old, he outlined a method for driving ships by means of water-powered screws. In 1795, Bramah designed the first hydraulic pump, laid out on principles first worked out by the French physicist Blaise Pascal (1623–1662) a century and a half earlier. The hydraulic pump consists of two pistons connected to each other through a pipe. One of the pistons has a large diameter and the other, a smaller diameter. The whole system is made watertight.

When pressure is applied on the smaller piston, it is transmitted equally throughout the pump and into the larger piston. The total *force* in the two pistons, however, is very different because the force on an object depends on both pressure and the area on which the pressure works: $F = p \times A$. Any force exerted on the smaller piston, then, is able to produce a much larger force in the larger piston.

Bramah's hydraulic pump is still in wide use today. It forms the basis of hydraulic car lifts, hydraulic braking systems, and hydraulic devices used to lift large pieces into place for new buildings, bridges, and other structures.

In addition to the hydraulic press, Bramah invented machines for numbering bank notes, pumping beer, manufacturing paper, making nibs for pens, uprooting trees, ringing doorbells, and aerating water. During his lifetime, the two inventions for which he was best known were a flushing water closet and some of the best locks ever made. In fact, one of the locks he invented was so effective that it was not until thirty years after his death that a locksmith, working for fifty-one hours, was able to pick the lock.

Further Reading

McNeil, Ian. *Joseph Bramah: A Century of Invention, 1749–1851*. Newton Abbot, UK: David & Charles, 1968.

Buoyancy

See Archimedes; Hydraulics, Hydrodynamics, and Hydrostatics

C

Cairns, John, Jr. (1923–)

Cairns is a prolific researcher, teacher, and writer in the field of limnology and environmental biology. He is the author of more than a thousand publications, including forty-five books and nearly five hundred scientific papers. He has been awarded the United Nations Environmental Programme Medal in the field of Environmental Restoration and Sustainability and was elected in 1993 as a foreign member of the Linnean Society of London, a restricted association of 50 members. In 1991, Cairns was elected to the National Academy of Sciences.

Personal History

John Cairns Jr. was born on 8 May 1923 in Conshohocken, Pennsylvania. He entered Pennsylvania State University in 1940, but took a leave of absence when World War II broke out. He served in the U.S. Navy from 1942 until 1946, after which he returned to school at Swarthmore College. He received his A.B. degree in biology from Swarthmore in 1947 before earning his M.S. and Ph.D. degrees from the University of Pennsylvania in 1949 and 1953, respectively.

Major Accomplishments

While still a graduate student at Penn State, Cairns accepted a position as assistant curator at Philadelphia's Academy of Natural Sciences. One of his first assignments involved a study of river biology, a study that took him across North America and to a number of other countries. As part of this research, he carried out a study of the effects of pollution on marine organisms in the Conestoga River in Pennsylvania, a study that led to his master's degree. The study also reinforced his growing interest in the study of environmental pollution, a relatively new field of research at the time.

Much of Cairns's later research was focused on the concept that the extent and effects of water pollution can be understood by studying its effects on organisms that live in the water. As a result of this work, Cairns also became interested in developing methods for restoring ecosystems that had already been damaged by pollution, a field known as *restoration ecology.*

In 1966, Cairns resigned his position with the Academy of Natural Sciences to accept a post as professor of zoology at the University of Kansas. Two years later, he moved to the Virginia Institute of Technology (VIT; now the Virginia Polytechnic Institute and State University), where he became research professor of zoology. He later was appointed director of VIT's Center for Environmental and Hazardous Materials Studies and University Distinguished Professor.

Canals

A canal is an artificial waterway built for navigation, irrigation, water supply, drainage, or the diversion of water for a hydroelectric dam. This entry deals primarily with the use of canals for navigation and water supply. For the use of canals for irrigation, *see* Irrigation and Aqueducts.

Construction

Canals are usually built by digging a trench in the ground with heavy construction equipment. Canals used for the delivery of water may be lined with concrete, plastic, or some other material to avoid seepage of water into the ground. They may also be covered to prevent evaporation of water into the air. Shipping canals are generally simpler in design, consisting primarily of an open trench deep enough to permit passage of the types of ships that will use the canal.

Canal construction is very much limited and regulated by the physical terrain. Canals can be used most efficiently for agricultural irrigation, for example, in regions where the land is relatively flat. The cost of constructing bridges, viaducts, tunnels, and other structures to carry canals through mountainous terrain often limits their use in other than relatively flat land.

When canals do have to traverse surfaces with different vertical heights, engineers use locks, inclined planes, or other devices to move water from one level to another. A *lock* is a section of the canal that can be closed off by gates at both ends. A ship enters the lock at one level, and water is pumped into or out of the lock. The ship is then able to leave the lock at the opposite end at a new water level. Inclined planes can be used to move water in a canal from a higher point to a lower point, but the reverse

effect can be achieved only by using pumps to move the water uphill.

Canal History

The first canals were probably little more than simple trenches built by the early Sumerians and Babylonians to bring water from distant sources to cities. One of the first navigation canals was built in 510 B.C. by order of King Darius I of Persia. The canal joined the Nile River to the Red Sea.

Canal building was a well-advanced science in China by the sixth century B.C. Probably the finest example of the art was the Grand Canal, said to be the oldest and largest man-made waterway in the world. It was started during the Zhou dynasty in 486 B.C., and was extended under the Emperor Yangdi of the Sui dynasty between A.D. 605 and 610.

The canal runs north and south, from Beijing in the north to Hangzhou in the south. It contains twenty-four locks and sixty bridges. For many centuries, it was poorly maintained and fell into serious disrepair. In 1958, however, a program to renew and rebuild the canal was begun. Today, it is again regarded as one of the most important artificial waterways in the world.

Where topography permitted, canals eventually became a very important means of transportation. They were often less expensive to build and maintain than roads. In England, for example, canals played an important role in the Industrial Revolution. They provided a relatively inexpensive way of bringing wood, coal, and raw materials to factories, where they were converted to products that could then be delivered to market by means of canals. It was only with the successes of the Industrial Revolution and the development of steam-powered transportation that canal traffic began to decline in importance.

Some Historic Canals

In many parts of the world, canals have been replaced by roads, railway lines, sophisticated water-delivery systems, and other structures that work more efficiently. However, canals still have many important functions in certain parts of the world. For example, the Garagum Canal in Turkmenistan is one of the longest artificial waterways in the world. It brings water from the Amu Dar'ya River to the city of Krasnovodsk on the Caspian Sea, a distance of 1,400 kilometers (870 mi). The water provided by the canal is used both for irrigation and as the primary public water supply for many towns along its route.

Probably the most important canals in the world by some standards are those that join two large bodies of water. Such canals reduce, often enormously, the distance that must be traveled between those two bodies of water. The Kiel Canal, across the base of the Jutland Peninsula, is an example. It was built in 1895 to reduce the time needed to travel between the Baltic and North Seas.

The Suez Canal

Two other memorable engineering feats were the construction of the Suez Canal in 1869 and the Panama Canal in 1914. Humans had dreamed for centuries about the construction of an artificial waterway between the Red and Mediterranean Seas. For all practical purposes, there was no way to travel between these two bodies of water without such a connection. Canals like the one built under Darius made it possible for ships to travel from the Red Sea to the Nile River and then to the Mediterranean. But a more direct route between the two seas was also recognized as a preferable choice.

The impetus for the modern Suez Canal came from Napoleon's engineers, who were looking for a more direct route from Europe to India. They thought the differences in elevation along the proposed route were too severe for a canal to be built, however. Indeed, it was not until 1859 that an engineer was willing to take on the challenges surrounding the construction of such a waterway.

It was in that year that French engineer Ferdinand-Marie de Lesseps began work on the canal. Construction continued for nearly a decade and was finally completed in 1867. The canal was officially declared open on 17 November 1869 and, six years later, was purchased by Great Britain, which operated the waterway for almost the next century. In 1956, Egypt nationalized the canal and evicted the British operators. In the decades that followed, the canal was often at the center of conflicts among Arabs, Jews, and westerners. Today, it is again operating normally, one of the most important artificial waterways in the world.

The Panama Canal

The story surrounding the construction of the Panama Canal is at least as complex and intriguing as that concerning the Suez Canal. The earliest mention of a canal across the isthmus of Panama can be traced to 1534, when Charles I ordered a survey of the area with a view toward building such a waterway. Charles envisioned the canal as a way of saving the thousands of miles of travel around the Cape of Good Hope that was then necessary to get from the eastern shores to the western shores of North America.

Charles's survey came to naught for a variety of reasons. The equipment needed and the engineering skills required were clearly not yet available. In addition, the terrible health and sanitation conditions present in the isthmus created a whole set of problems with which scientists of the time were ill-equipped to deal. Among these problems was the widespread occurrence of yellow fever, a disease that made working in the area nearly impossible.

In fact, it was not until 1880 that an effort to construct a cross-isthmus canal was initiated by a French engineering team. After

**Coast and Geodetic Survey ship *Explorer OSS28* passing through the Panama Canal, 1964.
*NOAA Ship Collection, NOAA.***

twenty years, the effort failed because of a combination of financial and health problems. In 1903, the United States signed an agreement with Panama to take over construction of the canal. The U.S. Corps of Engineers began construction in 1907 and completed the work in 1914. In large measure, the success of the operation was due to the success of Colonel William C. Gorgas in bringing yellow fever under control.

Control over the operation of the canal rested with the United States until 2000. In that year, under the terms of an agreement signed in 1977, ownership of the canal reverted to the government of Panama.

Urban Canals

Canals are found in urban areas throughout the world. They have been constructed in a variety of ways and serve many different purposes. Perhaps the best known urban canals are those in Venice, Italy. These canals are formed by bodies of natural water that originally separated the 120 islands that make up the city and that have, over the years, been shaped and adapted to form the system of "water-streets" on which the city depends for most of its transportation needs. The city contains 177 canals in all, the largest being the Grand Canal, a waterway that extends about 3 kilometers (2 mi) in length and divides the city into two nearly equal parts.

Bangkok, the capitol of Thailand, was also once covered with a complex system of canals, known as *klongs*. In fact, many westerners have referred to Bangkok as the Venice of the East. The first klong was built in the eighteenth century by King Rama I. It was constructed across a bed in the Caho Phraya River to form an island for defense of the king's palace. Over time, additional klongs were added, forming a honeycomb-like network of canals that served as the city's primary means of transportation. Over the past century, many of the original klongs have been filled in and covered with modern streets and highways.

Many urban canal systems were originally built for the purpose of drainage, but now serve a variety of other functions. In southeast Florida, for example, the land is traversed by a complex system of primary, secondary, and tertiary drainage canals. These canals were originally constructed to carry away rainwater from the flat land around the area, especially that which makes up the Everglades. At the simplest level, the system consists of thousands of drainage ditches spread out through hundreds of neighborhoods. These ditches drain into larger, or *secondary*, canals, which in turn drain into even larger, or *primary*, canals. The whole system is administered by the South Florida Water Management District and the U.S. Corps of Engineers. In addition to serving as a powerful water-management tool, the canal system provides opportunities for many forms of recreation, including boating, bird-watching, and fishing.

Canals are also used in urban areas to supply water from distant reservoirs. Many urban areas in Arizona, for example, are normally quite dry, too arid to sustain industry, commerce, and large residential populations. The development of cities such as Phoenix, Tempe, and Scottsdale, then, has depended on the construction of canal systems that bring water from distant mountain reservoirs. These canals are used not only for drinking water and other forms of consumption, but also to maintain the cities' green lawns; numerous golf courses; and boating, fishing, and other recreational opportunities. *See also* Barges

Further Reading

Hadfield, Charles. *World Canals: Inland Navigation Past and Present.* New York: Facts on File, 1986.

Payne P. S. Robert. *The Canal Builders: The Story of Canal Engineers through the Ages.* New York: Macmillan Company, 1959.

Canoeing

See Aquatic Sports, Canoeing

Capillary Action

Capillary action (or capillarity) is the tendency of a liquid to rise or fall in tubes of small diameter. The phenomenon gets its name from the Latin term *capillus*, meaning "hair."

The process of capillarity can be thought of as a two-step process. The first step involves the force of adhesion—a force of attraction between unlike molecules. Consider the case of a thin glass tube containing water. A significant force of adhesion exists between water molecules and glass molecules. This force of adhesion causes water molecules to be pulled upward on the outer edge of the water's surface. This effect produces the lens-shaped appearance, called a *meniscus*, visible on the surface of water whenever it is placed in a glass container.

The second step in capillary action involves the process of cohesion—the attraction of like molecules of water for each other. Cohesion is responsible for the phenomenon known as *surface tension*, in which the surface of water appears to behave as if it were covered with a skin. The skin-like effect is produced because water molecules on the surface pull strongly on each other.

In the case of capillarity, surface tension tends to pull the mass of water in the glass tube upward to a height determined by the outer edge of the surface. In effect, the combination of adhesion and cohesion act to pull water to a higher level in the glass tube then normal.

The amount of capillary action that occurs in this example depends largely on the diameter of the glass tube. In wide tubes, the force of adhesion is less able to overcome the gravitational forces (the mass) of the water. In narrower tubes, the adhesive forces may be strong enough to raise the small amount (small mass) of water in the tube. Roughly speaking, the narrower the tube, the more significant will be the capillary effect.

The effect of capillary action differs widely depending on the liquid involved and the tube in which it is contained. The force between alcohol and glass molecules, for example, is less than that between water and glass. As a result, capillary action is less significant for alcohol in glass than it is for water in glass.

Capillary action can also result in the depression of a liquid in a tube. In the case of mercury, for example, the cohesive force between mercury molecules is much greater than any adhesive force between mercury and glass molecules. In this case, a negative (convex) meniscus is formed, and the surface of the liquid is depressed in the tube.

Capillary action has a number of important applications. For example, molten wax rises in a candle through tiny fibers in the candlewick. Sponges can be used to absorb water because they contain tiny capillary tubes through which water can rise. The movement of capillary action through masonry and concrete is an important problem because it may weaken the material, destroy adhesives, and damage painted surfaces and other coverings on the material.

Capillarity is also an important phenomenon in agriculture. As soil dries out, it tends to form tiny vertical cracks through which subsurface moisture can escape and evaporate. In regions with adequate rainfall,

this phenomenon has no practical importance for farmers; but in very dry areas, this loss of moisture can be significant. One way to deal with the problem is to cultivate the soil on a regular basis, breaking up the vertical cracks and reducing the rate of evaporation. This agricultural technique, known as *dry farming*, has become an important method of increasing crop production in arid regions of the world. *See also* Surface Tension

Further Reading

Dregne, H. E., and W. O. Willis, eds. *Dryland Agriculture*. Madison, WI: American Society of Agronomy, 1983.

Carson, Rachel (1907–1964)

Rachel Carson was one of the most effective popular writers about science during the mid–twentieth century. Her most famous book, *Silent Spring*, is sometimes credited as being the most important force in initiating the modern environmental movement.

Personal History

Carson was born in Springdale, Pennsylvania, on 27 May 1907. She became interested in writing at an early age and had one of her stories, "A Battle in the Clouds," accepted by *St. Nicholas* magazine when she was only ten. In 1925, Carson entered the Pennsylvania College for Women (now Chatham College), planning to become a professional writer. Her plans changed, however, after her exposure to a first year biology class. She found that she was fascinated by the subject and decided to combine her two major interests—writing and biology—into a career of science writing. She eventually earned her bachelor's degree in zoology in 1928.

After graduation, Carson spent a summer at the Marine Biology Laboratory in Woods Hole, Massachusetts (now the

A 1951 photo of biologist and author Rachel Carson. *Bettmann/Corbis.*

Woods Hole Oceanographic Institute), before beginning graduate studies at Johns Hopkins University. At Woods Hole, Carson was introduced to the study of marine biology, a subject in which she would remain interested for the rest of her life.

Carson received her master's degree from Johns Hopkins in 1932, at the height of the Great Depression. She was able to find part-time teaching assignments at the University of Maryland and Johns Hopkins, but was just barely able to survive financially. Her problems were made more difficult in 1935 when her father died, leaving her responsible for her mother, who was in poor health.

Major Accomplishments

At this point in her life, Carson was fortunate enough to find a more lucrative position as a science writer at the U.S. Bureau of Fisheries. Her job there was to prepare radio scripts and to edit a variety of government publications dealing with fisheries. She was also able to do some freelance writing on the side, and in 1937, she published her first major article, "Undersea," in *Atlantic Monthly* magazine.

The article attracted the attention of an editor at Simon & Schuster, who encouraged Carson to write a book on the topic of her *Atlantic Monthly* article. Carson did so, and the book, *Under the Sea Wind*, appeared in 1941. Unfortunately, the book was published just as World War II was beginning, and it was never a commercial success.

Carson remained at her job with the Bureau of Fisheries (then renamed the Fish and Wildlife Service), where she eventually became chief editor of the agency. Her experience with *Under the Sea Wind*, however, convinced her to try her luck with another book. That book, *The Sea around Us*, was published in 1951 and achieved much greater success. It was chosen as an alternative selection by the Book of the Month Club and remained on the *New York Times* best-seller list for eighty-six weeks. The book also won Carson a National Book Award and honorary doctoral degrees from Chatham College and Oberlin College. Perhaps most important, the financial success of *The Sea around Us* allowed Carson to retire from government service and work full-time on her writing.

Carson's next book, *The Edge of the Sea*, was also a great critical and financial success. She also devoted her time to a variety of other projects, including a script for the *Omnibus* television series titled "Something about the Sky."

Silent Spring

For many years, Carson had been interested in and concerned about the widespread use of pesticides in agriculture and other fields. She had become convinced that these chemicals were making dramatic changes in many ecosystems, and she decided to write a book on the topic. That book, *Silent Spring*, was published in 1962 by Houghton Mifflin. The book argued that the indiscriminate use of pesticides was an ill-advised effort by humans to bring nature under our control, an effort that was causing irreparable harm to nature and that would, in the end, turn out to be fruitless as organisms adapted to the chemicals being used against them.

Carson's book was received with hostility by chemical companies and some reviewers, but eventually won over the hearts and minds of many legislators and the general public. She received many awards for the book, including the Audobon Medal, the Cullen Medal of the American Geographical Society, and a special award from the Izaak Walton League of America. She died of heart failure on 14 April 1964 at the age of 56.

Further Reading

Brooks, Paul. *The House of Life: Rachel Carson at Work.* New York: Houghton Mifflin, 1972.

Lear, Linda J. *Rachel Carson: Witness for Nature.* New York: Henry Holt, 1998.

Cavendish, Henry (1731–1810)

Henry Cavendish was the first person to describe the true chemical nature of water. Prior to Cavendish's research, many people thought of water as a basic and fundamental substance, a chemical element. This idea—developed by the ancient Greeks—had come down through the ages largely unchanged, into the early eighteenth century.

Personal History

Henry Cavendish was born in Nice, France, on 10 October 1731. His birth occurred in France because his mother was hoping to improve her poor health with a visit to the Riviera. As it happens, she was unsuccessful in this hope and died when Henry was two years old.

Cavendish is one of the most eccentric characters in all of the history of chemistry. He was extraordinarily shy and was unable to earn a college degree because of his inability to face his professors during examinations. Later in life, he made every possible effort to avoid contact with other people. He is said to have been able, with the greatest difficulty, to carry on a simple conversation with one man at a time, but never with a woman. He used written notes to communicate with his female servants for even the simplest of needs, such as the preparation of dinner. He even had a separate entrance built onto his house so that he could come and go without seeing other people.

Cavendish died in London on 24 February 1810 at the age of 78.

Major Accomplishments

For all his lack of social polish, Cavendish was one of the most brilliant scientific minds of his time. He devised new and ingenious experiments and made astute interpretations of the results of those experiments.

The work for which he is best known was carried out in the mid-1760s. In that research, he treated a variety of metals with acid and observed the production of a flammable gas. That gas was *hydrogen*. Today, we would represent the type of reaction Cavendish saw as follows, where Fe is the symbol for iron metal and HCl, the symbol for hydrochloric acid:

$$2Fe + 6HCl \rightarrow 2FeCl_3 + 3H_2$$

Cavendish continued to study the properties of hydrogen gas in succeeding years. In 1784, he burned hydrogen gas in a closed cylinder containing oxygen and found that a substance was formed that had "no taste nor smell, and which left no sensible sediment when evaporated to dryness; neither did it yield any pungent smell during the evaporation; in short, it seemed pure water." Cavendish later went on to determine the composition of water and found that it consisted of two parts hydrogen and one part oxygen by volume, a discovery that led to its current chemical formula of H_2O.

Cavendish made a number of other important contributions to chemistry and physics. For example, he first determined the chemical composition of nitric acid, and he found a way to determine with considerable accuracy the value of the gravitational constant.

Further Reading

Berry, Arthur John. *Henry Cavendish: His Life and Scientific Work*. London: Hutchinson, 1960.
Jungnickel, Christa, and Russell McCormmach. *Cavendish*. Philadelphia: American Philosophical Society, 1996.

Caves

A cave is a naturally occurring open space inside the earth. Caves can be formed by a variety of mechanisms. For example, rivers or streams may cut into rock, hollowing it out and forming a cave. Winds can have a similar effect when they blow strongly and persistently against fragile rock, such as sandstone. Such caves are known as *aeolian caves*. Ocean waves can erode the faces of cliffs and other rocky shorelines, cutting out *sea caves*, whose entrances may be underwater at least part of the time.

Ice caves are formed by running water. As the ice in a glacier melts, it may run down through cracks and crevices in the glacier, increasing the size of those fissures and eventually forming a cave. *Talus caves* are formed by rockslides in which boulders collect and form an enclosed underground space. *Lava tubes* are formed when lava flows begin to cool and harden. The tube is created when hot lava continues to flow through and eventually empties out of a surrounding tube that has already cooled and solidified.

Solution Caves

The most common type of cave is a *solution cave*, formed when groundwater dissolves rock located beneath the earth's surface. That process begins when rainwater dissolves carbon dioxide (CO_2) from the atmosphere as it falls to the ground. When water dissolves carbon dioxide, a chemical change takes place in which the weak acid, carbonic acid (H_2CO_3), is formed:

$$H_2O + CO_2 \rightleftharpoons H_2CO_3$$

The double arrow ($\rightleftharpoons$) above indicates that carbonic acid is unstable and decomposes rather easily, changing back into water and carbon dioxide.

As rainwater sinks into the ground, it becomes groundwater and slowly moves deeper into the earth. During that process, it absorbs more carbon dioxide from the ground, increasing the amount of carbonic acid that is formed.

Carbonic acid is a weak acid, not strong enough to dissolve most kinds of rock or mineral. There are a few exceptions to that statement, however, the most important of which is limestone, which consists of calcium carbonate ($CaCO_3$). When carbonic acid comes into contact with limestone, a chemical reaction occurs in which the limestone is converted into a more water-soluble form, calcium bicarbonate ($Ca(HCO_3)_2$):

$$H_2CO_3 + CaCO_3 \rightleftharpoons Ca(HCO_3)_2$$

Again, the double arrow means that calcium bicarbonate is not very stable and breaks down relatively easily into calcium carbonate and carbonic acid.

As groundwater containing carbonic acid flows through cracks in limestone, it dissolves small amounts of calcium carbonate, carrying away the soluble calcium bicarbonate that is formed. Over long periods of time, those cracks become larger and larger until a cave has been formed. The size and shape of the cave produced are determined by a number of factors, including the time over which water has been flowing through the rock, the size and shape of the cracks, and earth movements in the surrounding area.

Speleothems

During the evolution of a cave, the sequence of chemical reactions described above may be reversed. For example, suppose that a drop of groundwater has soaked through the earth and become suspended on the roof of a cave. Over time, the carbonic acid in that drop of water will tend to give up its carbon dioxide to the air:

$$H_2CO_3 \rightleftharpoons H_2O + CO_2$$

As the acidity (amount of carbonic acid) of the water droplet decreases, it will be able to hold less dissolved calcium bicarbonate. The calcium bicarbonate is converted back to calcium carbonate (limestone), which is then deposited on the roof of the cave:

$$Ca(HCO_3)_2 \rightleftharpoons CaCO_3 + H_2CO_3$$

The limestone formed in this process is known as *flowstone*.

When this process takes place not with a single drop of water but with a stream of water, a mass of limestone known as a *speleothem* is formed. Many types of speleothems are possible, depending on the way in which water flows through a cave. For example, if the water simply flows from the roof of a cave, drop by drop (as in the example above), an increasingly long icicle-like formation known as a *stalactite* is formed. If some of that water drops onto the floor of the cave, the "icicle" may grow upward rather than (or in addition to) downward. The floor-based "icicle" is a *stalagmite*. Under other circumstances, twisting, hollow tubes of flowstone, known as *helictites*, may be produced. And some of the most massive speleothems are formed when water runs down the side of a cave, leaving a mass of precipitated calcium carbonate behind in the form of a *curtain*, or *drape*.

Cave Life

Cave life can be categorized into one of three divisions: troglobites, troglophiles, and trogloxenes. Troglobites (*troglo-* = "cave"; *-bios* = "life") are organisms that spend their entire lives inside caves. They have adapted to conditions of near or complete darkness, relatively constant temperature, and other physical surroundings found in a cave. Many such organisms are blind because they have no need for vision in caves where little or no light penetrates. Some examples of troglobites are blind cave fish; cave shrimp; certain cave salamanders; and some species of isopods, amphipods, and millipedes.

Troglophiles (*-phile* = "loving") are organisms that can live within or outside of caves, but tend to prefer the former. They include cave crickets and certain types of beatles, cockroaches, millipedes, fish, and salamanders.

Trogloxenes (*-xenos* = "guest") are organisms that travel back and forth regularly between the interior of a cave and the outside world. Although a host of species, ranging from rats and raccoons to humans, belong to this category, the best known trogloxenes may be bats, which tend to spend daylight hours at rest in caves and evening hours hunting in the outside world.

Caves and Humans

Caves have played an important role in the study of human history. Because temperature and humidity tend to remain relatively constant in caves, artifacts left by our earliest ancestors are often well preserved. From these artifacts, archaeologists have been able to deduce a number of interesting facts about cave dwellers going back many thousands of years. Artifacts that have contributed to this knowledge include pottery, baskets, tools, clothing, food, feces, mummies, rock carvings (petroglyphs), mud drawings (mud glyphs), and cave paintings.

As an example, some of the best preserved artifacts are those found in caves near the town of Altamira in northern Spain. These artifacts date to a period between 11,000 and 19,000 years ago and provide a host of information about the daily lives of the Magdalenian people who lived in the area at the time.

Famous Caves

A few of the most famous caves in the world, with their most distinctive feature, include the following:

Aggtelek Caves (Hungary and Slovakia): a collection of more than seven hundred caves, one of which contains a stalactite 13 meters (43 ft) long and a stalagmite 32.7 meters (107 ft) in length, the largest in the world
Carlsbad Caves (New Mexico): the largest cave system formed by hydrogen sulfide (which reacts with water to form sulfuric acid) rather than carbonic acid
Fingal's Cave (Coast of Scotland): a massive sea cave that extends about 60 meters (200 ft) inland
Mammoth Cave (Kentucky): the world's largest cave system, extending over a distance of more than 550 kilometers (340 mi)
Peak Cavern (Derbyshire, United Kingdom): Largest cave entrance in the world, 20 meters (66 ft) high and 60 meters (200 ft) wide

Further Reading

Jackson, Donald Dale, and the editors of Time-Life Books. *Underground Worlds.* Alexandria, VA: Time-Life Books, 1982.
Middleton, John. *The Underground Atlas: A Gazetteer of the World's Cave Regions.* London: Hale, 1986.
Taylor, Michael Ray, and Ronald C. Kerbo. *Caves: Exploring Hidden Realms.* Washington, DC: National Geographic Society, 2001.

Center for Marine Conservation

The Center for Marine Conservation (CMC) was founded in 1989 for the purpose of aiding in the protection of ocean environments and conserving the global abundance and diversity of marine life. The organization works through science-based advocacy, research, and public education to promote citizen participation in its programs to protect the oceans and marine life.

CMC's activities fall into one of two major categories: Marine Ecosystem Protection and Marine Wildlife and Fisheries Conservation. Each of these categories is, in turn, divided into other areas of interest and concern, as follows:

Marine Ecosystem Protection
  Marine Reserves
    Protecting Coral Reef Ecosystems
    Sambos (Florida) Ecological Reserves
  Marine Sanctuaries
    Florida Keys National Marine Sanctuary
    Northwest Straits National Marine Sanctuary
  Clean Oceans Campaign
    Clean Water Act Reauthorization
    Regional Clean Oceans Programs
    Water Quality Monitoring and Citizen Outreach
    Scientific Research and Public Education
  Caribbean Reef Campaign
Marine Wildlife and Fisheries Conservation
  Alaskan Seas Campaign
  Threatened and Endangered Marine Wildlife
    Sea Turtles
    Marine Mammals Protection Act
    Overfishing and Fisheries Conservation

The center publishes a number of educational and informational brochures, pamphlets, audiovisual materials, posters and stickers, and other materials, including Kid's Marine Debris Kit, Beach Cleanup Kit, Plastics Recycling and Degradability Kit, and Commercial Fishing Impacts Kit, brochures on recreational fishing, boat maintenance, and environmentally safe boat trips; curricula entitled, Save Our Seas and "Turning the Tide on Trash"; and video entitle. "Trashing the Oceans and Port of Newport" and "Troubled Water, Plastics in the Marine Environment."

Further Information

Center for Marine Conservation
1725 DeSales Street, N.W., Suite 600
Washington, DC 20036
Telephone: (202) 429-5609
Fax: (202) 872-0619
URL: http://www.cmc-ocean.org
e-mail: cmc@dccmc.org

Clean Water Act of 1972

The Clean Water Act of 1972 is the most comprehensive piece of legislation passed by the U.S. Congress dealing with the issue of water pollution in U.S. rivers, lakes, streams, oceanfronts, and other waterways. The law was preceded by older and usually less-effective legislation, such as the Rivers and Harbors Act of 1899 (also called the Refuse Act of 1899), the Water Pollution Control Act of 1948, and the Water Quality Act of 1965.

The intent of the Rivers and Harbors Act of 1899 was to ensure that rivers and streams were maintained in a navigable condition by limiting the amounts and types of wastes that could be dumped into them. Until the 1970s, it was arguably the strongest piece of legislation available to the federal government to control pollution in the nation's waterways. The 1948 and 1965 acts offered funds and assistance to states for cleaning up their waterways, and charged them with setting standards for interstate waterways. But neither act was very effective in dealing with the growing problem of pollution in rivers, streams, lakes, and other waterways.

The Clean Water Act was originally passed as the Water Pollution Control Act of 1972. It was amended in 1977, at which time its name was changed to the Clean Water Act, and it was reauthorized in 1987. The act has been supplemented by other water-quality-related legislation, such as the Marine Protection Research and Sanctuaries Act of 1972, the Safe Drinking Water Act of 1974, the Toxic Substances Control Act of 1976, the Ocean Dumping Ban Act of 1988, the Shore Protection Act of 1998, and the Coastal Zone Act Reauthorization Amendments of 1990.

The original act declared that it was the intent of Congress to "restore and maintain the chemical, physical, and biological integrity of the Nation's waters." To achieve this objective, Congress established the following goals and policies:

- elimination of the discharge of pollutants into all navigable waters by 1985;

- provision for a water-quality level that would provide for the protection and propagation of fish, shellfish, and wildlife and for recreation in and on the water by 1983;

- prohibition of the discharge of toxic pollutants into waterways;

- provision of financial assistance for the construction of new waste-treatment facilities;

- assistance in the development of area-wide waste-treatment planning and facilities;

- encouragement and support for research and development projects on the development of new technology for the reduction of pollution in the nation's waterways;

- development and implementation of programs for the reduction and eventual elimination of non-point sources of pollution.

To carry out these objectives, Congress authorized funding to state and local programs designed to reduce water pollution; directed that a permitting process be developed to reduce the amount of pollutants released into waterways; charged states and tribes with the creation of specific water-quality standards; and established standards for the release of pollutants by industrial and municipal facilities.

In its own survey of the first twenty-five years of the Clean Water Act, the U.S. Environmental Protection Agency pointed to a number of accomplishments resulting from the act. The portion of the nation's waters safe for fishing and swimming, it pointed out, had increased from one-third to two-thirds. Wetland losses had decreased from an average of about 460,000 acres per year to about 70,000–90,000 acres annually. The number of Americans served by sewage treatment plants had increased from 85 million to 173 million. And the amount of soil erosion had dropped from 2.25 billion tons per year to about one billion tons per year ("Brief History of the Clean Water Act," http://www.epa.gov/owow/cwa/history.htm). *See also* Water Pollution

Further Reading

Clean Water Act. U.S. Code. Title 33, chap. 26. It is available online at ⟨http://www4.law.cornell.edu/uscode/unframed/33/ch26.html⟩.

Climate

See Clouds

Clouds

A cloud is a visible mass of water in the form of tiny water droplets of ice crystals. Clouds are an essential part of the hydrological cycle of any planet, including Earth. Clouds form when water evaporates from the earth's surface, rises into the atmosphere as a vapor, and condenses to a liquid or solid state when it reaches a point in the atmosphere where the temperature is less than the dew point. Dew point is defined in meteorology as the temperature at which air is saturated with water vapor. Any further cooling of air at its dew point normally results in the condensation of water from its vapor state.

Condensation of Water Vapor

An adequate supply of moisture in the air is not sufficient for the formation of a cloud. Tiny particles, called *condensation nuclei*, must also be present. Condensation nuclei provide surfaces on which water vapor is able to condense. Without condensation nuclei, the relative humidity of air can easily exceed 100 percent. Such conditions are rare—if not nonexistent—in practical circumstances because so many different materials can act as condensation nuclei. Tiny particles of carbon from forest fires, volcanoes, and anthropogenic activities; crystals of sulfate and nitrate compounds produced by both natural processes and human activities; and sea salt tossed into the atmosphere from ocean waves are examples of common condensation nuclei.

The process of cloud formation begins when water evaporates from oceans, lakes, rivers, streams, other bodies of water, and the earth's surface. As water vapor rises in the atmosphere, it begins to cool for two reasons. First, the higher something rises in the troposphere, the cooler the surrounding air becomes. Second, atmospheric pressure decreases the higher something rises above the earth's surface. As atmosphere pressure decreases, both air and the water vapor it contains expand. The process of expansion itself cools both air and water vapor.

At some point, called the *lifting condensation level*, the air and water vapor have cooled to the dew point, permitting condensation to occur. At that level of the atmosphere, condensation occurs quite rapidly, resulting in the formation of a very large number of tiny water droplets with diameters in the range of 10–20 μm (micrometers; millionths of a meter). This suspension of water droplets in air constitutes a cloud that is visible from the earth's surface.

Clouds may consist of either water droplets or tiny crystals of ice, depending on the temperature of the surrounding atmosphere and the presence or absence of freezing nuclei. *Freezing nuclei* are tiny particles whose geometric form is roughly similar to that of an ice crystal. In the absence of freezing nuclei, water may condense as droplets of water rather than ice crystals. The factor that determines the form in which water condenses is the surrounding temperature. At temperatures above 0°C, the water in clouds will be in the form of liquid water droplets. At temperatures between about 0°C and −10°C, water tends to exist in the form of *super-cooled water*—that is, water that has been cooled below its normal condensation point (0°C) but that remains in a liquid state. Between −10°C and −20°C, water exists as a mixture of water droplets and ice crystals. And below −20°C, water tends to occur most commonly in the form of ice crystals.

Types of Clouds

Meteorologists today classify clouds into one of ten major categories. The basic principles for this system were first suggested in 1803 by an English pharmacist and

Ten Major Cloud Types

Cloud Type	Category	Characteristics
cirrus	high	thin, wispy, very white, often with a slight curl; called "mare's tail"; composed of ice crystals
cirrocumulus	high	thin, white, with a rippled or wavy appearance; called a "mackerel sky"; made of ice crystals
cirrostratus	high	thin, white, and veil-like; may cover the sun or moon to produce a halo-like effect
altocumulus	middle	elliptical, puffy balls of white to gray clouds; may occur as collections of individual ball-like structures
altostratus	middle	sheetlike or layered, white to dark gray in color; sun or moon may be just barely visible through the cloud
stratocumulus	low	large, puffy, grayish clouds that may overlap each other to produce a rolling effect
stratus	low	dark gray, low-lying clouds that appear just above the earth's surface; may be associated with mist or drizzle
nimbostratus	low	generally shapeless dark gray clouds that are one of the primary sources of precipitation
cumulus	vertical	dense, white, billowy clouds with flat bases that reach to altitudes of 6,000 m (20,000 ft) or more
cumulonimbus	vertical	similar to cumulus clouds, but darker and more likely to produce severe storms; top may be flat, with an anvil-shaped head

amateur meteorologist Luke Howard. The modern system of cloud classification takes into account two factors: the altitude at which the cloud is found and its general shape. Most clouds can be described as *high* (more than 6,000 m [20,000 ft] above the earth's surface), *middle* (2,000–6,000 [6,500–20,000 ft] m in altitude), *low* (less than 2,000 [4,800 ft] m above sea level), and *vertical* (with a base close to the earth's surface and a top high in the atmosphere).

The list above summarizes important features of the ten major cloud types.

Many clouds fall into the neat categories in the table above, but some do not. Meteorologists have assigned names for these clouds that fit their particular shape and form. Some examples of these "unusual" clouds include:

fractus clouds, which appear to be broken apart into separate segments
castellanus clouds, with towering peaks that make them look like castles
lenticular clouds, with shapes like convex lenses
uncinus clouds, named for their hooklike shapes.

Clouds have long been used as an element in forecasting the weather. For example, mariners and farmers have known for centuries that the presence of a mackerel sky (cirrocumulus clouds) was a harbinger of bad weather. They learned that a squall line with heavy showers could be expected within a few hours.

Meteorologists now understand why clouds can be a reasonably good aid to weather forecasting. Clouds often form at the boundary between a cold front and a warm front, and the relative movements of such fronts generally produce certain types of clouds and certain types of weather. Some cloud types and the weather conditions with which they may be associated are the following:

cirrus: may presage the arrival of a distant storm that will arrive in 24–36 hours
cirrocumulus: precursor of a warm front; may develop into cirrostratus, a common source of rain
altocumulus: may precede the approach of severe weather and thunderstorms
altostratus: if becoming more intense, may portend the arrival of long-lasting rain or snow
nimbostratus: often accompanies constant and heavy precipitation
stratocumulus: generally associated with fair weather
stratus: often the source of light rain or drizzle or modest snow falls
cumulus: usually accompanied by fair weather unless they turn into cumulonimbus, in which case they are likely to produce thunderstorms, heavy rain, and even hail.

Fog

In terms of composition, fog is identical to a cloud—a mass of tiny droplets of water suspended in air. The major difference is that fog lies close to the earth's surface, while clouds are suspended a few hundred or thousands of meters into the atmosphere. Also, the manner in which fog is formed differs from the way clouds develop. In the case of clouds, water vapor typically condenses into water droplets as air masses rise higher in the atmosphere. In the case of fog, condensation occurs at or just above ground level.

Various forms of fog can be classified on the basis of both the manner of their formation and their general characteristics. For example, *radiation fog* forms when the ground and air just above it give off heat, an event that usually occurs at night. In such a case, the cooling trend may be sufficient to

November 1968 cloudscape above Tabuaeran (formerly Fanning Island) in the South Pacific nation of Kirbati. *Dr. James P. McVey, NOAA Sea Grant Program, NOAA.*

cause the adjacent air to reach its dew point, causing the condensation of water vapor onto grass, the ground, or other objects close to ground level. When winds are calm, a radiation fog tends to be relatively thin, immobile, and patchy. Radiation fogs are generally fairly dense and tend to drain downhill, filling valleys. Radiation fog usually dissipates in the morning when the sun begins to warm the earth. Heat from the earth, in turn, warms the lower boundary of the fog, causing the dew point to rise and water to evaporate to water vapor.

Advection fog occurs when warm moist air flows over cool ground. The ground absorbs heat from the air passing over it, reducing its temperature to the dew point. At that point, water vapor from the air begins to condense on ground-level objects, forming an advection fog. Because advection fogs

are caused by air movements, they are more likely than radiation fog to be pushed upward into the atmosphere. Such fogs may extend more than 500 meters (1,600 ft) upward and remain over extended periods of time. Geographic and topographic factors can be significant in the development of advection fogs. At Cape Disappointment in Washington State, for example, warm, moist ocean winds constantly blow over cool land on the cape, causing persistent, dense fogs. On average, the cape experiences more than one hundred days a year of advection fog.

Another form of fog caused by air movement is *upslope fog*, which develops when wind blows up a steep slope. As air moves upward on the slope, it expands, releases heat, and becomes cooler. In some cases, this sequence of events may cause the air temperature in the air mass to drop to its dew point, causing fog to form along the slope. Air masses moving up the sides of the Rocky Mountains may produce this effect.

The opposite of an advection fog is a *steam fog*, a phenomenon that occurs when cool air moves over a warm surface, such as warm water. In this instance, water may evaporate into the cold air above it, increasing the humidity to the dew point. If this happens, water vapor may recondense to form a layer of fog just above the surface of the water. Perhaps the most pronounced example of steam fog occurs in the Arctic and Antarctic regions. During winter, water temperatures in these regions stay significantly warmer than the air that blows over them from adjacent land masses. Huge amounts of steam fog may be produced, resulting in the condition known as *arctic sea smoke*.

Yet another form of fog, *frontal fog*, is produced when a warm air front collides with and rises up and over a cold front. When that happens, rain produced in the warm front may evaporate as it falls through the cold air mass, producing a cloud (fog) just above the earth's surface.

Clouds and Climate Change

Clouds may play an important role in climate change, about which many scientists, politicians, and ordinary citizens are now very concerned. On the one hand, an increase in annual global temperatures would cause an increase in the rate of evaporation from lakes and oceans. The water vapor thus formed would become part of the atmosphere and, eventually, condense into clouds. Therefore, the warmer the earth becomes, the more clouds should develop.

But clouds play an important role in determining climate. A certain fraction of the sunlight reaching the troposphere is reflected back into space by clouds. The more clouds there are, the more sunlight will be reflected, and the cooler the earth will become.

But the effect of clouds on weather is not that simple. Some clouds are bright and tend to reflect solar radiation. Others are dark and tend to absorb radiation. High and low clouds also behave differently in terms of their response to solar radiation. In addition, clouds over tropical areas have different properties and different effects from those over middle and higher latitudes.

Because it is difficult to predict the types of clouds that might be produced as the result of atmospheric warming, the nature of the clouds' influence on weather patterns is unclear. Some experts believe that cloud development might actually be a self-correcting mechanism that will automatically prevent the earth's temperature from rising.

Extraterrestrial Clouds

Astronomers have discovered the presence of cloudlike objects on planets other than Earth. Venus, for example, is forever enshrouded in heavy clouds that make it impossible for researchers to observe the planet's surface directly. However, Venusian clouds are not made of water, but of sulfuric acid. They appear to be distributed in three levels, at altitudes of about 48, 58, and 65 kilometers (30, 36, and 40 miles) above the planet's surface.

Similar cloudlike phenomena have been observed on Jovian planets and their satellites. For example, Jupiter has a complex mass of clouds that appear cream-colored from Earth, but turn out to be red, orange, and tan when viewed by telescope. Water may make up some part of these clouds, but they almost certainly consist of hydrogen, helium, ammonia, and methane. Jupiter's sister planet, Saturn, has similar formations above its surface. Astronomers believe that the whirling white masses they see may be huge storms in the Saturnian atmosphere. One of the most interesting recent discoveries about Senturian clouds was announced in late 2000. Astronomers found clouds on Saturn's satellite, Titan, that look and behave very much like cloud systems here on Earth.

Probably the only planet to have Earth-like clouds is Mars, which appears to have two types of cloud systems. One may consist of water and carbon dioxide, which condenses from the atmosphere during the cold evenings on the planet and around the poles during its winter season. The second cloud system seems to be made of fine dust particles stirred up by periodic storms on the planet's surface. *See also* Hydrologic Cycle; Ice; Precipitation

Further Reading

Hamblyn, Richard. *The Invention of Clouds: How an Unknown Meteorologist Forged the Language of the Skies*. New York: Farrar, Strauss & Giroux, 2001.

United States Weather Bureau. *Cloud Forms According to the International System of Classification*. Washington, DC: Government Printing Office, 1938.

World Meteorological Organization. *International Cloud Atlas*. 2d ed. Geneva: Secretariat of the World Meteorological Organization, 1975.

Colwell, Rita Rossi (1934–)

Rita Rossi Colwell is probably best known today for her work in the field of marine biotechnology, the search for ways of using materials found naturally in the oceans for medical, industrial, and other applications. She has also been interested in the study of the relationship between marine organisms and various types of water pollution. For example, in 2000 she reported on methods for using satellites to monitor the location and spread of the bacterium that causes cholera (*Vibrio cholerae*) in drinking water sources in Third World nations.

Personal Life

Rita Rossi was born in Beverly, Massachusetts, on 23 November 1934. She earned her bachelor's degree in bacteriology (1956) and her master's degree in genetics (1958) at Purdue University, and her Ph.D. in marine biology from the University of Washington in 1962. She married Jack Colwell, a fellow student at Purdue, in 1956. In 1964, Colwell accepted an appointment at Georgetown University, where she remained until 1972. In that year, she became professor of microbiology at the University of Maryland. During the early 1990s, Colwell was instrumental in the founding of the University of Maryland Biotechnology Institute, of which she was president from 1991 to 1998. She was appointed the eleventh director of the National Science Foundation (and the first woman director) in 1998, a post she currently holds.

Professional Recognition

Colwell has been awarded fourteen honorary degrees, the 1985 Fisher Award of the American Society for Microbiology, the 1990 Gold Medal Award of the International Institute of Biotechnology, and the 1993 Phi Kappa Phi National Scholar Award. She has authored or co-authored sixteen books, including *Estuarine Microbial Ecology* (University of South Carolina Press, 1973), *Biotechnology of Marine Polysaccharides* (Hemisphere Press, 1985), and *Biomolecular Data: A Resource in Transition* (Oxford University Press, 1989).

Cook, James (1728–1779)

Cook is often regarded as the greatest explorer and navigator since Magellan. He was born into an agricultural family in the village of Marton, Yorkshire, on 27 October

1728. After working as an apprentice on commercial sailing ships, he joined the Royal Navy in 1755 and served in the French and Indian War of the late 1750s and early 1760s. Cook then spent several years studying the coasts of northeastern Canada. He carried out many ocean-depth soundings and geographical surveys in the waters around Newfoundland and Labrador, and at the mouth of the St. Lawrence River.

In 1768, Cook received a commission from the Royal Society to captain the sailing ship *Endeavour* on a voyage to the newly discovered island of Tahiti. The main purpose of the trip was to observe the transit of Venus, an astronomical phenomenon of considerable interest and importance to scientists. While on this voyage, Cook discovered the Admiralty and Society Islands, explored the eastern coast of Australia, and circumnavigated New Zealand. He was the first foreigner to find that the country actually consisted of two separate islands.

Cook set out on a second expedition to the Pacific Ocean in 1772. On this voyage, he was searching for the possible existence of a southern continent beyond both South America and Australia. He traveled as far as the Antarctic Circle and proved that no such continent existed, at least as far north as he had traveled. As a result of this voyage, which ended in 1775, scholars gained a relatively complete and accurate description of the southern Pacific Ocean and the lands bordering it.

A year after returning to England, Cook launched a third voyage to the Pacific. On this trip, he discovered the Hawaiian Islands and sailed as far north as the Alaskan and Siberian coastlines. On his way home, he stopped once more in Hawaii, where he was set upon during a brawl with some natives and killed. His death occurred in Kealakekua Bay on 14 February 1779. According to legend, Cook was eaten by natives of the island, who were, at the time, cannibals.

In addition to his outstanding work in cartography and hydrography, Cook carried out some important studies on the role of vitamins in maintaining good health. On his second voyage, he tested the proposal originally made by Scottish physician James Lind (1716–1794) that scurvy, a disease endemic among sailors, could be prevented by incorporating an adequate supply of citrus fruits (which we now know to provide vitamin C) into one's diet. *See also* Oceanography

Further Reading

Beaglehole, J. C. *The Life of Captain James Cook*. Stanford, CA: Stanford University Press, 1992.

———, ed. *The Journals of Captain Cook*. New York: Penguin, Penguin Classics, 2000.

Hough, Richard. *Captain James Cook*. New York: W. W. Norton, 1997.

Cooling Water

Cooling water is water that is used as a heat-exchange medium in some industrial, chemical, power-generating, air-conditioning, and other facilities. A heat-exchange medium is a material that is used to transfer heat from one point to another. Water is one of the most common heat-exchange mediums because it is abundant, is relatively inexpensive, and has a high heat capacity—the ability to absorb a large amount of heat per unit of mass.

By far the largest use of cooling water is in the production of electrical power. According to the Environmental Protection Agency, the average steam electric power plant in the United States uses about 700 million gallons of cooling water per day. Such plants account for about 92.4 percent of all cooling water used in the United States, about 71,172 bgy (billion gallons per year). Other important consumers of cooling water include the chemicals and allied products industry (3.6 percent of all cooling water; 2,797 bgy), the primary metals industry (1.7 percent; 1,312 bgy), the petroleum and coal products industry (0.8 percent; 590 bgy), and paper mills and related paper plants (0.7 percent; 534 bgy).

Steam Electric Power Plants

Electricity is generated in a fossil-fueled steam power plant when coal, oil, or gas is burned. Heat from the burning fuel is used to boil water and produce steam. That steam is then used to drive turbines which, in turn, operate electric generators. Nuclear power plants operate on essentially the same principle except that the heat needed to make steam comes from controlled nuclear reactions in the plant's reactor core.

Steam that has passed through a turbine must be cooled and converted back to liquid water, which can then be reused in the plant to generate more steam. Cooling water is used to convert spent steam back to liquid water.

At one time, cooling water was simply taken from a lake or river near the power plant, used to condense steam, and then released back into the river or lake. The returned water contained more heat than it did when removed from the lake or river, of course, and also contained new impurities, such as dissolved solids and microorganisms from the plant's water system. When freshwater was abundant and environmental concerns minimal, this system was widely used.

Cooling Towers

In recent decades, however, traditional ways of using cooling water have become unacceptable in most cases, because of both the increasing scarcity of large amounts of freshwater and the potential environmental risks posed by recycled cooling water. In this context, the use of *cooling towers* has become much more popular. A cooling tower is a tall structure through which hot water passes and gives up its heat.

Heated cooling water loses its heat in a cooling tower by two mechanisms. First, some

heat passes directly from the hot water into the air because of a difference in temperature of the two media. The amount of heat lost in this way depends on the season (that is, the surrounding air temperature), but tends to be a relatively small amount, of the order of about 20 percent of the total heat loss.

Second, a much larger amount of heat is lost from warm cooling water by the process of evaporation. The used cooling water is exposed to the air in a cooling tower, allowing some of it to evaporate and be lost to the atmosphere as water vapor. The reason for this procedure is that water has a very high heat of evaporation, the amount of heat required to change a unit mass of liquid water into a gas. For each pound of water lost by evaporation, 1,000 Btu of heat are removed from the cooling water. Of course, water lost by evaporation must be replaced from a source other than the plant's cooling system, such as a nearby lake.

Cooling towers also provide a convenient way to monitor the quality of water being used in a plant. Chemicals can be added at some point within the tower to remove dissolved solids that cause corrosion of pipes, to kill potentially harmful microorganisms, to maintain the proper acidity of the cooling water, and to make other necessary adjustments in water quality.

Environmental Considerations

The use of cooling towers has solved many, but not all, of the economic and environmental problems associated with the traditional system of removing and returning cooling water directly from and to a natural body of water. One remaining environmental issue is that water vapors released from a cooling tower may, under certain meteorological conditions, condense to form fog, which, in turn, can result in dangerous driving or flying conditions.

Also, cooling water released by the older, direct in-and-out process can some-

The cooling towers of Three Mile Island nuclear power plant near Harrisburg, Pennsylvania, shown here shortly after the 1979 accident that shut down the facility. *Bettmann/Corbis.*

times have some positive environmental effects, as in the operation of commercial aquaculture farms, in irrigation systems where the growing season may be extended by the use of heated water, and in heat transfer systems by which homes and other buildings can be heated by spent cooling water.

Further Reading

Cheremisinoff, Nicholas P., and Paul N. Cheremisinoff. *Cooling Towers: Selection, Design, and Practice.* Ann Arbor, MI: Ann Arbor Science Publishers, 1981.

Hill, G. B., E. J. Pring, and Peter D. Osborn. *Cooling Towers: Principles and Practices.* London: Butterworth-Heinemann, 1990.

The Coral Reef Alliance

The Coral Reef Alliance (CORAL) works with the dive industry, governmental agencies, local communities, and other organizations to promote coral reef conservation around the world. Its goal is to protect and manage coral reefs, establish marine parks, fund conservation efforts, and promote public awareness of coral reefs.

Some examples of the Alliance's work include the following:

• grants to local organizations for the conservation and protection of coral reefs, such as emergency relief for reefs struck by hurricanes and the purchase of outboard motors for boats used to patrol marine-protected areas;

• funding and expertise for the creation of coral reef parks;

• development of partnerships with the dive industry to promote responsible diving and marine conservation;

• production of an educational slide presentation, underwater photo exhibition, and the alliance's website to raise public awareness about coral reef issues.

CORAL's website provides a search function on a wide variety of reef-related issues and links to a number of other organizations worldwide interested in issues of coral reef conservation and protection.

Further Information

The Coral Reef Alliance
2014 Shattuck Avenue
Berkeley, CA 94704
Telephone: (510) 848-0110 or (888) CORAL-REEF
Fax: (510) 848-3720
URL: http://www.coralreefalliance.org
e-mail: info@coral.org

Cousteau, Jacques-Yves (1910–1997)

Jacques Cousteau is best known for his studies of and popular writings about the marine world. During his lifetime, he invented some important devices for the study of ocean life and wrote a number of widely acclaimed books, television programs, and feature-length films about the ocean.

Personal Life

Jacques-Yves Cousteau was born on 11 June 1910 in St. André-de-Cubzac, France. He spent his early years traveling throughout Europe and North America with his family. His father was a legal advisor whose work took him to a variety of countries. As a young teenager, Cousteau lived and went to school in New York City and spent summers at Lake Harvey in Vermont. It was during these summer vacations that Cousteau first became interested in and skilled at deepwater diving.

Back in France, Cousteau was an indifferent student who was expelled from high

Jacques Cousteau (background) with his son Philippe (at the wheel). *Bettmann/Corbis.*

school for throwing rocks through the school's windows. In addition to his interest in diving, Cousteau developed an interest in photography and is said to have owned one of the first home-movie cameras sold in France.

In 1930, Cousteau was admitted to the French Naval Academy, from which he graduated two years later with a degree in engineering. His academic work had improved significantly since his rock-throwing days, and he graduated second in his class. After graduation, he entered naval aviation school with plans to become a pilot. Those plans were dashed, however, when he was involved in a car accident in which his right arm was paralyzed.

That accident turned out to be a blessing in disguise. To regain full use of his arm, Cousteau began an intensive program of swimming in Le Mourillon Bay. It was his daily exercise in the bay that sparked an interest in marine studies, an experience that was to transform his life. "Sometimes we are lucky enough to know that our lives have been changed," he once wrote, "to discard the old, embrace the new, and run headlong down an immutable course. It happened to me at Le Mourillon on that summer's day, when my eyes were opened on the sea" (quoted in McMurray, 414).

Major Accomplishments

By the late 1930s, Cousteau had begun to combine his two primary interests: photography and deep-sea diving. He tried to develop methods for taking both still and motion pictures underwater. One of the problems he encountered, however, was the lack of adequate diving equipment. At the time, the only two options available were goggle diving and deep-sea diving suits. Goggle diving was not very satisfactory because one could not go very far beneath the surface. And diving suits were of limited value because they were so large, heavy, and cumbersome. In addition, one's movements were limited because of the air line that connected the diver to the mother ship overhead.

By 1943, Cousteau and a colleague, Emile Gagnon, had developed a new type of diving device, which they called an *aqualung*. The invention was later renamed the *scuba* by the U.S. Navy, standing for self-contained underwater breathing apparatus. The aqualung consisted of a tank of compressed air, a regulator to supply air to the diver, and a breathing tube through which the air was delivered. The invention not only provided Cousteau with a new method for underwater photography but also became useful as a way of locating and disabling German mines during World War II.

Cousteau's reputation as a marine researcher was made in the early 1950s after he purchased a converted U.S. minesweeper, had it refurbished as a research vessel, and named the ship *Calypso*. Cousteau took *Calypso* on an extended expedition to the Red Sea, where he made a number of important discoveries—including new plant and animal species and volcanic basins on the sea bottom. In February 1952, Cousteau made one of the discoveries for which he is best known—a sunken Roman ship filled with great treasures off the coast of southern France.

The voyages of *Calypso* provided Cousteau with the information on which he based his first book, *The Silent World*. The book was eventually translated into twenty languages and sold more than five million copies worldwide. The book was eventually made into a film that won a Palme d'Or at the 1956 Cannes Festival and an Academy Award in 1957.

Cousteau had long been interested in devices that would allow researchers to remain underwater for extended periods of time. In the 1950s, he and his colleagues invented a small submarine called a *diving saucer*, in which researchers could work for a few hours at a time. A decade later, he had improved that invention with a device known as *Conshelf*. A Conshelf station was a self-contained, underwater base in which researchers could live and work for days or weeks at a time. The first Conshelf station was built on the floor of the Mediterranean Sea, 10 meters (33 feet) below water level. Two men remained in the station for a week. A year later, an improved version of the station was built on the floor of the Red Sea at a comparable depth. A crew of five men remained in the station for a month.

By the late 1960s, Cousteau had begun to realize the potential of television for educating the general public about the marine world. He produced a series of programs in 1968 called *The Undersea World of Jacques*

Cousteau, a series that was to run for eight seasons. In 1970, he produced another series of twelve one-hour programs under the title *The Undersea Odyssey of the "Calypso."* In all, Cousteau made more than a hundred documentary films, winning him ten Emmys and three Oscars.

Cousteau was long an active spokesman for protection of the world's oceans. In 1970, he formalized his concerns in this area by founding the Cousteau Society, which now has its home base in Bridgeport, Connecticut. For his work in support of the world's oceans, he has received a number of awards and honors, including the Grand Croix dans l'Ordre National du Merite from the French government and the U.S. Presidential Medal of Freedom, both in 1987. Cousteau died on 25 June 1997. *See also* Scuba

Further Reading

Crawford, Tom. "Jacques Cousteau." In *Notable Twentieth Century Scientists*, edited by Emily S. McMurray, vol. 1, 413–417. New York: Gale Research, 1995.

Madsen, Axel. *Cousteau: An Unauthorized Biography*. New York: Beaufort Books, 1986.

Munson, Richard. *Cousteau: The Captain and His World*. New York: Morrow, 1989.

Crinaeae

See Nymphs

D

Dams

A dam is a structure built across a waterway to prevent the flow of water, thereby creating a lake or large reservoir of water that can be used for a variety of purposes. The primary function of dams is to control flooding; provide a water supply for industrial, commercial, agricultural, or some other use; and supply water for the generation of hydroelectric power.

History

The earliest dams may have been constructed in Mesopotamia about 6000 B.C. Although it is difficult to say with any certainty, it appears that the function of these dams was to provide water for irrigation of crops. Over the next 1,500 years, the Sumerians developed a complex system of dams and irrigation canals used to manage water from the Tigris and Euphrates Rivers.

By about 3500 B.C., the Egyptians had also become dam builders. The primary function of these dams was probably to control the flooding of the Nile River, a problem that has plagued Egyptian society for more than five millennia. One of the earliest dams was constructed near the modern-day city of Helwan. The dam appears to have collapsed during the first flooding following its construction, but remnants of its remaining underlying structure can still be observed today.

Dams were being built almost everywhere in the world by the beginning of the Christian era. For example, sophisticated structures began to appear in Sri Lanka as early as 400 B.C. One of the largest of these was built in A.D. 460, a structure 34 meters (110 ft) in height. The dam remained the world's highest dam for more than 1,000 years after.

The pace of dam construction in Great Britain and the rest of Europe was accelerated with the arrival of the Industrial Revolution in the nineteenth century. By 1900, Great Britain had more dams in operation than all of the rest of the world combined.

The developing interest in hydroelectric power in the 1930s provided yet another impetus for dam construction. It was during this period that some of the world's largest dams were begun, including the Hoover Dam (completed in 1936), the Grand Coulee (completed in 1942), and the Fort Peck (completed in 1940) in the United States, and the Afsluitdijk in the Netherlands (completed in 1932). By the end of World War II, large dams were being built in nearly every part of the world.

The largest dam in existence today is the Syncrude Tailings structure in Canada, with a volume of 540,000 cubic meters (708,000 cu yd). Other large dams around the world are the Chapetón and Pati in Argentina, the Tarbela in Pakistan, the Kambaratinsk in Kyrgyzstan, the Lower Usuma in Nigeria, and the Cipasang in Indonesia.

Today, the largest number of dams is to be found in China, with over 19,000 such structures. The countries with the next largest numbers of dams include the United States (5,500), the countries of the former Soviet Union, Japan, and India. For a variety of reasons, the rate of dam construction has been declining over the past half century, especially in more developed nations. That number has fallen from an average of about 1,000 new large dams per year in the 1950s to about 260 per year in the mid-1990s ("About Rivers and Dams"). At the present time, the greatest level of dam construction is occurring in less developed parts of the world, including China, Turkey, and South Korea.

Types

Dams can be classified according to the material from which they are made and their structural style. One of the most common types of dams is the *arch dam*, so named because it is built in the shape of an arc that is curved upstream. With this structure, the force of water in the lake is transmitted to dam abutments, the parts of the dam that are connected to the surrounding rock. Arch dams are commonly built of concrete in narrow canyons with steep walls of stable rock. Examples of arch dams in the United States include the Flaming Gorge in Utah, the Glen Canyon in Arizona, and the Hungry Horse in Montana.

Buttress dams have upstream sides that slope at angles of about 45 degrees and are supported by strong buttresses on the downstream side of the dam. The force of water in the lake pushing on the dam wall is transmitted to the buttresses and thence to the ground or surrounding rock. Buttress dams can take a variety of forms, such as arch buttress, flat slab or slab buttress, multiple-arch buttress, and solid-head buttress. The Bartlett in Arizona, Lake Tahoe in California, and Link River in Oregon are examples of buttress dams in the United States.

Gravity dams are constructed of concrete, masonry, or some other very heavy material. The weight of the material acts as a counterforce against the pressure of stored water in the lake or reservoir. The effectiveness of the dam material is increased by making the dam very wide at the bottom and providing a slight arclike curve. Gravity dams are often used to control floods and are built so that water can flow over their top without washing away the dam. As with buttress dams, gravity dams can take many forms, such as arch gravity, curved gravity, and hollow gravity. Examples of gravity dams in the United States are the Altus in Oregon, Angostura in South Dakota, and Elephant Butte in New Mexico.

Embankment dams are made from any type of excavated natural material and are named according to the materials used. For example, *earth dams* consist of more than 50 percent earthy material less than 3 inches in diameter. Other types of embankment dams are rockfill, rolled fill, zoned, and hydraulic fill.

Environmental Issues

During the heyday of dam construction in the mid-twentieth century, relatively little attention was paid to possible environmental consequences of these structures. Engineers, politicians, and ordinary citizens were generally delighted that it had become possible to control floods and provide abundant water for many uses, including power generation and irrigation.

More recently, questions have been asked about the environmental impact of dams on the areas in which they are constructed. One obvious effect is on the flora and fauna that live in the dammed river. Changing the flow of water in a river by damming it changes water conditions both above and below the dam, altering the plant and animal life that may be able to live there.

Dams also change the pattern of erosion above and below the dam itself. Perhaps the most dramatic example of such changes is the Aswan High Dam, built on the Nile River in the 1960s. The purpose of the dam was to control the unpredictable flooding of the Nile, which it has been effective in doing. However, the dam has also prevented the flow of silt and sediments down the Nile into the populated regions of Egypt. This release of silt and sediments had made the land around the Nile one of the richest and most agriculturally productive in the world for more than five millennia. As a result of the dam's construction, no new soil is being delivered to Egypt's farmers and, for the first time in human history, they are having to use fertilizers on their crops.

There are also concerns about the displacement of human populations when large dams are built. According to one estimate, anywhere between thirty and sixty million people have lost their homes to make way for the construction of new dams. The same estimate claims that about two million people every year are still being displaced for dam construction, the majority of them in China and India ("About Rivers and Dams"). *See also* Energy from Water; Floods; Reservoirs

Further Reading

"About Rivers and Dams," http://www.irn.org/basics/dams.shtml.

Dorcey, Anthony H. J., ed. *Large Dams: Learning from the Past, Looking at the Future.* New York: World Bank, 1997.

International Water Tribunal. *Dams.* Utrecht: International Books, 1994.

World Commission on Dams. *Dams and Development: A New Framework for Decision-Making.* London: Earthscan, 2000.

Further Information

United States Society on Dams
1616 Seventeenth Street, Suite 483
Denver, CO 80202
Telephone: (303) 628-5430
Fax: (303) 628-5431
URL: http://www2.private1.com/~uscold
e-mail: stephens@ussdams.org

The International Journal on Hydropower and Dams
123 Westmead Road
Sutton, Surrey, SM1 4JH
United Kingdom
Telephone: +44 (0) 20 8643 5133
Fax: +44 (0) 20 8643 8200
URL: http://business.virgin.net/hydropower.dams
e-mail: subs@hydropower-dams.com

Deliquescence

See Hydrate

Deluge

See Flood Legends

Desalination

Desalination is the process by which dissolved salts are removed from salt water or brackish water to produce water that can be used for human consumption. Desalination is also known as *desalinization* or *desalting*. In many parts of the world, obtaining an adequate supply of freshwater is an important ongoing problem. About 97 percent of the planet's waters constitute the oceans, and this water contains too many dissolved solids to be used for drinking, cooking, or other domestic and industrial purposes.

The term *saline water* is generally defined as water that contains more than 1,000 ppm (parts per million) of dissolved solids. Saline water is further categorized as *brackish water* (1,000 to 4,000 ppm dissolved solids), *salted water* (4,000 to 18,000 ppm dissolved solids), and *seawater* (18,000 to 35,000 ppm dissolved solids). Desalination procedures can produce water with a concentration of anywhere from 1 to 1,000 ppm dissolved solids, depending on the method used for purification.

Growth of Desalination Plants

Although the principles on which desalination is based are simple and well known, the process has become commercially viable only within the past half century. One impetus to studies on desalination was World War II, during which military troops were sometimes based in areas where freshwater was difficult to obtain. The United States government became actively involved in research on desalination to deal with this problem, a commitment that continued after the war. The Department of the Interior's Office of Saline Water (later the Office of Water Research and Technology) supported much of the research that was to lead to present-day desalination systems.

As of 2001, more than 7,500 desalination plants were in operation worldwide. About 60 percent of these plants are located in the Middle East, where freshwater is scarce and national governments can often afford the cost of building and operating such plants. The world's largest desalination plant is in Saudi Arabia, a plant that produces 485,000 m^3 (cubic meters of water), or 128 mgd (million gallons per day). About 12 percent of the world's desalination capacity is located in the Western Hemisphere, with most plants located in the Caribbean and Florida ("Seawater Desalination," http://www.coastal.ca.gov/desalrpt/dchap1.html).

Worldwide desalination capacity is now estimated at about 22.7 million m^3 (6 billion gal) per day, an increase of about 70 percent over the preceding decade. Ten countries, primarily in the Middle East and North Africa, account for 75 percent of this capacity. Saudi Arabia is the world's largest producer of desalinated water, accounting for nearly a quarter of the world's capacity. The United States ranks second after Saudi Arabia, with about 16 percent of the world's capacity. (International Desalination Association, 1998).

Desalination Technologies

Most desalination technologies can be classified into one of two categories: thermal or membrane processes. In thermal technologies, saline water is heated to vaporize the water component of the mixture. The steam formed is then condensed to produce pure water. The salts present in the saline water are then removed in a more concentrated saline solution. In membrane technologies, saline water is passed through a membrane so as to "filter out" pure water, leaving behind a saline solution more concentrated than the original solution. These two types of technologies account for about 86 percent of the world's desalination capacity. The remaining 14 percent makes use of other methods, such as electrodialysis and vapor compression techniques.

Thermal Technologies

Thermal technologies can be subdivided into three different methods: (1) multi-stage flash distillation, (2) multi-effect distillation, and (3) vapor compression distillation. In all three methods, the general principle is to convert some of the water in a saline solution into vapor, which can then be condensed and removed as pure water. All methods operate at reduced air pressure because water boils at a lower temperature at reduced pressures. For example, the boil-

A desalination plant in Oman. *Bojan Breceli/Corbis.*

ing point of water at normal atmospheric pressure (760 mm Hg) is 100°C (212°F), but at half that pressure (380 mm Hg), its boiling point is only 80.9°C (178°F). By using this approach, such plants significantly reduce the amount of heat energy needed to boil water and, thus, the cost of operating the plant.

A *multi-stage flash distillation* (MSFD) facility consists of two major parts: a brine heater and a *stage*—a chamber in which saline water is heated and steam produced. A pipe carrying saline water passes through the upper part of the stage and into the brine heater. In the brine heater, the pipe carrying the saline water is heated by steam from an external boiler. The heated saline water is then fed back into the stage, where it empties out of the inlet pipe into the chamber itself. At the reduced pressure within the stage, the hot saline water begins to boil—

that is, it "flashes." Steam formed during boiling rises in the chamber and heats the inlet pipe through which it passed earlier. As the steam gives up its heat to the inlet pipe, it condenses to form pure water, which is captured and drawn off at the bottom of the stage.

In fact, an MSFD plant, as its name suggests, contains more than one stage. Purified water from the first stage passes into a second stage, where the process that occurred in stage one is repeated. The water is further purified as it gives up more heat to the inlet pipe at the top of the chamber. An MSFD plant may have anywhere from 15 to 25 stages. The final products of the technology are nearly pure water and a saline solution with a much higher concentration of salts than the original raw material.

In a *multi-effect distillation* (MED) plant, saline water is fed into the plant in a long pipe that empties into a chamber known as an *effect*. The effect is heated by a pipe carrying steam from an external boiler and is maintained at reduced pressure. When the saline water empties into the effect, it begins to boil immediately, as in an MSFD plant. The steam formed rises in the effect, warming the inlet pipe carrying incoming saline water and condensing to form pure water.

As in the MSFD method, an MED plant consists of more than one effect, giving the technology its name. The inlet pipe carrying saline water empties not into a single effect but into a series of effects. In each of the chambers, the same process occurs, with flash boiling of water in the saline solution producing heat that warms the incoming saline solution and pure water, condensed from the steam. A typical MED plant has from 8 to 16 effects.

A *vapor compression distillation* (VCD) plant is often used for small facilities, such as resorts or industries where freshwater is scarce. It may also be used in conjunction with other types of desalination plants, such as the MSFD or MED processes described above. In the VCD process, saline water enters the plant through a long pipe and is sprayed into a chamber that is heated at reduced pressure. Water in the saline solution boils and is removed through a pipe at the top of the chamber. The steam is then condensed and the hot water is passed through the original chamber in an outlet pipe. As the hot water passes through the chamber, it gives off heat, which is used to vaporize the next batch of water sprayed into the chamber.

Membrane Technologies

Osmosis is a process by which charged particles (ions) pass through a semipermeable membrane. A *semipermeable membrane* is one that allows particles of one size to pass through, but not those of a larger size. If two solutions of unequal concentrations are separated by a membrane, the net movement of particles is such that the more concentrated solution becomes less concentrated, and the less concentrated solution becomes more concentrated. After a period of time, the concentrations of both solutions on either side of the membrane are equal.

The purpose of desalination, however, is to produce the opposite result, making a saline solution more concentrated by removing and collecting pure water from it. Two methods have been developed to achieve this objective—electrodialysis and reverse osmosis.

In *electrodialysis*, saline water is passed through a chamber that contains hollow cylindrical tubes covered with thin membranes. Half the tubes carry a positive electrical charge and half, a negative electrical charge. As salt water passes through the chamber, positively charged ions in the solution are attracted to the negatively charged tubes and are removed from solution. Negatively charged ions are attracted to the positively charged tubes and are also removed from the solution. The positively and negatively charged tubes are arranged in alternate positions in the chamber so that saline water flows past one type of tube first, then the other type of tube. The space between each pair of charged cylinders through which the saline water flows is known as a *cell*. A typical electrodialysis plant contains several hundred cells.

A *reverse osmosis* plant is very simple in concept. Saline water under high pressure is forced into a chamber containing a large membrane. The membrane used must be carefully selected so as to permit the passage of water molecules but not charged ions in the saline solution. The pressure forces water molecules through the membrane, leaving behind a more concentrated saline solution than the original. The pure water is collected on the opposite side of the membrane, while the more concentrated saline solution is drawn off at the bottom. As with other desalination processes, the separation process described here may be repeated in more than one chamber, further purifying the water produced each time.

Other Technologies

Other methods for desalinizing water have also been developed, although none has been as commercially successful as those described above. One method which might be familiar to the general public is the solar still. Many young people are taught to make solar stills in Boy Scouts, Girl Scouts, school science classes, and other settings.

A solar still consists of three parts: a pan to hold saline water, a pan of larger diameter in which to place the first pan, and a transparent covering for the two pans. Clear glass or plastic is often used for the covering. When this arrangement is placed in direct sunlight, energy from the sun's rays heats water in the saline solution and causes it to evaporate. Water vapor rises and condenses on the covering, collecting into droplets, and running down the inside of the covering. Liquid water then drops into the outer pan in the form of pure water.

Although the concept of a solar still is simple and the energy needed to operate it (sunlight) is free, this technology has not found wide commercial use. The major problem is the cost of constructing and maintaining the necessary facility. In addition, the still operates effectively only in places where significant amounts of sunlight can be reliably depended on.

As the world's freshwater problems become more difficult, interest in desalination is likely to increase. During the last decade of the twentieth century, the industry began to develop a more sophisticated professional structure with the creation of the International Desalination Association as well as national and regional desalination organizations in Europe, the United States, India, Japan, Pakistan, and the Gulf nations. In the United States, desalination research is supported by both government and private agencies, such as the American Water Works Association, the U.S. Bureau of Reclamation, and Canada's National Water Research Institute. *See also* Distillation; Salinity

Further Reading

Balaban, Miriam, ed. *Desalination and Water Re-Use: Proceedings of the Twelfth International Symposium*. Rugby, UK: Institution of Chemical Engineers, 1991.

Buros, O. K. *The ABCs of Desalting*, 2d ed. Topsfield, MA: International Desalination Association, 1999.

Buros, O. K., et al. *The USAID Desalination Manual*. Englewood Cliffs, NJ: IDEA Publications, 1982.

"Desalination of Water and Sea-Water," http://www.world-wide-water.com/Desal.html.

International Desalination Association. *1998 10A Worldwide Desalting Plants Inventory—Report No. 15*. Topsfield, MA: IDA, 1998.

Further Information

International Desalination Association
P.O. Box 387
Topsfield, MA 01983
Telephone: (978) 887-0410
Fax: (978) 887-0411
URL: http://www.ida.bm
e-mail: idalpab@ix.netcom.com

Deuterium Oxide

See Heavy Water

Distillation

Distillation is the process by which two or more substances are separated from each other by means of evaporation and condensation. For example, pure water can be separated from solid impurities dissolved in it, or two liquids of different boiling points (such as water and alcohol) can be separated from each other more or less completely.

History

Philosophers as early as the second century B.C. in China appear to have understood the principles of distillation. They discussed the possibility of separating the "light" from the "heavy" and the "pure" from the "dross." But evidence for the existence of actual distillation apparatus appears no earlier than the ninth century A.D., where it is described in manuscripts by Mary the Jewess of Alexandria. By the eighth century, distillation had become a relatively common procedure among Arabic alchemists, especially in the preparation of perfumes. Only a few centuries later, it was a standard procedure used by alchemists in their efforts to transmit impure substances into more pure forms.

Procedure

Modern distillation equipment can take a variety of forms. In one of the most common of those forms, the liquid to be purified is placed into a round flask with a tall neck. Extending from the top of the neck is a sidearm attached to a condensing tube. The condensing tube consists of a long tube surrounded by a second tube of larger diameter. Water flows into the bottom of the outer *cooling tube* and out the top.

Suppose that one wishes to obtain pure water from seawater by means of distillation. The seawater is placed in the distilling flask and heated. At some point, the water begins to boil. The steam thus formed rises in the distilling flask and passes out of the sidearm into the condensing tube. Dissolved solids in the seawater remain behind in the distilling flask. As steam passes into the condensing tube, it gives up heat to the cooling water and

condenses to pure water. The pure water is collected at the end of the condensing tube.

Suppose that one wishes to separate water from another liquid, such as alcohol, with which it has been mixed. Again, the mixture (alcohol and water) is placed in the distilling flask and heated. In this case, the liquid with the lower boiling point, alcohol, begins to boil first. The alcohol changes into a vapor, passes out the sidearm, condenses in the condensing tube, and is collected at the bottom of the tube. When all of the alcohol has boiled off, nearly pure water remains in the distilling flask.

In some cases, the distilling process—such as that involving a water-alcohol mixture—is more complex than described here. For example, water and ethyl alcohol form an *azeotropic mixture* consisting of 4.4 percent water and 95.6 percent alcohol that cannot be further separated by distillation.

With appropriate adjustments in the distillation apparatus and procedure, it is possible to separate complex mixtures of gases, solids, and liquids of different boiling points.

In theory, distillation provides an attractive method of obtaining pure water from an impure source, such as seawater or polluted water. In practice, however, the energy costs of generating enough heat to boil water are often prohibitive. Distillation is used, nonetheless, for a great variety of industrial processes in which the separation and purification of substances is required. *See also* Desalination

Further Reading

"Distillation Water Treatment," http://www .msue.msu.edu/msue/imp/mod02/01500609 .html.

Distillation is described and discussed in most introductory high school and college chemistry laboratory manuals.

Diving

See Aquatic Sports, Diving

A 1977 photo of King Faria, an 80-year-old water witch from San Rafael, California, and his willow divining rod. *Bettmann/Corbis.*

Dowsing

Dowsing is an ancient art of locating hidden objects, such as water or precious metals, by the use of an instrument known as a *divining rod*. It is probably best known to the general public as a method for finding underground sources of water. Some individuals, companies, and municipalities have hired dowsers to locate areas in which they should dig new water wells.

The origin of dowsing is unknown, although references to its use can be found in some very old manuscripts. For example, the German mineralogist Georgius Agricola describes in his classic work *De re metallica* how the use of a forked hazelwood branch can be used to locate deposits of silver.

Adherents of dowsing claim that the art is not a special gift possessed by certain individuals but a "supernormal" intuition that can be developed with training and practice. Some people are said to be able to choose the best place to dig a well or look for metallic ore simply by using their well-honed senses. Other people may need some sort of mechanical assistance, such as a divining rod.

Most classical references refer to the use of a branched stick for dowsing. Traditionally, the two forked ends of the stick are held in the dowser's two hands, with the single branch extended horizontally above the ground. When water or some other natural resource is detected, the protruding branch of the stick dips downward.

Today, the divining rods used by dowsers may be made of a variety of materials, including various types of wood or metal. Many dowsers prefer to make their own divining rods, although professionally made rods can also be purchased. One such rod, the Cameron Aurameter, is said to serve five different functions, including a direction finder, a gravimeter, and a highly selective divining rod with various subfunctions.

For many dowsers, the art offers an opportunity to learn more about their own natural powers and can be used for a large variety of purposes, including finding underground water, lost valuables, and missing persons or pets, or detecting energy released by crystals, blessed objects, or emanations from human auras. Dowsing has also been proposed as a method for detecting microscopic flaws in metallic structures, detecting radioactive emanations from beneath the earth's surface, and testing the purity of foods.

Disagreement exists as to the efficacy of dowsing techniques. Some research studies have failed to show that dowsing is an effective means for finding water. For example, a 1959 study showed that 56 percent of all county agricultural extension agents queried expressed a disbelief in the efficacy of the practice. Another 20 percent said that they believed in the practice, while 24 percent said that they were open-minded about its effectiveness. A 1977 pamphlet on dowsing published by the U.S. Geological Survey concluded that there was little scientific support for the efficacy of dowsing.

But other studies have produced contrary conclusions. A 1995 study sponsored by the German government found that dowsing was far more effective than would have been predicted by chance. In Sri Lanka, for example, 96 percent of dowser predictions were correct.

Today, most scientists probably reject the efficacy of dowsing as a method for locating water or other buried objects. However, the practice continues to have its adherents, and dowsers are still relied on by at least some proportion of the population as dependable predictors of water sources.

Further Reading

Baum, Joseph L. *The Beginner's Handbook of Dowsing.* New York: Corwn Publishers, 1974.

Discover Dowsing, http://www.dowsing .com/Dowsing.

Vogt, Evon Z., and Ray Hyman. *Water Witching U.S.A.* Chicago: University of Chicago Press, 1959.

Drought

The National Weather Service (NWS) defines a drought as "a period of abnormally dry weather which persists long enough to produce a serious hydrologic imbalance [such as] crop damage, water supply shortage, etc. The severity of the drought depends upon the degree of moisture deficiency, the duration and the size of the affected area" (http://www.nws.noaa.gov/om/drought.htm).

Definitions

A drought can be defined in various ways. From a meteorological standpoint, a drought can be thought of as a period of time during which rainfall is significantly less than normal over a number of months or years. Low rainfall in a desert would probably not be considered a drought, since the amount of precipitation in desert areas is normally low. But low rainfall in the American Midwest over a period of three years would be abnormal and might be considered a drought.

A drought can also be defined based on its effect on agriculture. If there is not enough moisture in the soil for crops to grow normally, a drought may be said to have occurred. From a hydrological standpoint, a reduction in the amount of surface and subsurface water can also be defined as a drought.

Finally, a disruption of normal social, economic, and other human activities as the result of reduced precipitation can result in drought conditions.

Indices

Scientists, farmers, government workers, and others use a variety of measures to determine whether a drought has occurred, how serious the drought is, and whether a drought is likely to occur in the future. The most common drought index used in the United States is the Palmer Drought Severity Index (PDSI), developed by Wayne C. Palmer in the early 1960s. The PDSI is based on measurements of current precipitation, air temperature, and soil moisture, as well as historical measures of these conditions. The PDSI has numerical values ranging from −4.0 and less (extreme drought conditions) to −0.49 to +0.49 (normal conditions) to +4.0 or more (extremely wet conditions). The PDSI is most useful in determining long-term droughts over relatively small areas. It is used by the U.S. government to determine when drought-relief programs should be put into effect.

The Crop Moisture Index (CMI) is a second drought index developed by Palmer in 1968. The CMI is designed to express the amount of moisture available over short-term periods in crop-producing areas of the nation.

A more recent drought index is the Standardized Precipitation Index (SPI) developed by Tom McKee and his colleagues at Colorado State University in 1993. The SPI was originally designed to predict the likelihood of a drought over three-, six-, twelve-, twenty-four-, and forty-eight-month time periods. It is less complex than the PDSI and is primarily based on measurements of water availability, such as precipitation, stream flow, groundwater resources, and stored reserves of water. The SPI is widely used by planners who need to be prepared for future droughts. The index has values ranging from −2.0 and less (for extremely dry conditions) to +2.0 and more (for extremely wet conditions). A major problem with the index is that the data on which it is based are constantly changing and may make long-term projections somewhat unreliable.

Other drought indices have been developed by various federal and state agencies to predict the likelihood of droughts in specific geographical regions.

Impacts

Droughts may have both positive and negative impacts, although the latter tend to be far more extensive and devastating. These negative impacts can be classified as economic, environmental, and social. Examples of economic impacts include the following:

- losses from crop, dairy, and livestock production
- losses from timber production due to forest fires
- losses from fishery production because of damage to rivers, streams, and lakes
- losses to farmers, who may eventually have to declare bankruptcy
- losses to recreational and tourism industries
- losses to electricity-generating companies because of reduced river flows
- disruption of water supplies
- revenue losses to federal, state, and local governments as a result of reduced crop productions and other economic losses
- increased costs of water supply and transportation
- increased land prices

Environmental effects that may result from drought conditions include the following:

- damage to plant and animal species
- loss of forests and other types of flora
- loss of wetlands
- loss of biodiversity
- erosion of soil by wind and water

- diminished water and air quality
- damage to aesthetic quality of the land

Social impacts of a drought include the following:

- shortages of food and water
- increased risk because of fires
- increased political conflicts over water use and water-related issues
- reduction in recreational activities
- loss of cultural and aesthetic sites
- increased mental and physical stress
- increased health problems, such as respiratory disorders
- changes in quality of life and lifestyle

Not all drought impacts are negative, however. In many cases, one person's loss is another person's gain. For example, researchers William E. Riebsame, Stanley A. Changnon, Jr., and Thomas R. Karl found that there were a number of "winners" in the devastating 1987–1989 drought that struck many parts of the United States. Among those winners were the following:

- agricultural producers in areas not affected by the drought
- railroads, who took over much of the shipping that had previously been carried by barges
- water-producing companies, such as well-drillers and weather modification companies
- electricity-generating companies, for whom demand increased as people and businesses depended more and more on air conditioning

Historic Droughts

Droughts are a never-ending part of many human societies. During the period between 1999 and 2001, for example, devastating droughts were being reported in many countries around the world. For example:

- Ethiopia experienced a drought in 1984–85 that killed as many as a million people
- A 70 percent reduction in rainfall in Jordan in 1999 cut cereal harvests by more than 40 percent and pushed unemployment rates to 25 percent
- Drought conditions in Afghanistan and Pakistan in 2000 were "the worst in memory" according to the international charitable organization, Save the Children
- Half a million people in Mongolia—about 20 percent of the population—faced starvation in 2000 in the worst drought in 30 years
- Water levels in the Tigris and Euphrates Rivers fell to such low levels in 1999 that one could walk across them in some places, leading to the worst drought in Iraq in 50 years

The United States endured three particularly serious drought periods during the twentieth century. The first, and perhaps the most famous in American history, was the so-called Dust Bowl of the 1930s. This historic event was the result of two crucial factors: an extended period of low precipitation in the midwestern and southern plains, and careless farming practices that denuded vast reaches of agricultural land from protective vegetation. The drought covered some parts of the central plains for the better part of the decade, although it was most severe in localized regions in 1934, 1936, and 1939–1940. The Dust Bowl got its name from huge dust storms that swept across immense parts of the region, blowing away valuable topsoil and

Dried up irrigation ditch in Rio Bravo, Mexico, during the drought of 1996. *Jack Dykinga, USDA.*

leaving wastelands behind. At times, the storms were so severe that the sun was blotted out for days at a time.

Tens of thousands of residents of the plains, their livelihood destroyed by the drought, fled west in an effort to find new lives. The social disruption by this mass emigration was unlike anything ever seen in the nation. At one point in 1936, the Los Angeles chief of police assigned 125 officers to patrol the state's borders between Arizona and Oregon to prevent "undesirable" immigrants from the Dust Bowl from entering the city.

The second great drought of the century occurred in the 1950s as the result of a combination of unusually high temperatures and low precipitation. The drought began in the southwestern states in 1950 and gradually spread eastward as far as Kansas and Nebraska. It reached its most severe level in 1956 when the PDSI reached the lowest point ever measured. During the drought, crop losses reached 50 percent in many regions.

A third drought spread across the midwestern plains in the period between 1987 and 1989. It was the most expensive drought—indeed, the costliest natural disaster—in U.S. history. Energy, water, agricultural, and other costs were estimated at $39 billion in the United States, and comparable losses in western Canada reached $1.8 billion. The drought began in the western states in 1987 and spread across the northern midwestern states, where it had its greatest effect.

Further Reading

Von Kotze, Astrid, and Allan Holloway. *Living with Drought: Drought Mitigation for Sus-*

tainable Livelihoods. Bourton-on-Dunsmore, UK: Intermediate Technology, 1999.

Wilhite, Donald A. *Drought Assessment, Management, and Planning: Theory and Case Studies*. Dordrecht: Kluwer Academic, 1993.

————, ed. *Drought: A Global Assessment*. London: Routledge, 2000.

Further Information

National Drought Mitigation Center
University of Nebraska—Lincoln
236 L. W. Chase Hall
P.O. Box 830749
Lincoln, NE 68583-0749
Telephone: (402) 472-6707
Fax: (402) 472-6614
URL: http://enso.unl.edu/ndmc/center
e-mail: dwilhitel@unl.edu

NOAA's Drought Information Center
14th Street & Constitution Avenue, NW
Room 6013
Washington, DC 20230
Telephone: (202) 482-6090
Fax: (202) 482-3154
URL: http://www.drought.noaa.gov
e-mail: answers@noaa.gov

Dry Farming

See Capillary Action

E

Earle, Sylvia A. (1935–)

Sylvia A. Earle has established herself as one of the world's foremost marine biologists and oceanographers. She was one of the first researchers to use the equipment developed by Jacques Cousteau in 1943 for the study of marine organisms. In 1970, she led a team of researchers who lived for two weeks in an underwater chamber in the Tektite Project. Although the project was a success, Earle felt that the most work could be accomplished by a completely same-sex team. As a result, she was chosen to lead a team of researchers—this time, four other women—in the Tektite II project.

Personal History

Sylvia A. (Reade) Earle was born in Gibbstown, New Jersey, on 30 August 1935. Her family moved to Clearwater, Florida, in 1948, where Earle became fascinated with the Gulf of Mexico and marine studies in particular. She attended Florida State University, from which she received her bachelor's degree in 1955. She was then awarded a master's degree (1956) and a doctoral degree (1966) from Duke University. After receiving her Ph.D., she spent a year as resident director of the Cape Haze Marine Laboratory in Sarasota, Florida. She then became a Radcliffe International Scholar at

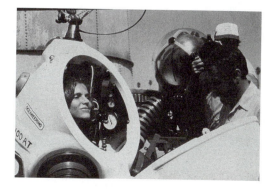

Dr. Sylvia Earle prepares to dive in a JIM suit. *OAR/National Undersea Research Program (NURP).*

Radcliffe College and research fellow at Harvard University's Farlow Herbarium. Earle continued as research fellow and associate at Harvard until 1981.

Major Accomplishments

In 1976, Earle accepted an appointment as curator of phycology (the study of algae) at the California Academy of Sciences and, concurrently, as a fellow in botany at the Natural History Museum of the University of California at Berkeley. In 1981, Earle and her former husband, Graham Hawkes, founded Deep Ocean Engineering for the purpose of designing and building equipment for working underwater and in other hazardous conditions. One of the company's first products was a one-person submersible vessel called *Deep Rover*.

In 1990, Earle was appointed chief scientist at the National Oceanic and Atmospheric Administration, the first woman to hold that position. She held that post for less than two years, after which she returned to Deep Ocean Engineering.

Of her more than 6,000 hours underwater, Earle's most memorable dive may have been one she took in 1979 off the island of Oahu in Hawaii. She descended to a depth of 380 meters (1,250 ft), where pressures reach more than 40 kilograms per square centimeter (600 lb per sq. in.). At the time, the dive was the deepest ever performed without a tether to the surface.

Earle has been awarded eleven honorary doctorates and has received numerous honors, including the 1980 Explorers Club Lowell Thomas Award, the 1981 Order of the Golden Ark from the Prince of the Netherlands, the 1989 New England Aquarium David B. Stone Medal, the 1990 Society of Women Geographers Gold Medal, the 1991 Golden Plate Award of the American Academy of Achievement, the 1992 Director's Award of the Natural Resources Council, the 1996 Lindbergh Award, the 1998 John M. Olguin Marine Environment Award, and the 1999 Ding Darling Conservation Medal. She has written more than a hundred publications on marine science and technology, including the books *Sea Change* (1995) and *Wild Ocean* (1999). Her children's books, *Hello Fish, Sea Critters*, and *DIVE!*, have won five awards for excellence.

Further Reading

"Sylvia A. Earle," http://www.doer-INC.com/saeprofile2001.html.

Sylvia Earle, Ph.D., Biography, http://www.achievement.org/autodoc/page/ear0bio-1.

Earth, Air, Fire, and Water

Earth, air, fire, and water are four materials commonly mentioned as elements by the ancient Greeks. The term *element* refers to any form of matter that is basic and fundamental. Today, scientists believe that all matter in the universe is composed of various combinations of about one hundred chemical elements.

The concept of an *elementary material* goes far back into history. For example, a Chinese document called the *Shu Ching*, or "Book of records," from the Eastern Chou dynasty (722–221 B.C.) mentions "five things," or "five movers," thought to be responsible for the composition of all objects in the natural world. Those five things were water, fire, wood, metal, and earth.

Greek Science

It was not until about the sixth century B.C., however, that the concept of an element was developed to any substantial extent. At about that time, a revolution occurred among Greek scholars with regard to the way they thought about the natural world. Prior to that time, the universe was generally thought to have been created by and be under the control of gods, goddesses, spirits, and other supernatural beings. The only efforts that humans made to "study" nature was to try to learn what it was the gods wanted and how humans might influence the decisions that the gods made.

Beginning in the sixth century B.C., a group of Greek thinkers began to ask a totally different kind of question. Their focus shifted from the gods and goddesses who created and ran the universe to the universe itself. What was there about matter and energy, the earth and the sky, living and nonliving objects, that humans could understand? they asked. As the great historian of science, W. P. D. Wightman, put it, this shift in thinking produced the first truly speculative investigations about nature.

Those speculations were very different from the type of thinking scientists do today. Greek philosophers were primarily interested in reasoning out the correct answers to questions about nature. They rarely carried out experiments to see whether their ideas were valid. Still, the attitude of objectivity with which they approached their analyses was largely unique in human history up to that time and formed an important basis for the way scientists think about the world today.

The Elements

One basic premise about the universe to which nearly all Greek thinkers subscribed was the principle of simplicity. When we look around us in the world, we see a multiplicity of various objects and events. The world often seems confused, complex, and even chaotic.

But that condition cannot be fundamentally true, the Greek scholars said. Beneath all the complexity and confusion, there are a few fundamental and basic materials and concepts. Interestingly enough, this drive to find simple explanations for the complexity of the real world continues to drive the vast majority of scientific research today. Scientists continue to believe that, in its most basic form, nature follows a few very simple laws. It is the job of scientists to discover what those laws are.

And so, among the Greeks, there began a search for the fundamentals of matter, the handful of substances from which everything else is made. This exercise was, of course, primarily an exercise in rational thought. One scholar might logically, suggest air as one of the elements because it is so fundamental to all forms of life. Another scholar might suggest fire as an element because of its remarkable ability to change solid materials into vaporous substances, which can then be conveyed to the gods.

The first known discussion of elements in Greek history is found in the writings attributed to the philosopher Thales of Miletus. Thales was thought to have lived from 640–546 B.C., although we have almost no firsthand information about his life. Almost everything we know about Thales has come to us from secondary or tertiary sources. However, many historians of science believe that we know enough to recognize Thales's contributions to thought. Wightman, in fact, calls Thales "the father of speculative science," if not "the founder of natural philosophy, and through that . . . of natural science" (Wightman, 10, 9).

Thales believed that water was the single fundamental substance from which all other materials in the world were created and on which the earth still floats in the form of an immense solid disk. He said that water can evaporate and condense in a variety of different ways, and each of these results in the formation of a different substance. Thales was never very clear as to exactly how untold numbers of real substances could be formed from water, but that was probably not very important. What *was* important was the question he asked: What is there in nature that is truly elemental?

Other Views of the Elements

Over the next three hundred years, the nature of elements was discussed in great detail by all of the great thinkers of the time. Each philosopher developed his own view as to which substances were elemental. Some, like Thales, chose water, but others preferred other substances. For example, Anaximenes (560–500 B.C.) declared that air was the one and only elemental substance. He believed that air could condense to form water or earth or rarefy to make fire.

Another philosopher, Heraclitus (536–470 B.C.), taught that fire was the one fundamental element. He made his choice based on a more general philosophy that the only reality in the world is change. Nothing remains constant, he taught, and all is in a state of flux. In such a case, what could be more representative of and fundamental to the real world than fire, which is always being born, changing, and dying?

The Four Elements

Over time, philosophers developed more complex and sophisticated views of the nature of elements. Matter, they said, was probably made of two, three, or more elementary substances, which could be combined in a variety of ways. For example, Empedocles (490–430 B.C.) was among the first scholars to list earth, air, fire, and water as the four basic elements of matter. He said that these four materials could combine and separate in an endless number of ways, but they always remained fundamentally the same. "Only commingling takes place," he wrote, "and the separation of the commingled." Out of this commingling and separation, he said, arose the "illusion" of a physical world (Van Melsen, 23).

The doctrine of the four elements was brought to its highest level of development by Aristotle (384–322 B.C.). He took the analysis of matter one step back by saying that earth, air, fire, and water were not, themselves, fundamental but were manifestations of various combinations of four other fundamental properties: heat, cold, dryness, and moistness. Thus, he taught that heat and moistness combined to produce air; heat and dryness, to produce fire; dryness and cold, to produce earth; and cold and moistness to produce water.

As with so much else he wrote, Aristotle's theory of the elements came to dominate much of scholarly thinking about the natural world for nearly 2,000 years. For example, the eminent physician Galen of Pergamon (A.D. 129–199) taught that the human body contains four *humors*: blood,

phlegm, yellow bile, and black bile. Illness was caused, Galen said, by an imbalance of these four properties, which resulted in an excess or deficiency of one of the humors.

With the spread of Christianity and the disparagement of worldly knowledge in the first few centuries after the death of Christ, the development of scientific thought largely came to an end in the Western world and was not revived until the early Renaissance. When this new breed of scientists began to think about the nature of elements, earth, air, fire, and water were no longer part of their theories.

Further Reading

Bynum, W. F., E. J. Browne, and Roy Porter. *Dictionary of the History of Science*, 117–19. Princeton, NJ: Princeton University Press, 1981.

Van Melsen, Andrew G. *From Atomos to Atom.* Pittsburgh: Duquesne University Press, 1952.

Wightman, W. P. D. *The Growth of Scientific Ideas*, chapter 1. New Haven, CT: Yale University Press, 1951.

Electrolysis of Water

Electrolysis is the decomposition (breaking down) of a compound into its component parts by means of an electric current. The electrolysis of water results in the formation of hydrogen gas and oxygen gas. The primary commercial significance of water electrolysis is the preparation of oxygen and hydrogen gas.

Mechanism

The electrons needed to bring about electrolysis are provided by a battery or some other source of DC (direct current) electricity. The electrons travel from the source through wires until they reach the reaction cell, which contains the substance to be electrolyzed (in this discussion, water).

The electrons flow into the substance through one end of the conducting wire—the *cathode*, or negative *electrode*. They flow through the substance in the cell to the opposite electrode—the *anode*, or positive electrode. Electrons then travel out of the anode, into another wire, and back to the source of electricity. The system of source, wires, electrodes, and conducting substance is the *electric circuit*.

Electrolysis is possible only with substances that conduct an electric current in the liquid state or when dissolved in water. Such substances are called *electrolytes*. For example, when sodium chloride (NaCl) is dissolved in water, it dissociates (breaks apart) into positively charged sodium ions (Na^+) and negatively charged chloride ions (Cl^-). During electrolysis, positively charged sodium ions migrate through the water solution to the cathode, while negatively charged chloride ions migrate to the anode.

Water is not normally considered to be an electrolyte because it ionizes to such a small degree. At room temperature, only one water molecule out of 10,000,000 ionizes to produce hydrogen ions and hydroxide ions:

$$H_2O \equiv H^+ + OH^-$$

For the electrolysis of water to proceed efficiently, a few drops of an acid is added to the electrolytic cell. The presence of additional hydrogen ions from the acid increases the rate at which electrolysis occurs.

Charged particles (*ions*) from the substance undergo chemical changes when they reach electrodes of the opposite electrical charge. For example, sodium ions take on an electron from the cathode to become neutral sodium atoms:

$$Na^+ + e^- \rightarrow Na^0$$

At the same time, chloride ions give up electrons at the anode to become neutral chlorine atoms:

$$Cl^- - 1e^- \rightarrow Cl^0$$

(In actual practice, the electrolysis of an aqueous solution of sodium chloride is somewhat more complex than indicated here.)

Decomposition of Water

When water is electrolyzed, hydrogen ions travel to the cathode, take on electrons, and become hydrogen atoms:

$$H^+ + e^- \rightarrow H^0$$

Two hydrogen atoms then combine with each other to form a hydrogen molecule, which escapes from the reaction cell as hydrogen gas:

$$H^0 + H^0 \rightarrow H_2 \, (g)$$

Meanwhile, at the anode, a somewhat more complicated reaction takes place. Each water molecule gives up two electrons, forming a single oxygen atom and two hydrogen ions:

$$H_2O - 2e^- \rightarrow O^0 + 2H^+$$

Two oxygen atoms then combine with each other to form an oxygen molecule, which escapes from the reaction cell as oxygen gas:

$$O^0 + O^0 \rightarrow O_2 \, (g)$$

The overall chemical reaction that occurs within the cell, then, is a combination of the two reactions that occur at each electrode:

$$2H_2O \rightarrow 2H_2 + O_2$$

Quantitative Aspects

The quantity of a substance produced during electrolysis depends on a number of factors. The fundamental quantitative laws of electrolysis were first announced in 1832 by the great English physicist and chemist Michael Faraday (1791–1867). Faraday said that the amount of a substance released during electrolysis depends on three factors: (1) the atomic mass of the substance, (2) its oxidation number (valence), and (3) the amount of electricity that passes through the electrolytic cell.

The fundamental relationship between electric current and mass of substance produced is expressed in a unit known as the *Faraday constant*, whose value is 96,485 coulombs per mole of electrons. A *coulomb* (C) is equal to the flow of 1 ampere of electric current for one second. Faraday's law says that the flow of 96,485 C of electricity will release one mole of electrons or their equivalent (such as one mole of hydrogen ions [H^+]).

Practical Implications

Electrolysis is a simple and straightforward process. Many high school science students have seen the reaction performed in one of their classes. The reaction would appear to have some practical value as a source for either hydrogen gas or oxygen gas, or both.

Such is not the case, however. The cost of the electricity needed to carry out electrolysis on a commercial scale is too great to justify this method for preparing the two elements. Oxygen can be obtained much more cheaply by the fractional distillation of air, and hydrogen is prepared quite inexpensively by treating petroleum products at high temperature. A small amount of hydrogen is obtained as a by-product of the electrolysis of sodium chloride solutions in the manufacture of sodium hydroxide and chlorine. But this amount is small compared to that obtained from all other methods of producing the gas.

Elements

See Earth, Air, Fire and Water

Emulsion

An emulsion is a mixture of two liquids that are insoluble in each other. For example, oil and water do not dissolve in each other, but a mixture of the two can be made by adding a third substance, an *emulsifying agent*, and shaking the mixture vigorously. Emulsions are a special class of mixtures known as *dispersions*, which consist of two substances—one suspended in the other. Other examples of dispersions are mist, fog, clouds, gelatin, milk, mayonnaise, and paint.

Emulsions consist of two components, one of which (the *dispersed*, or *internal*, phase) is usually finely divided and uniformly distributed in the other (the *continuous*, or *external*, phase). The two are held in suspension by the emulsifying agent, also known as a *surface active agent* or *surfactant*.

Forming an Emulsion

Emulsifying agents consist of molecules with one important property: one end is attracted to one of the two liquids in the emulsion, while the other end is attracted to the other liquid. Thus, the emulsifying agent molecule acts as a bridge that links together molecules of the two liquids that would otherwise not be attracted to each other.

Probably the most common emulsifying agent is soap. A soap molecule is a long, spaghetti-shaped molecule, one end of which is organic and the other, ionic. The organic end of a soap molecule is attracted to other organic molecules, such as fats and oils. The ionic end is attracted to molecules that carry an electrical charge, such as water.

Soap is used to clean dirty objects because soap molecules form a bond between the greasy surface of dirt particles and the water used for washing. When the soap-dirt-water mixture is agitated, an emulsion is formed among the three materials. During the process of agitation, the dispersed phase is broken apart into very tiny particles, each of which is surrounded by many emulsifying-agent molecules. When the wash water is thrown out, the emulsion carries away dirt particles bonded to the water, leaving behind a cleaner object.

Properties of an Emulsion

Two important properties of most emulsions are opacity and instability. The physical appearance of the emulsion depends on the size of the dispersed particles of which the emulsion is made. If the dispersed particles are less than 0.05 μm (micrometers) in diameter, they do not scatter light effectively, and the emulsion appears to be transparent.

When the dispersed particles are larger, however, they are able to scatter light, producing an opaque mixture that may range in color from gray to blue to milky white. Generally speaking, an emulsion appears to be grayish and semitransparent if its dispersed particles are between 0.05 and 0.1 μm in diameter; bluish white if the particles are between 0.1 and 1.0 μm in diameter; and milky white if the particles are greater than 1.0 μm in diameter. This property makes it possible to estimate the size of dispersed particles in an emulsion simply by looking at it.

Emulsions are also unstable. That is, over some period of time, the two phases of which they are made tend to separate from each other. An emulsion of oil and water with soap as an emulsifying agent will, after some time, separate into two layers, oil on top of water. The time required for separation is variable, depending on the components of the emulsion and the way it was made. Emulsions with relatively large dispersed particles tend to be less stable than those with small dispersed particles.

Uses

Emulsions provide an attractive method for dispensing substances that would be difficult to use under other circumstances. For example, many people use conditioning products containing oil on their hair. In principle, one might just apply the oil directly to one's hair, but the hair would turn out to be sticky and difficult to comb. By making an emulsion of the oil, it can be added in much smaller amounts to provide sheen and texture without making the hair impossible to manage.

Emulsions also offer a way of administering precise amounts of a substance. For example, nasal sprays are used to deliver a small, precise amount of medication to a person's nasal passages to prevent allergic reactions. The amount of medication needed is so small (a few milligrams) that there is no effective way to administer it undiluted to the nose. But an emulsion can be made with a known amount of medication as the dispersed phase in a known amount of water as the continuous phase. A single spray, then, is known to produce a small but exact amount of medication.

Emulsions can also be used to apply a material to a surface more easily than is possible with other methods. One of the most common examples is paint. The component of paint that provides surface protection and/or color consists of a thick organic liquid that cannot be applied to a surface with a brush or by any other direct method. But an emulsion of the protective/coloring component (as the dispersed phased) and water, alcohol, or some other liquid (as the continuous phase) can be made to allow the paint to be spread on a surface easily.

Because of properties like these, emulsions are used in a variety of industries, such as:

Drugs and pharmaceuticals. Emulsions make it possible to provide small, specific doses as well as to administer drugs that are unpleasant to taste, toxic in larger quantities, or more readily absorbed by the body. For example, the drug amphotericin B is used as an emulsion to reduce its overall toxicity and decrease pain when it is administered. Some barbiturates are provided as emulsions to provide a time-release effect of their anesthetic action.

Foods. Many foods occur naturally as emulsions or are prepared in the form of emulsions. Perhaps the best known example of a natural emulsion is milk. Some prepared foods that consist of emulsions include butter, peanut butter, ice cream, salad dressings, and whipped dessert toppings. Emulsions make foods taste better because they blend two or more flavors. They may also make foods easier to package and store and may add a desirable texture to foods.

Polymerization. Polymerization is a chemical reaction by which individual small molecules (monomers) are joined to each other to make very large, complex molecules (polymers). Polymers have a great many applications in modern society, perhaps the best known of which may be a variety of plastic materials. The physical and chemical conditions under which polymerization occurs are very important in determining the type of polymer produced. Chemists have learned how to control the types of monomers used, the temperature at which the reaction occurs, catalysts that can be added, and the physical state of the monomers to get specific types of polymers with exactly the right kinds of properties.

Emulsions are sometimes used during polymerization. The monomer is added to a mixture of soap and water and agitated. Some of the monomer is broken apart into the dispersed phase, while some remains in large droplets in the soapy water. A chemical is added to the emulsion to initiate the polymerization reaction, which begins with monomers in the dispersed phase. These monomers locate and react with individual monomer molecules in the larger droplets. The dispersed particles grow larger and larger as they react with more and more monomer molecules from the droplets. Since the polymerization reactions in the emulsion all begin with individual monomer molecules in the dispersed phase, the rate and nature of the polymerization reaction can be carefully controlled.

Agricultural and dairy operations. Farmers now use many different types of pesticides to control insects, fungi, worms, and other pests. It is important that these chemicals be sprayed on crops or animals at exactly the right dosage. Since the chemicals used are often insoluble in water, they cannot simply be made into solution and sprayed. Instead, they are converted to emulsions, which can be administered in controlled dosages.

Asphalt emulsion. Asphalt has a great many industrial and commercial applications, but it is difficult to use because it is so thick and viscous. One way to make it easier to use is to make a water-asphalt emulsion, which can then be applied to a surface. After a period of time, the water from the emulsion evaporates, leaving behind the asphalt covering. Asphalt emulsions are now widely used in road construction and maintenance and as a protective covering on walls, floors, and other surfaces.

Wax emulsions. There are many instances in which a wax covering on a surface is desirable. For example, fruits and vegetables are often picked long before they are actually sold to a consumer. Some method must be found to prevent the product from beginning to decay before it can be marketed. One way to do so is to cover the surface of the fruit or vegetable with a thin wax covering that is safe for humans to eat.

The problem, of course, is how to apply the wax directly. It would have to be melted and brushed or painted on, an unreasonable approach. However, it is possible to make a wax-water emulsion into which the fruit or vegetable can be immersed or that can be sprayed on the product. By controlling the composition of the emulsion and the rate of application, a wax covering of any desired thickness can be provided.

Wax emulsions can also be used as floor coverings, sealants to protect wood form moisture and decay, and surface coverings to protect from abrasion and wear.

Silicone emulsions. Silicone is a very popular kind of polymer that can be prepared in many physical states. As an oil, it has a host of industrial and other applications. For example, it is used to cover a painted surface to prevent the paint from breaking down. As with other oils, silicones are not easily applied to a surface directly, but can be converted into an emulsion and then sprayed or brushed on a surface. When the water in the emulsion evaporates, the silicone covering is left behind.

Silicone emulsions are also used as a defoaming agent. For example, they are used in the pulp and paper industry to reduce foaming during the preparation of pulp. They are also used as defoaming agents in the manufacture of textiles, the preparation of pharmaceuticals, and the refining of oil.

Further Reading

Becher, Paul, ed. *Encyclopedia of Emulsion Technology.* New York: M. Dekker, 1983.

Mittal, K. L., and Promod Kumar. *Emulsions, Foams, and Thin Films.* New York: Marcel Dekker, 2000.

Energy from Water

The kinetic energy of moving water can be captured and converted to useful energy through a variety of mechanisms. One of the oldest of these mechanisms is the water-wheel, thought to have been invented by the ancient Greeks or early Norsemen. Water-wheels are used to convert the energy of running water into some other form of mechanical energy needed to operate simple devices, such as grinding wheels, saws, and ore crushers.

In modern times, the most important function of water power has been to produce electricity. The four most common systems for generating electrical power from moving water are hydroelectric dams, ocean thermal energy conversion (OTEC) devices, tidal power systems, and ocean wave and current strategies.

Hydroelectric Power

The amount of power (P) generated by a hydroelectric plant depends on two factors: the rate at which water flows through the dam (f) and the vertical distance through which it falls, also called the *head* (d). A useful equation in determining the approximate amount of power that can be obtained from a dam is the following:

$$P \ (in \ kilowatts) = 5.9 \times f \ (meters/second) \times d \ (meters)$$

Types of Plants

Hydroelectric facilities are generally classified into one of two major categories—*low-head* and *high-head* plants—depending on the height through which water falls. Low-head plants operate using the energy provided by a rapidly running river or stream or with low dams. In such facilities, water falls through no more than a few meters. The first electricity-generating hydroelectric dam built in the United States was a low-head dam constructed on the Fox River in Appleton, Wisconsin, in 1882. The Fox River plant used a waterwheel to provide the water needed to operate the system.

Because of the limited height through which water falls in low-head dams, they are able to provide only modest amounts of electrical power, usually in the range of a few kilowatts—enough energy to operate a small plant. For larger power needs, high-head dams, capable of generating thousands of megawatts of electricity, are needed.

The critical component of a high-head hydroelectric plant is a very large body of water produced by damming a river or stream. The height of the dam can be more than a hundred meters, providing a very large drop for water passing through it. For example, the largest hydroelectric plant in the world, at Itaipu, Brazil, is 225 meters (738 ft) in height.

A common variation of the basic high-head plant is a pumped-storage facility. In such a plant, there are two water reservoirs—the upper reservoir created by the dam, and a lower reservoir. During periods of high demand, some water flowing through the dam is diverted into the lower reservoir. Then, during periods of low demand, some of the water in that lower reservoir is pumped back up into the upper reservoir, where it later passes through the dam once more.

All types of hydroelectric plants produce electricity by the same general principle. Water flowing through the facility is directed against the blades of a turbine, causing it to rotate. The rotation of the turbine, in turn, is used to activate a generator, from which electrical current is produced.

In addition to their much greater power capacity, high-head plants have another important advantage over low-head plants. The latter provide energy at the whim of nature. If river flows are low, little electricity can be generated. In high-head plants, the availability of very large reservoirs of water behind the dam ensure that a constant flow of electrical current will always be available.

Geographic Locations

The value of hydroelectric plants as a source of power varies widely throughout the United States and around the world, depending on the availability of water. In the United States in 1990, for example, states such as Alaska, Delaware, Hawaii, Iowa, Kansas, Mississippi, New Jersey, and Ohio get less than 0.1 percent of their energy from hydroelectric sources. By contrast, Oregon and Washington receive more than 80 percent of their electricity from hydroelectric plants, largely because of the large number of rapidly running rivers in the area. Overall, 10.7 percent of the electricity generated in the United States comes from hydroelectric sources ("Percent of a state's total power production . . . ").

Similar patterns hold for the rest of the world. In Africa, for example, hydroelectric power is important in only five nations: Egypt, Ghana, Mozambique, Zaire, and Zambia. The lack of hydroelectric power in other parts of the continent reflect not only the absence of rapidly flowing rivers but also the financial reserves necessary to construct the large dams needed for such facilities.

By contrast, hydroelectric power is an important source of electricity in nations that do have adequate water resources as well as the financial capability of building the needed structures. In Quebec, Canada, for example, 96 percent of all electric power is generated by hydroelectric plants. And in Mexico, more than a third of all electricity comes from such plants ("Renewable Energy").

The largest hydroelectric facility now in existence is the Itaipú plant on the Paraná River that separates Brazil and Paraguay. The plant was built in 1984 and has a rated capacity of 12,600 megawatts. The largest hydroelectric plant in the United States is the Grand Coulee facility in Washington State, with a rated capacity of 6,494 megawatts of electric power. Other large hydroelectric plants are the Krasnoyarsk plant in Russia (6,096 MW), La Grande 2 plant in Canada (5,328 MW), and the Churchill Falls plant, also in Canada (5,225 MW).

Advantages and Disadvantages

Probably the most important advantage of hydroelectric power plants is that they use a renewable source of energy. The water needed to operate such plants comes from rainfall and is available without charge in rivers and streams. The only cost in operating plants is the equipment needed to hold and move water through the facilities. Also, there is little or no pollution produced by the plants.

The most serious disadvantage of such plants is the large amount of land required for water storage. When the plants are first built, agricultural and forest land may have to be flooded and, in some cases, human communities may have to be abandoned to make room for the artificial lake being created. Some questions are also being raised about the decay of vegetation drowned by such lakes. Studies suggest that this decay may release significant amounts of greenhouse gases into the atmosphere. Finally, the presence of dams on waterways may have important impact on fish living in the river and on the movement of silt through the waterway.

Ocean Thermal Energy Conversion (OTEC)

The OTEC system of generating electricity depends on the fact that ocean waters are divided into approximately two vertical layers. The upper layer, extending to a depth of about 100 meters, is heated sufficiently by the sun to give it a temperature from 10°C to 25°C warmer than the lower (deeper) layer. This difference varies greatly according to latitude and is largest in the tropics.

The presence of a warm upper layer in the ocean suggests that it should be possible to build heat energy that can be used to power an electric generator. Engineers have devised two systems by which this heat can be utilized—the open-cycle OTEC and the closed-cycle OTEC. The open-cycle OTEC is the simpler system and uses warm seawater as the working fluid in the system. The warm surface water is introduced into a chamber in which a partial vacuum is maintained. The seawater evaporates in the vacuum, producing a low-pressure steam that drives a turbine. The turbine, in turn, drives an electric generator. After passing through the turbine, the steam is condensed and returned to the ocean.

In the closed-cycle OTEC, ammonia—or some other low-boiling-point liquid—is used as the working fluid. Warm seawater is used to evaporate liquid ammonia, which is then used to drive the turbine and generator. After passing through the turbine, the ammonia is cooled by cool water from a deeper layer of the ocean. Both warm and cool water are then returned to the ocean.

The concept of ocean thermal energy conversion was first introduced by the French engineer Jacques-Arsène d'Arsonval in 1881. His ideas were further developed by his student, Georges Claude, who built the first operating OTEC system. In the succeeding century, only one working model of OTEC has been successfully operating, that constructed by the Natural Energy Laboratory of Hawaii. Proponents of OTEC point to a number of its advantages: it is nonpolluting, uses a renewable energy source, is always available, and produces potable water as a side-effect of its operation. Critics point out that the initial cost of constructing an electricity generating plant on the ocean's surface is enormous, possibly more than could ever be covered from the sale of electricity. In addition, the efficiency of an

OTEC plant is unlikely to exceed about 5 percent, a very low value compared to almost any other source of electricity.

Tidal Power

The regular, periodic tidal flow of ocean water onto and away from the land is another possible method for using water to generate energy. The idea of tidal power has been around for at least two centuries. During the 1800s in the United States, a number of small communities used the flow of tides as a way of generating electricity for their own needs. The tides were dependable and a free source of energy, and the equipment needed to make them operate for human purposes was inexpensive and easy to build. As other sources of electricity became available, however, tidal power plants disappeared from the marketplace.

The simplest method of building a tidal-power-generating plant is to construct a dam across the entrance of an estuary into which the tide flows. The dam is called a *barrage* and is designed to allow water to flow beneath it and into the estuary at high tide. As the tide reaches its highest point, gates are dropped from the barrage, preventing water from flowing back into the ocean. Instead, the water is diverted through an opening into the barrage, where it is used to operate a turbine and generator.

Within the past decade, an alternative method of using the tides to generate power has been developed. Tall towers are placed on the bottom of the mouth of the estuary. Windmill-type propeller blades are attached to the towers below the surface of the water. As water rushes into the estuary during high tide, it pushes against the blades, causing them to rotate and operate a turbine and generator.

Tidal power is potentially an important source of energy in areas that abut the ocean and experience significant tidal changes. But the cost of building such systems is still too great in most cases for them to compete viably with other sources. At the present time, only two commercially active tidal power stations are in operation. One is a 240-megawatt plant at La Rance in Brittany, France, and the other is a 16-megawatt plant at Annapolis Royal, Nova Scotia, Canada.

Ocean Waves and Currents

In theory, the movement of ocean waters in currents and waves is an enormous potential source of energy. According to one estimate, such water movement could produce 2–3 million megawatts of energy, equivalent to the output of three thousand large power plants ("Energy Fact Sheet: Ocean Energy Systems"). The problem, thus far, has been to find mechanical devices that can capture this energy and convert it into electrical energy or other useful forms of energy.

Many experimental devices have been suggested for this purpose. Some devices are anchored along the shore, designed in the shape of a funnel through which water flows into a large reservoir. The stored water is then allowed to flow back into the ocean through a turbine that drives a generator. Other devices designed to float on the ocean can capture wave energy at a distance from the shore. The so-called *Salter Duck* is an example of such a device. The Salter Duck consists of a cam-shaped floating tank attached to one or more shafts anchored to the ocean bottom. Waves cause the Salter Duck to rock back and forth, forcing a fluid through a turbine. In experimental tests, Salter Ducks have been able to convert 80 percent of wave energy into electricity.

Oscillating wave columns are another method for capturing the energy of waves and ocean currents. These devices consist of long hollow columns filled with air. As ocean water is forced into the columns by wave motion, the air is compressed. The compressed air can then be used to drive a turbine. Prototype examples of oscillating wave columns have been used successfully in navigation buoys. The compressed air in such buoys is used not to generate electricity but to blow a whistle as a warning signal.

As with other forms of water-generated energy, the devices described here have the advantages of using a free, plentiful resource (water) without producing pollutants. Designing machines that can withstand storms and transmit electricity to land-based stations is, however, still a major problem. *See also* Dams; Waterwheels

Further Reading

Avery, William H., Chih Wu, and John P. Craven. *Renewable Energy from the Ocean: A Guide to OTEC*. New York: Oxford University Press, 1994.

Energy Fact sheets, http://www.iclei.org/efacts.

"Percent of a state's total power production that came from hydroelectric power plants in 1990," http://ga.water.usgs.gov/edu/tables/maphypctoftotpower.html.

"Renewable Energy," http://www.eia.doe.gov/oiaf/ie095/hydro.html.

Seymour, Richard J., ed. *Ocean Energy Recovery: The State of the Art*. New York: American Society of Civil Engineers, 1992.

Tidal Power: Symposium Proceedings. New York: American Society of Civil Engineers, 1987.

Zagar, A., ed. *Hydropower: Recent Developments*. New York: American Society of Civil Engineers, 1985.

Erosion by Water

See Rivers and Streams; Water Waves

Estuaries

An estuary is a partially enclosed body of water along a coastline, where freshwater from a river or stream flows into the ocean. The characteristic feature of an estuary is that it contains a mixture of freshwater (from the river or stream) and salt water (from the ocean). The estuary is separated from the ocean itself by a peninsula, barrier island, reef, or salt marsh, or by some other type of obstruction.

Estuaries are called by a variety of names, such as bay, harbor, inlet, lagoon, or sound. The largest estuary in the United States is Chesapeake Bay, while other well-known estuaries are Boston Harbor, Galveston Bay, Indian (Florida) River, New York/New Jersey Harbor, Puget Sound, San Francisco Bay, and Tampa Bay.

Benefits of Estuaries

To the naive observer, an estuary may appear to be an unproductive area that can be dredged, filled, or otherwise developed at no loss to humans or the rest of the natural world. The new highway, shopping mall, housing development, or other project built on old estuary land may appear to provide greater benefits to humans that did the original body of water.

Biological Productivity

But appearances are deceiving. An estuary is one of the most productive ecosystems found anywhere on Earth. The explanation for this productivity lies in the fact that rivers and streams that empty into an estuary carry with them silt and nutrients scoured and dissolved from upland areas. When the moving water in these rivers and streams reaches the estuary, it slows down and deposits the suspended solids that it has been carrying. Also, the dissolved nutrients are contained within the water that fills the estuary. These nutrients make possible an incredibly rich environment for the growth of marine plants and animals, and other organisms that prey on them.

The primary level of biological producers in an estuary consists of phytoplankton (free-floating aquatic plants) and marsh and sea grasses, which grow abundantly in the estuary's nutrient-rich waters. These plants

Aerial photo of the Ashepoo-Combahee-Edisto (ACE) Basin National Estuarine Research Reserve, Ashe Island, South Carolina. *NOAA National Estuarine Research Reserve Collection, NOAA.*

serve as food for a host of marine consumers, including many species of fish, crabs and lobsters, clams and other shellfish, marine worms, wading marsh birds, birds of prey, waterfowl, migrating birds, and reptiles. By one estimate, estuaries are the source of more than 75 percent of all commercial fish caught in the United States, and 80–90 percent of all recreational fish caught. They accounted for nearly $2 billion in income for commercial fisheries in 1990 ("About Estuaries," 2).

Estuaries, in fact, are one of the three most biologically productive ecosystems in the world (the other two being swamps and marshes, and tropical rain forests). The typical estuary produces about 9,000 kcal/m²/yr (kilocalories of energy per square meter per year). By comparison, the average piece of agricultural land produces about a third as

much, and the open ocean, less than a tenth as much (Miller, 1961).

Other Benefits

But estuaries have other values to human life as well. For example, they are an important mechanism for flood control. They act as giant sponges that collect and hold water carried into them by swollen rivers and streams, thus protecting homes and other buildings on nearby coastal areas. Estuaries also act as buffer against ocean storms that bring high waves onto the shore. Estuaries may also act as a natural filtering system, removing pollutants carried into them from rivers and streams.

Finally, estuaries are an important recreational facility. People use estuaries for swimming, boating, fishing, and many other marine activities. According to a 1990 survey conducted by the National Safety Council's Environmental Health Center, more than 180 million Americans visit ocean and bay beaches, generating $8 to $12 billion in income for the tourist industry ("About Estuaries," 2).

Threats to Estuaries

Two factors have accounted for the rising threat to estuaries, both in the United States and around the world. The first factor is the seeming uselessness of estuaries, mentioned above. The second factor is the planet's growing population, particularly in oceanfront areas. In the United States, for example, more than half of the nation's population now lives near estuaries. Development has led to the destruction or damage of about half of the country's estuaries, primarily by dredging and filling. In the two-decade period between 1947 and 1967, the nation's most populous state, California, lost two-thirds of its estuaries (Miller, 390). Our largest estuary, Chesapeake Bay, was so badly polluted that an emergency recovery program was launched to restore its health.

The National Estuary Program of the U.S. Environmental Protection Agency has identified seven major threats to the world's estuaries: nutrient overloading, pathogens, toxic chemicals habitat loss and degradation, introduced species, alteration of natural flow regimes, and declines in fish and wildlife populations. These factors do not operate independently, but may influence and interact with each other.

Nutrient overloading refers to the accumulation of nitrogen and phosphorus in waters carried into an estuary. Moderate levels of these nutrients are essential to plant and animal growth, but excessive amounts can create a variety of problems. For example, they may cause diseases in fish, produce algae blooms, and lead to a reduction in dissolved oxygen. When oxygen levels fall below about 2 ppm (parts per million), many aquatic species begin to die. Excessive levels of nutrients are found in rivers and streams because of runoff from lawns, agricultural and dairy lands, damaged or leaky septic systems, inadequate sewage treatment plants, groundwater discharges, and atmospheric deposition from power plants and vehicle emissions.

Pathogens enter rivers and streams through urban and agricultural runoff, damaged or leaky septic systems, overflow from sewer systems, illegal discharges from sewage plant systems, waste from recreational vehicles or campers, and waste from pets and other animals, both domestic and wild. Pathogens in estuarine waters can cause illness and death both in aquatic life and in humans who use those waters. The presence of pathogens in the water is a common cause for the closing of beaches and prohibition of shellfishing activities.

Toxic chemicals present similar threats to the health of aquatic organisms and humans. Such chemicals include heavy metals (such as cadmium and mercury), organic chemicals (such as PCBs and PAHs), and pesticides. These chemicals get into rivers and streams through industrial discharges, runoff from urban streets and sewers, drainage from mined areas, and atmospheric deposition.

Habitat loss and degradation occur when the area around an estuary is developed for new housing or an industrial or commercial purpose. Other forms of development that can prove harmful include conversion of land to agricultural purposes and construction of highways, dams, and marinas. Such construction may completely destroy an estuary, or it can result in the loss of food, cover, breeding and nursing areas, or migratory corridors for resident and transient animals.

When humans alter an estuary, they may bring with them *introduced species* that can drastically alter the region's ecosystem. These new species often compete successfully with native species, changing (and sometimes damaging or destroying) the natural food web in the estuary. As an example, more than two hundred nonindigenous species have been introduced into the San Francisco Bay, accounting for 40 to 100 percent of resident species in various parts of the bay.

Alteration of natural flow regimes can also harm an estuary. The species who live in an area have adapted over long periods of time to water with certain specific characteristics, such as amount of salinity, temperature, and pH (acidity). If the flow of water into an estuary is disrupted, those characteristics may change, and species may not be able to adapt and survive in new conditions. As an example, the flow of water into the Indian River Lagoon in Florida has doubled over time because of extensive drainage systems constructed upstream. This increase in flow has reduced salinity in the estuary, leading to changes in the marine environment in the lagoon.

Any one or some combination of the above factors can also lead to *declines in fish and wildlife populations*. Another important factor is human involvement, in the form of overfishing. In the Delaware Bay, for example, the taking of horseshoe crabs for bait by fishermen has been implicated in a serious decline in the population of that shellfish.

National Estuary Program

In 1987, the U.S. Congress responded to growing concerns about the fate of American estuaries by creating the National Estuary Program (NEP) as part of the 1987 Amendments to the Clean Water Act. The charge to the NEP was to "identify, restore and protect estuaries along the coasts of the United States."

The structure of the NEP is somewhat different from that of most other government agencies in that its primary goal is to encourage local communities to take responsibility for the care of their own estuarine environments. Programs in twenty-eight regions have now been created. They include estuarine environments such as Puget Sound, Columbia River, Morro Bay, and Santa Monica in the West Coast region; Corpus Christi Bay, Barataria-Terrebonne, and Sarasota Bay in the Gulf of Mexico region; Indian River, Delaware Estuary, and Albermarle/Pamlico in the Southeast region; and Casco Bay, Buzzards Bay, and Long Island Sound in the Northeast region.

In each region, local committees consist of representatives from federal, state, and local governmental agencies, business leaders, researchers, educators, and private citizens. These committees work together to identify problems in the local estuary, to develop specific plans of action to deal with these problems, and to create a formal management plan to restore and protect the estuary. *See also* Limnology; Wetlands

Further Reading

"About Estuaries," http://www.epa.gov/owow/estuaries/about1.htm.

Dyer, K. R. *Estuaries: A Physical Introduction.* 2d ed. New York: John Wiley, 1997.

Miller, Q. Tyler, Jr. *Living in the Environment,* Belmont, CA: Wadsworth Publishing, 1985.

Nordstrom, Karl F., and Charles T. Rowan, eds. *Estuarine Shores: Evolution, Environments, and Human Alterations.* New York: Wiley, 1996.

Further Information

Coastal Management Branch
Environmental Protection Agency (4504F)
Ariel Rios Building
1200 Pennsylvania Ave., NW
Washington, DC 20460
Telephone: (202) 260-6502
Fax: (202) 260-9960
URL: http://www.epa.gov/owow/estuaries
e-mail: OW-GENERAL@epamail.epa.gov

Evaporation

Evaporation is the process by which a substance changes from a liquid to a vapor. Some substances, such as ice, change from the solid state directly to the vapor state, a process known as *sublimation*.

Dynamics

Consider a sealed container of water. The molecules of which the water is made move about in a variety of speeds, some faster and some more slowly. The kinetic energy of any one molecule is a function of its velocity, and can be expressed as:

$$K.E. = \tfrac{1}{2}mv^2$$

where m is the mass of the molecule and v, its velocity.

Some molecules of water with greater-than-average energy may be able to overcome the attraction of other molecules adjacent to them and escape from the liquid into the space above it. When a container is first filled with water and speed, the rate of escape (the *rate of evaporation*) will be high. Over time, however, that rate begins to decrease. As more water molecules escape into the space above the liquid, they begin to collide with each other, some rebounding back into the liquid. At some point, the rate at which water molecules escape from the liquid is equal to the rate at which they are returning to the liquid. The pressure of water molecules in the space above the liquid at that point is said to be the *water vapor pressure*.

Evaporation results in a decrease in the temperature of the liquid that remains. Molecules that escape from the liquid tend to have higher kinetic energy than those left behind. As they leave, the average kinetic energy (i.e., the temperature) of the remaining liquid is reduced. This effect has wide applications in everyday life. For example,

refrigerators produce a cooling effect because a liquid with a low boiling point is allowed to evaporate, taking heat out of the refrigerator and vaporizing the liquid.

Factors Affecting Rate of Evaporation

The rate at which evaporation occurs depends primarily on three factors: the nature of the substance, its surface area, and the temperature. Substances whose molecules do not form intermolecular bonds with each other (including many organic liquids) tend to evaporate rather easily. Substances that do form such bonds tend to evaporate more slowly. In the case of water, hydrogen bonds among adjacent water molecules in the liquid state tend to hold these molecules together, creating a relatively high energy barrier that molecules must overcome before they can escape from the liquid.

Liquids with a large surface area tend to evaporate more quickly than those with a smaller surface area. The more molecules there are exposed to the atmosphere, the greater the likelihood that one will escape. Thus, a liter of water will evaporate more quickly if placed in a shallow pan than in a tall cylinder.

The rate at which molecules escape from a liquid (i.e., evaporate) can be increased by increasing their energy—that is, by raising the temperature of the liquid. As a rule of thumb, an increase in temperature of about 10°C approximately doubles the rate of evaporation. At some point, the vapor pressure of the liquid is equal to or just greater than atmospheric pressure, and liquid molecules escape freely into the atmosphere. At this point, the liquid is said to have reached its *boiling point*.

Latent Heat of Vaporization

The amount of energy needed to vaporize a given mass of any substance is known as the *latent heat of vaporization* of that substance. The latent heat of vaporization of liquid water is about 540 cal/g (calories per gram; 2,256 kJ/kg [kilojoules per kilogram]). By comparison, the latent heat of vaporization of ethanol (ethyl alcohol) is 204 cal/g (1,780 kJ/kg), and of methanol (methyl alcohol), 263 cal/g (1,072 kJ/kg).

Liquids that have been evaporated remain in the vapor state until enough energy has been removed to slow down their molecules and convert the vapor back to a liquid. The amount of energy required for this change is equal to the latent heat of vaporization. That is, if 540 calories of heat is removed from a single gram of water vapor, all of that water will condense in the form of liquid water.

Practical Applications

The evaporation of water has many practical applications. It plays an essential role in the water cycle, where it constitutes one of the primary mechanisms by which liquid water stored on the earth's surface returns to the atmosphere.

Living organisms also depend on the process of evaporation. In plants, water is pulled out of the soil through root hairs and root cell into stems and leaves, and then out of the leaves by means of an evaporative process known as *transpiration*. During transpiration, large quantities of liquid water escape from tiny openings in leaves known as *stomata* and into the atmosphere in the form of water vapor.

Many chemical and industrial operations rely on the process of evaporation. For example, solid crystalline sugar is obtained from sugar beets, sugar cane, or maple sugar by evaporation. In the case of sugar beets and sugar cane, the beets or cane are first cleaned, chopped, and then treated with hot water, which dissolves the sugar they contain. The sugar solution obtained from all three of these natural products is then heated in an evaporator. Water in the liquid vaporizes, leaving behind the solid sugar.

Evaporators are also used to reduce the volume of waste liquids that must be disposed of in some operations. The liquids themselves often consist of toxic or hazardous liquids mixed with water. When treated in an evaporator, the water portion of the mixture is removed and released as pure water, while the remaining wastes—now much reduced in volume—can be disposed of more easily and at less cost.

Evaporative cooling devices are offered as an alternative to traditional air-conditioning systems. In such devices, a stream of relatively warm air is passed through a filter containing water. As the water evaporates, it removes heat from the area, producing a cooling effect.

Evaporation is also an important method of food preservation, perhaps the first ever developed by humans. The bacteria that cause disease are unable to live in the absence of moisture, so drying foods is a good method for preserving them. The original method for dehydrating foods may simply have been to expose them to the sun for a period of time. Solar energy caused water in the foods to evaporate, leaving behind dried foods that could be eaten at some later time. Today, a variety of home devices are available for dehydrating foods. Although now less important than freezing, drying, and other methods of preservation, dehydrating foods is still a significant commercial process of food preservation. *See also* Evapotranspiration; Hydrologic Cycle; Transpiration

Further Reading

Brutsaert, Wilfried. *Evaporation into the Atmosphere: Theory, History, and Applications.* Dordrecht: Reidel, 1982.

Jones, Frank E. *Evaporation of Water: With Emphasis on Applications and Measurements.* Chelsea, MI: Lewis Publishers, 1992.

Evapotranspiration

Evapotranspiration is the process by which water is lost from the earth's surface (evaporation) and from the leaves and stems of plants (transpiration). The amount of evapotranspiration that occurs in an area depends on a number of factors, including the net solar energy received; the surface area of lakes, rivers, and other bodies of water; wind speed; amount, type, and density of vegetation; type (reflectivity) of land surface; depth of plant roots; amount and availability of soil moisture; and season of the year.

The most important of these factors is solar radiation, which explains why the rate of evapotranspiration differs by season, time of day, amount of cloud cover, and latitude. In the United States, for example, the southwestern region receives a significantly larger amount of solar radiation per day than does any other part of the nation. The rate of evapotranspiration in states such as Arizona, New Mexico, southern California, and west Texas, therefore, is higher than elsewhere in the nation. In these states, nearly 100 percent of the moisture that falls as precipitation returns to the atmosphere by evapotranspiration. By contrast, much of the precipitation that falls in the northeastern states runs off into lakes, rivers, and the ocean, and only about 40 percent returns to the atmosphere by means of evapotranspiration. Nationwide, the average loss of water to the atmosphere by evapotranspiration is about 67 percent.

Knowledge of evapotranspiration patterns is very important to farmers and others who use irrigation water. The rate at which evapotranspiration occurs varies widely throughout the growing season, so that the amount of irrigation water that should be applied varies comparably. Irrigation water supplied during times of low evapotranspiration is essentially wasted because it cannot be used by crops. *See also* Aquatic Plants and Animals; Evaporation; Transpiration

Ewing, Maurice (1906–1974)

Maurice Ewing is perhaps best know for his research on ocean bottoms. Over a period of more than ten years, Ewing led and participated in more than fifty expeditions to study ocean bottoms, providing more comprehensive and complete data on this part of the planet than had ever been obtained.

Personal Life

Maurice Ewing was born in Lockney, Texas, on 12 May 1906. He attended Rice Institute (now Rice University), from which he received his B.A. (1926), M.A. (1927), and Ph.D. (1931). He taught physics at the University of Pittsburgh from 1929 to 1930; physics, geology, and geophysics at Lehigh University from 1930 to 1944; and geology at Columbia University from 1944 to 1972. He was also the first director of Columbia's Lamont-Doherty Geological Observatory. Upon his retirement from Columbia, Ewing was named head of the Division of Earth and Planetary Sciences at the University of Texas Medical Branch at Galveston. He held that post until his death in Galveston on 4 May 1974.

Major Accomplishments

Prior to Ewing's studies of the ocean bottom, most of our knowledge of Earth's seafloor came from simple plumb lines dropped from ships to the ocean floor. These methods produced little detailed information about the shape of the ocean bottom.

After World War II, Ewing and his colleagues adapted a variety of tools and technologies to enhance their study of the seafloor. For example, he devised underwater cameras with which to observe the ocean bottom directly, developed methods for taking core samples from the seafloor, designed ultrasound equipment to study floor topography, and adapted seismic studies to analyze the interior structure of the ocean bottom. Largely as a result of Ewing's work, scientists learned that the seafloor has a complex and varied structure not unlike that of dry land. It contains mountain ranges, deep valleys, flat-topped hills, and underwater volcanoes. In 1956, Ewing showed that the now-famous mid–Atlantic ridge mountain chain actually extends far beyond that ocean itself and into the Indian, Pacific, and Antarctic Oceans. In fact, the ridge forms a continuous range that stretches around the planet.

Further Reading

Wertenbaker, William. *The Floor of the Sea: Maurice Ewing and the Search to Understand the Earth.* Boston: Little, Brown & Company, 1974.

Extraterrestrial Water

Extraterrestrial water is water that occurs anywhere in the universe outside Earth's atmosphere. Although there is currently no evidence that water exists anywhere in the universe other than Earth's solar system, the possibility of its presence in other planetary system does exist.

For example, astronomers announced in late 1999 that they had found six new planets orbiting distant stars. That brought to twenty-eight the number of planets known to exist outside the solar system. Based on the planets' distances from their own stars, astronomers hypothesized that at least five of the six new discoveries might contain water. If so, those planets could contain some form of life.

The Moon

Scientists have long speculated about the possible presence of water on the Moon,

Mars, and Venus. One reason for this interest is that water is one of two essential conditions necessary for the existence of life as we know it, the other being oxygen. If a solar body contains both of these factors, it is then theoretically possible that life in some form now survives on that body or may have been present at some time in the past.

Certain features on the surface of the Moon have an appearance similar to bodies of water on Earth. As a result, these features were long ago given names appropriate to such bodies of water, names such as the Ocean of Storms, the Sea of Rains, the Sea of Serenity, the Sea of Tranquility, the Sea of Crises, and the Sea of Fertility.

With the development of good telescopes, however, it soon became obvious that the flat, dark surfaces originally identified as bodies of water were in fact not. Instead, they appeared to be extensive flat plains of rocky material with reflective properties very different from the surfaces around them.

This view was confirmed when astronauts in the Apollo Space Program brought back rock samples from the surface of the Moon. Analysis of those rocks showed a complete absence of water in any form whatsoever. It is now evident that any water that may have been present on the Moon was lost very early in its history.

Venus

The presence of water on both Mars and Venus has also been long suspected. Both planets are somewhat similar to Earth and hold at least the possibility of possessing some water on their surfaces or in their interiors. Data collected by space probes to both planets over the last three decades have now clarified the water situation for both planets.

In the case of Venus, there appears to be little or no free water in the planet's atmo-

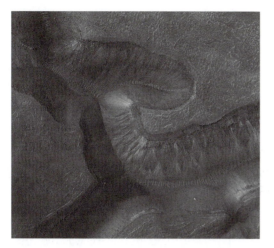

Taken in January 2000, this Mars Global Surveyor camera view of the Gorgonum Chaos region of Mars shows gullies that may have been formed by seeping groundwater. *NASA/JPL/Malin Space Science Systems.*

sphere or on its surface. Scientists now believe that any water present on Venus was lost long ago as the result of a "runaway greenhouse event" on the planet. The following scenario seems to be a likely explanation:

Huge oceans of water may have been present on the planet's surface following its creation. However, intense heat from the Sun would have caused the evaporation of most of that water. As water evaporated, it rose into the Venusian atmosphere, forming large masses of clouds. These clouds produced a "greenhouse effect" on the planet, raising the surface temperature even further. Eventually, all of the water on the planet's surface would have evaporated and formed even more clouds in the planet's atmosphere.

Water in the planet's atmosphere would have been exposed to solar radiation, however, capable of breaking down water molecules into hydrogen and oxygen atoms. The planet's gravitational attraction is not sufficient, however, to hold these atoms, and they drifted away into space.

Mars

The situation for Mars is, of course, very different from that for both Venus and Earth. Mars is far enough from the Sun that a "runaway greenhouse event" would have been very unlikely if not impossible. As with the Moon, Mars long seemed to be a likely candidate for holding water on its surface. In fact, some astronomers believed that certain features on the face of the planet could be explained by assuming that intelligent beings had once lived on the planet and built "canals" on its surface. Again, improved methods of observation long ago showed that no extensive bodies of liquid water are present on the planet.

Still, the question remained as to whether smaller amounts of water might be present beneath the planet's surface or in the form of ice elsewhere on the planet. The answers to those questions have been and are still being collected as the result of space probes sent to the planet over the past three decades.

Currently, scientists believe that large amounts of water may be stored in ice caps at the planet's poles. They think that the planet may once have been warm enough to permit that water to exist in liquid form. A number of features on the planet's surface suggest that giant floods may have swept across certain portions of the planet's surface, leaving behind geological features similar to those found on Earth.

There may also be small amounts of water in the Martian atmosphere. At most, that water could produce no more than a few micrometers of precipitation if it were to fall to the Martian surface as rain or snow.

Some evidence suggests that the upper few centimeters of the Martian surface may also contain some adsorbed water. Again, the actual amount of water present is probably very small.

Given these three possible water sources on the planet, some scientists are encouraged in their belief that life of some kind may well have existed on the planet in the past, even if it can no longer be found.

Comets and Meteoroids

Comets and meteoroids are both small bodies that travel through the space between the planets and, in some cases, beyond the solar system itself. Comets consist primarily of a dense solid nucleus no more than a few kilometers in diameter. As a comet approaches the Sun, some of the material in the nucleus is evaporated, forming a trailing tail pointing away from the Sun.

Comets are frequently referred to as dirty snowballs or dirty icebergs. These terms reflect the composition of the comet nucleus, primarily frozen gases and solid materials. Recent space probes have been able to identify the most common of these materials. They are frozen methane (CH_4), ammonia (NH_3), and water.

Meteoroids are usually described as large chunks of rock or metal that occasionally enter Earth's atmosphere. In 1999, the first evidence for the existence of water in meteoroids was obtained. A group of boys playing basketball in the town of Monahans, Texas, saw a small meteoroid strike the ground. When they brought it to a scientific laboratory for analysis, very small amounts of water were detected in the meteoroid. The water was encased in salt crystals, suggesting that they might have been formed during the evaporation of some primeval sea millions of years ago. The Monahans discovery was the first concrete evidence found on Earth for the existence of extraterrestrial water. *See also* Clouds

Further Reading

Carr, Michael H. *Water on Mars*. New York: Oxford University Press, 1996.

"Mars   Water,"   http://humbabe.arc.nasa .gov/MarsToday/MarsWater.html.

F

Firefighting

Water is the single most common material used in fighting fires. The reason for this fact is determined by the nature of fire itself. For fire to occur, three conditions must be met: there must be (1) a fuel, or something to burn; (2) an oxidizing agent, or something to make the fuel burn (such as oxygen itself); and (3) enough heat to raise the fuel to its kindling temperature, the lowest temperature at which a fuel will burn.

One way to prevent fire is to eliminate all combustible materials, the fuel mentioned above. Since that step is seldom possible, the process of fire extinguishing usually depends on eliminating one or both of the other two factors needed for a fire: the oxidizer (usually oxygen in the air) and heat.

Water is used in extinguishing fires because it achieves both of these objectives at once. When sprayed on a burning fuel, water coats the surface of the fuel, preventing oxygen from reaching the burning material. In addition, the steam formed when water comes into contact with a hot fuel forms a protective, inert atmosphere that isolates the fuel from surrounding oxygen.

Finally, water absorbs heat from a burning fuel, reducing the temperature of the fuel below its kindling temperature. Water is an

Firefighters use a hose to fight a house fire in Mendham, New Jersey, in 1990. *Michael S. Yamashita/Corbis.*

especially efficient agent for this purpose because it has one of the highest heat capacities (amount of heat absorbed per unit mass of a substance) and highest heats of vaporization (amount of heat needed to change a unit mass of the substance into a vapor) of any substance.

Water cannot be used on all kinds of fires. Some fuels, for example, react with water with the release of hydrogen gas, which is highly combustible. Spraying water on a piece of burning sodium metal, for example, results in the production of hydrogen gas, which itself catches fire, only intensifying the problem for a firefighter. Electrical fires also require a nonaqueous extinguisher because water is decomposed by an electrical current, another reaction in which hydrogen gas is produced.

The general principle of using water to fight fires has not changed over thousands of years of human history, although the equipment used to deliver water to fires has evolved dramatically. Instead of using handheld buckets of water passed from person to person along a human bucket brigade, firefighters today use powerful hoses that deliver streams of water at rates of up to 20,000 gallons per minute over distances of hundreds of feet.

One recent advance in the use of water for firefighting has been the introduction of water droplets, or "water fog," as the extinguishing agent. Tiny droplets of water have been found to be more effective than larger streams of water in fighting fires because the droplets have a larger total surface area and, therefore, are able to coat more of the fuel and absorb more heat.

Further Reading

Friedman, Raymond. *Principles of Fire Protection Chemistry.* Quincy, MA: National Fire Protection Association, 1989.

Fish

See Aquaculture and Mariculture, Aquatic Plants and Animals

Fish Farms

See Aquaculture and Mariculture

Fishing

See Aquatic Sports, Fishing

Flood Legends

Flood legends are stories that describe how the earth was totally, or almost totally, covered by water, wiping out most of humankind. In most cases, the flood is said to have been caused by the displeasure of a god or spirit over the wicked ways of humans. The flood provides an opportunity for the human race to make a fresh start. Many westerners tend to associate the Great Flood described in the biblical book of Genesis with this event. However, nearly every culture that has been studied by anthropologists has a similar legend as part of its mythic history.

The Biblical Flood

The story of the Great Flood told in the Hebrew Bible can be found in Genesis, chapters 6 through 9. The story exists in two quite different forms, known as the Yahwist and Elohist versions. In both, the cause of the flood is attributed to God's anger at the wickedness of the human race. Genesis 6:6, for example, says that "the LORD was sorry that he had made man on the earth, and it grieved him to his heart." Seven verses later, he tells Noah that "I have determined to make an end of all flesh; for the earth is filled with violence through them; behold, I will destroy them with the earth."

Only Noah and his family are allowed to survive the flood. God instructs Noah to build an ark and to bring into it one pair of each kind of animal that lives on the earth. It is from Noah's family and the animal pairs that God intends to restock the earth, hoping for better results in his second try.

The flood lasts for 40 days (in the Yahwist version) or 150 days (in the Elohist version) before Noah, his family, and the animals are able to disembark for dry ground.

The Epic of Gilgamesh

The biblical story of the flood was almost certainly derived from an earlier Akkadian legend dating back to the second millennium B.C. In this legend, the gods have begun to despair for the future of humankind. The world is overpopulated, and people are reveling in sinful affairs. Just before the torrential rains begin, however, the god Ea appears to a man by the name of Utnapishtim in a dream. Ea warns Utnapishtim of the coming deluge and tells him to build a large boat that will hold all his family and two of every kind of living creature.

In contrast with the Genesis story, the rains last for only six days and six nights. At the end of that time, a dove sent out by Utnapishtim returns because it can find no dry land on which to land (as in Genesis). But a day later, a raven sent out does not return, indicating that dry land is in the vicinity. When they disembark, Utnapishtim and his family are blessed by the gods and given immortality.

The Epic of Gilgamesh is a heroic poem named after its hero. In one part of the poem, Gilgamesh attempts to become reconciled to the loss of his dear friend, Enkidu. He approaches Utnapishtim to discover the secret of immortality. At this point, Utnapishtim tells Gilgamesh how he was granted this gift as the result of his experience with the flood. He tells Gilgamesh that he too may have immortality by eating of a certain plant. Unfortunately, a snake eats the plant before Gilgamesh can reach it, and the poem's hero is forced to reconcile himself to his own mortality. The theme of the Gilgamesh poem is, therefore, consistent with that of the biblical story of Adam and Eve and the Great Flood in forcing humans to confront the limitations of their own lives.

Flood Legends in Other Cultures

The details of the flood story from various cultures differ widely, although the major story line and morals to be learned are often quite similar. In Borneo, for example, the story is told that a group of men found and killed a giant boa constrictor. When they cut up the snake and attempted to fry it, torrential rains began to fall. Before long, the whole earth was covered with water except for one small hill. The only living things left on the hill were a woman by the name of Dayan Raca, a dog, a rat, and a few other small animals.

The woman noticed that a vine rubbing against a tree produced fire, and she followed that example to make a flame herself. She was, thereby, the first human to make fire. Soon afterward, she mated with the firedrill used to make the fire and later gave birth to a son called Simpang Impang. With the birth of this child, the human race began to regenerate again.

Among the Yoruba tribe of Africa, the flood legend is closely tied to the creation myth, as it is told in the Bible. After a long description of the process by which dry land, humans, plants, and other animals are created, the Yoruba myth explains that Olokun, the ruler of the sky, was angry because she had not been consulted about the creation of the earth. She released the ocean's might, and water spread across the dry land, submerging nearly everything. The few people who were left prayed to the god Eshu for relief from the flood. Eshu demanded the sacrifice of gods in his honor, and when this was done, he cast a series of spells that caused the oceans to recede from the dry land.

One of the most famous flood legends is that of the Greeks. According to this legend, Zeus decided to travel around the earth among humans in disguise. The longer he traveled, the more distressed he became at the sinfulness he saw. When he returned to Olympus, he ordered the gods to unleash the rains on the earth, to flood all dry land and destroy the human race.

The Titan Prometheus, however, heard of Zeus's plan and warned his mortal son Deucalion in a dream. Prometheus told Deucalion to build a chest large enough to hold himself and his wife and to fill it with food and water. Deucalion did so and he and his wife sailed on the flood waters for nine days and nine nights. At the end of that time, Zeus had relented and ordered the rains to stop. The chest came to rest on top of Mount Parnassus. Upon the advice of Themis, a Titan goddess, Deucalion and his wife traveled across the land, throwing stones in all directions. The stones turned into humans, the first members of the restored human race.

Flood legends vary to some extent depending upon the physical characteristics of the area in which a culture is located. Among the Aymara in Bolivia, for example, the flood is a deluge of ice and snow. The snow god Kun has become angry at the arrogance of the human race and has decided to eliminate them from the earth. In the flood sent down by Kun, all humans are killed except for the Eagle Man, who becomes the father of a new race called the Paka Jakes. The people who live near Lake Titicaca today regard themselves as the descendants of the Paka Jakes.

Flood legends are common among Native American tribes as well. For example, the Apache legend says that the god Dios appeared to an old man and old woman telling them that rains would soon begin and last for forty days and forty nights. He told them to have all members of the tribe go to the top of one of four mountains in the area. He also warned them not to look at the flood or at the sky.

Once the rains began, very few people tried to reach the mountains. Some people ignored the warnings about looking to the sky or flood and were turned into birds (in the first case) or fish or frogs (in the second case). When the rains ended, only twenty-four people had survived. It is from these twenty-four people that the new Apache tribe was formed. The legend adds that another disaster will strike the world at the end of the millennium. This time, however, the earth will be destroyed by fire, not water.

Further Reading

Cohn, Norman Rufus Colin. *Noah's Flood: The Genesis Story in Western Thought*. New Haven, CT: Yale University Press, 1996.

Dailey, Stephanie, ed. *Myths from Mesopotamia: Creation, the Flood, Gilgamesh, and Others*. Oxford: Oxford University Press, 1998.

Heiel, Alexander. *The Gilgamesh Epic and Old Testament Parallels*. 2d ed. Chicago: University of Chicago Press, 1970.

Floods

A flood is the inundation by a body of water on land that is normally dry. Floods are most commonly caused when rivers or streams overflow their banks and submerge surrounding areas. Those areas are called *floodplains*. But floods can also be caused in other ways, as when a dam breaks and releases huge amounts of water in a short time into the riverbed below it or when unusually large tides or ocean waves sweep up over a coastal area.

Floods are a major source of financial loss and loss of human life around the world. Between 1970 and 1995, more than 318,000 people were killed and another 81 million left homeless by floods. Property damage resulting from floods in the period 1991–1995 was estimated at $200 billion worldwide, about 40 percent of the cost due to all forms of natural disasters ("Social Changes Increase Flood Costs").

Causes

Flooding is a normal part of the hydrological cycle. When rain or snow falls on the earth's surface, it usually experiences one of two fates. First, it may soak into the ground until it reaches an area already saturated with water, the *water table*. The rate at which water is transported to the water table depends on a number of factors, one of the most important of which is the nature of the soil. Water tends to pass rapidly through soil consisting of large particles, such as sandy soil, and more slowly through soil containing smaller particles, such as clayey soil. Soil that does not soak into the ground runs off into the nearest river, stream, lake, or other body of water.

Under normal circumstances, the rate at which water falls on the earth's surface is roughly equal to the rate at which it is carried away into the water table or into rivers, lakes, and other bodies of water. But occasions arise when the rate at which water comes into contact with the ground is greater than the rate at which it is carried away by one of these two mechanisms. Persistent and/or unusually heavy rains and the melting of snow are two of the most common causes of this temporary hydrological imbalance.

When water accumulates more rapidly on the earth's surface than it can be carried away, flooding may occur. Rivers and streams become swollen with volumes of water greater than they usually carry, overflow their banks, and inundate surrounding areas. Floods can cause enormous amounts of damage because of the mass and velocity of water they carry. As an indication of the power of flood waters, one writer has estimated that an inch of rain falling 1,000 feet over a one-square-mile area has the energy potential of 60,000 tons of TNT, or three times the force of the first atomic bomb dropped on Hiroshima (Clark 1982, 23).

Because of the energy they possess, floodwaters tend to scour huge amounts of rock, sand, soil, and other materials off their beds and banks and carry them downstream. When those waters overflow their banks or empty into a lake or the ocean, they slow down and deposit those eroded materials. When deposition occurs on dry land surrounding the river or stream, they form the flood plains that are characteristic features of the land surrounding mature rivers. The deposited materials are known as *alluvial fill*. Alluvial fills contain some of the richest and most agriculturally productive soils found anywhere in the world. When deposition occurs in a lake or the ocean, the deposit is known as a *delta*.

One type of flood, the *flash flood*, can pose unusual and severe effects. A flash flood is defined as a flood that occurs very quickly, often within a few hours of a heavy rainfall, and produces unusually large volumes of water. The threat posed by flash floods is that they often appear so quickly that residents of an area have little or no opportunity to protect themselves and their property against the wall of water they bring with them.

Floods are sometimes rated according to their severity. For example, a *hundred-year flood* is one of such severity that it can be expected only once every hundred years, while a *ten-year flood* is one that can reasonably be expected to appear once every ten years. Of course, nature is not thoughtful enough to schedule floods according to any system such as this one. The Mississippi River Valley, for example, experienced "hundred-year floods" in 1943, 1944, 1947, and 1951. The frequency of the floods did not detract, however, from the severity suggested by the terminology used to describe them.

Floods in Legend

It should hardly be surprising to learn that nearly all human cultures have flood legends in their histories. In the Judeo-Christian culture, that legend is sometimes known as the *Deluge*, and is described in Genesis, chapters 6 through 9. Many authorities believe that the story was adapted from an even earlier flood legend from ancient Babylonia, that of *Gilgamesh*. In that legend, a god named Enlil became angry with humans because they made so much noise that he couldn't sleep. To punish them, he released a huge flood that wiped out almost all living beings.

Among the Norse, a worldwide flood was blamed on the release of blood from an evil god who was killed by Odin and his two brothers. A Lithuanian myth tells that the god Pramzinas released a worldwide flood on humankind, then provided the means for survival of a few when he accidentally dropped a nutshell into the floodwaters.

Flood myths held (and hold) an important place in many cultures because they are so closely associated with very real catastrophic natural events. In 1929, for example, a team of British archaeologists excavating a region near the Euphrates River found an 8-foot-thick layer of sediment, suggesting that a massive flooding of the river had occurred in about 3500 B.C. A flood of that magnitude might well have suggested to people of the time that Enlil, or some other god, had decided to destroy the world as they knew it.

Historic Floods

There are stories in just about every culture in the world telling of huge floods that caused widespread death and destruction of property. Only in relatively modern times, however, have detailed records been kept of the damage caused by floods. One of the earliest of these reports dates to 1228, when an estimated 100,000 people in the Dutch district of Friesland were said to have been drowned in floodwaters. In 1642, a rebel leader named Li Tzu-cheng laid siege to the Chinese city of Kaifeng in his war against the leaders of the Ming dynasty. To bring the city to its knees, he tore down the dikes around the city and allowed the Yellow River to flood in. More than 300,000 residents of Kaifeng were said to have died in that flood.

The most destructive flood in the history of the United States, based on lives lost, was the Johnstown, Pennsylvania, flood of 1889. That flood occurred when the old South Fork Dam on the Little Conemaugh River broke up during heavy rains and released 20 million tons of water on the city. Over 2,200 people were killed out of a population of 30,000, and thousands more were injured. It took more than five years for the city to recover from the disaster.

Probably the greatest flood disaster to strike the United Kingdom occurred in August 1952 in the village of Lynmouth in Devonshire. Lynmouth lies at the confluence of the West and East Lyn Rivers, whose drainage basin includes about 40 square miles of high plateau. The first two weeks had seen a steady rain that slowly saturated the ground in the Lyn Rivers' watershed. Then, on August 15, a torrential rain struck the area, dropping up to nine inches in a twenty-four-hour period. The rivers became swollen and poured through Lynmouth at a rate of about 575 tons of water per second. At times, the face of the water at the village was 30 feet high. Within a matter of hours, ninety-three houses had been destroyed or damaged beyond repaid and at least thirty-four people had been killed. Officials called the flood a fifty thousand-year flood, one that might be expected no more frequently than once every fifty thousand years.

The costliest flood in U.S. history, in terms of property damage, was the 1993 flooding of the Missouri and Mississippi Rivers. Damage ranged from Pipe Stem Reservation in North Dakota to St. Louis on the Missouri and from Minneapolis to St. Louis on the Mississippi. The flooding occurred as the result of an unusual combination of atmospheric conditions. A warm front carrying moist air from the Gulf of Mexico collided with a cold front carrying dry air flowing south from Canada. The two fronts produced heavy rains that lasted over a period of weeks.

Over time, water levels in the Missouri, Mississippi, and their tributaries rose to record highs. At nearly a hundred observation points along the two major rivers, flood levels exceeded any previous record, many by more than six feet. At one point or another, flooding was reported at more than five hundred distinct observation points along the rivers. Discharge from the Mississippi into the Gulf of Mexico was higher than any measured in the preceding sixty-three years.

The floods were responsible for forty-eight deaths and property damage estimated at $18 billion. More than fifteen million acres of land were inundated and about 54,000 people were forced to evacuate their homes.

Many coastal areas around the world are subject to severe flooding from huge ocean waves, known as a *sea surge, storm surge,* or *storm wave*. These surges can be caused by a variety of atmospheric, oceanic, and geological conditions. In November 1970, for example, a tidal wave caused by cyclonic winds drove a huge mass of water across East Pakistan, killing 200,000 people. An additional 100,000 residents of the area disappeared and were never accounted for. In July 1998, three enormous tsunamis swept across Papua, New Guinea, destroying entire villages and killing at least 2,000 people. The tsunamis had been caused by earthquakes on the ocean floor.

Flood Control

Humans have been trying for thousands of years to control flooding. The ancient Egyptians, for example, constructed a series of levees that ran nearly a thousand kilometers (about 600 mi) along the left bank of the Nile River. The levees were designed to control the annual surge of the river onto the adjacent floodplains. Among the methods that have been developed for flood control are levees, dams and reservoirs, channels, and ecosystem restoration.

Levees are raised structures built along the banks of a river designed to hold a river or stream within its natural course. Levees are sometimes classified as *engineered* or *nonengineered* structures, depending on the sophistication with which they are built. A nonengineered levee may consist of nothing more complex than a large pile of dirt thrown up along the side of a river or stream. An engineered levee is built with greater attention to the kind of soil on which it is laid down, the kinds of materials from which it is built, the proper packing of material in the levee, and additional features, such as covering the river-facing side of the levee with some protective material. Most levees are also planted with grass or other plant material to reduce the amount of erosion they experience during flooding.

One of the largest levee systems in the world is that along the Mississippi River. The system was begun by French settlers in Louisiana in the early eighteenth century and consisted of modest banks about a meter (3 ft) high. Over time, the levee system was expanded and enlarged. Today, it covers a distance of about 1,600 kilometers (1,000 mi), with embankments as high as 7 meters (24 ft).

Many factors contribute to the failure of a levee. As waters rise during a flood, they may flow over the top of the levee, eroding material and eventually breaking through the levee. Water may also seep through or under the levee, undermining its structure and causing it to be washed away. The material of which the levee is made may also become so saturated that it begins to collapse under its own weight.

When levees begin to fail during a flood, nearby residents often try to shore them up with sandbags, used to raise the height of the levee or to fill in breaches in the levee. Indeed, pictures of people piling sandbags onto a levee to protect themselves from a flood are perhaps one of the most common connections many people today have with flooding.

The goal of *dams* and *reservoirs* is to even out the flow of water in a river or stream. When heavy rains cause an increase in river flow, the excess water is trapped behind a dam and held until it can be released during drier times. In some cases, the excess water is actually drained away through channels into holding reservoirs, from which it can later be released.

One of the most successful flood-control systems based on dams is that of the Tennessee Valley Authority (TVA). When it was established in 1933, TVA was authorized to solve a number of problems in the Tennessee River valley. The challenge it chose to confront first was that of flood control. In preceding decades, uncontrolled flooding had washed away much valuable farmland, eroded natural soils, and silted the rivers and streams in the Tennessee River drainage basin. As a result of the complex set of dams constructed in the valley over the years, TVA has now largely solved the problem of flooding that was once endemic in the area. In solving this problem, the authority was also able to address a number of other related issues, such as reforestation, erosion control, improved navigation, and electricity production.

Stream channelization is a method of flood control in which a river or stream channel is widened, deepened, cleared, strengthened, and/or made more straight. Stream channelization became popular in the mid-1950s when the 1954 Watershed Protection and Flood Prevention Act authorized the U.S. Soil Conservation Service to use the procedure both to drain wetlands that could be converted to farmland and to reduce flood damage. Over the next few decades, various government agencies (primarily the U.S. Corps of Engineers) pursued channelization with a fervor, such that more than 8,000 miles of rivers and streams were "channelized" in the first two decades after the 1954 law was passed.

Only after stream channelization had been widely employed throughout the United States were some of its harmful effects recognized. These included damage to fish and wildlife habitats, contamination of downstream lakes and reservoirs, increased upstream erosion, dramatic changes in the physical characteristics of rivers and their floodplains, and the necessity for sometimes significant changes in near-river infrastructure (such as the need for rebuilding bridges and roads in the area). As these problems have become better understood, the initial enthusiasm for stream channelization has diminished. In addition, as with other forms of flood control, channelization has been found to solve flooding problems in one area of a river or stream, but to amplify those problems farther downstream.

One of the most logical, and sometimes most successful, methods of flood control involves the *restoration of ecosystems*. Flooding often becomes a problem when humans degrade the physical environment to build new homes, shopping malls, factories, roads, and other structures. As trees, grass, and other vegetation are removed during construction for such projects, one of the best natural flood-control devices—plant roots—is also removed.

As of the late 1970s, between 3.5 and 5.5 million acres of floodplain land in the United States had been developed for urban use. More than 6,000 communities with populations of more than 2,500 could be found on these lands. Population was growing at a rate of between 1.5 percent and 2.5 percent annually in these communities (Faber). Under such circumstances, the pressure placed on natural flood-control systems was sometimes simply too great, and flooding became much more common in these areas.

Even governmental efforts to turn back the tide of floodplain development has not always been successful. For example, the U.S. Congress created the National Flood Insurance Program (NFIP) in 1968 for the purpose of relieving the increase cost to taxpayers of losses caused by flooding. In communities that agreed to adopt floodplain management and preservation programs, citizens were eligible to apply for federally guaranteed loans. While such loans did protect many individuals and companies from the devastating losses suffered due to flooding, they also made development on floodplains less risky than it had previously been.

An important step in the recognition of ecosystem restoration occurred after the 1993 floods on the Mississippi and Missouri Rivers. A task force within the U.S. Army Corps of Engineers came to the conclusion that restoring watersheds and wetlands was a more reliable method for protecting against future floods than was the raising of levees or the construction of more dams (Faber). This philosophy appears to have made an impression on the members of the U.S. Congress. In 1994, for the first time, Congress failed to reauthorize the Water Resources Development Act, under which the Corps of Engineers carried out many of its structural flood-prevention programs.

Still, progress in reducing the economic and human costs of flooding has been discouraging. Between the 1940s and 1990s, inflation-adjusted economic costs have skyrocketed from $1 billion to $5 billion. And climate scientists have concluded that the increase cannot be attributed to more severe weather patterns but to an improvement in lifestyles that has led to greater development in floodplains ("Social Changes Increase Flood Costs").

As an example, about 40 percent of the $6.8 billion paid between 1978 and 1995 under the National Flood Insurance Program went to policyholders who had previously received similar payments. NFIP was apparently not discouraging people from building in floodplains and other hazardous flood areas. In fact, according to one authority, the program was "underwriting development in the floodplains [and] throwing money at people to be foolish" (Garland). *See also* Dams; Flood Legends; Johnstown Flood of 1899; Reservoirs

Further Reading

"Basics of Flooding," http://www.floodplain.org/P-BASICS.HTM.

Clark, Champ. *Planet Earth: Flood*. Alexandria, VA: Time-Life Books, 1982.

Faber, Scott. "Managing Watersheds to Reduce Flood Losses," http://www.epa.gov/owow/watershed/proceed/faber.html.

Garland, Greg. "It Wasn't Supposed to Work This Way," http://www.theadvocate.com/library/flood/1019side.htm.

Miller, John B. *Floods: People at Risk, Strategies for Prevention*. Geneva: United Nations, 1997.

Smith, Keith, and Roy Ward. *Floods: Physical Processes and Human Impacts*. Chichester, NY: Wiley, 1998.

"Social Changes Increase Flood Costs," http://www.usatoday.com/weather/clisci/floods102400.htm.

Flotation

See Archimedes; Hydraulics, Hydrostatics, and Hydrodynamics; Surface Tension

Fly Casting

See Aquatic Sports, Fly Casting

Fog

See Clouds

Forbes, Edward (1815–1854)

Forbes was an early contributor to the young science of oceanography. He was particularly interested in marine life and was dubious about the contemporary belief that life existed only in the upper layers of the ocean. His most important contribution to the field of oceanography was to discredit that belief, partly by his discovery of a starfish from a depth of about 400 meters (1,300 ft.) in the Mediterranean.

Forbes was born in Douglas, on the Island of Man, on 12 February 1815. He studied medicine at the University of Edinburgh, but found that he was more interested in natural history. He became active in the British Association for the Advancement of Science and served as curator and paleontologist to the Geological Society of London. He also held two academic posts, as professor of natural history at Edinburgh and also at the Royal School of Mines in London.

His book *The Natural History of European Seas*, published in 1859 after his death, became one of the early classic textbooks in biological oceanography. Forbes died at the young age of 39 on 18 November 1854 in Wardic, Scotland.

Fountain of Youth

The Fountain of Youth was a mythical water source thought to confer eternal youth on any person who drank from it or bathed in it. The person most commonly associated with the search for this site was the Spanish explorer Juan Ponce de León (1460–1521). He was born in the province of León in northwestern Spain and, as a teenager, served in the Spanish army in its battles against the Moors. In 1493, Ponce de León joined Columbus on his second voyage to the Americas. He served with distinction in the conquest of natives in Hispaniola and was awarded with the charge of exploring and conquering the island of Borinquen (Puerto Rico). In 1509, he was appointed governor of the island.

While serving as governor, Ponce de León was told by the natives of a Fountain of Youth on the island of Bee Mee Nee (Bimini) in the Caribbean Sea. In 1512, he received a commission from King Ferdinand II to explore and settle the island. He lost his way on the journey, however, and landed on the North American mainland instead. He called the region where he came ashore La Florida, in honor of Easter Sunday, which, in Spanish, is *Pascua florida*.

It goes without saying that Ponce de León never found the Fountain of Youth. In fact, some authorities believe that he was less interested in the fountain itself than in the gold and silver that were said to be found in abundance near the fountain.

After continuing his exploration of Florida, Ponce de León returned to Spain, where he remained until 1521. In that year, he undertook a second expedition to Florida. After reaching his destination, he was wounded in an Indian attack and died soon after.

Further Reading

King, Ethel M. *The Fountain of Youth and Juan Ponce de León*. Brooklyn, NY: T. Guaus' Sons, 1963.

Sanz, V. Murga. *Juan Ponce de León*. 2d ed. Rio Piedras, PR: University of Puerto Rico Press, 1985.

Fountains

A fountain is a stream of water that flows naturally from underground or is constructed artificially and released through a system of pipes. The term refers both to the flow of water itself and to the structure through which the water flows. Fountains have served a variety of functions throughout history, ranging from a source of water for drinking, washing, and other purposes to architectural ornament to religious inspiration.

Archaeologists have discovered evidence for the existence of artificial fountains dating as early as about 4000 B.C. in ancient Persia. Other examples have been found in the remains of ancient Assyrian and Mesopotamian cities that date to about 3000 B.C. A fountain discovered near the Comel River in the region of ancient Assyria, for example, consists of a series of basins cut into the rock at various levels. Water from the river flowed into the upper basin and then into lower basins.

Probably the most important function of early fountains was to provide a supply of water for a community. As long as people congregated in relatively small villages, each family was probably able to obtain the water it needed for washing, cooking, and other needs by transporting it from the nearest river or lake. But as communities grew larger and more complex, it became essential for the city government to provide a water supply for its citizens. Water fountains solved this problem. In many locations, systems of pipes and channels were developed to carry water from a reliable source, such as a large river or lake, to the center of a city. There it was released in the form of a fountain, from which citizens could draw the water they needed.

But fountains probably took on other kinds of significance relatively early in human history. Some of the earliest fountains discovered are adorned with carvings and statues that suggest they may have been used for religious or mystical purposes. The spiritual significance assigned to water in many cultures may explain the supernatural roles assigned to fountains in many societies.

Among the early Greeks, for example, fountains were sometimes thought to be dwelling places for water gods and goddesses as well as other deities. The custom of throwing coins into a fountain for good luck probably arose in cultures where such acts were regarded as propitiation of the gods.

Fountains have been constructed in all parts of the world not only as reliable water sources but also as elements of architectural design. Many of the great private residences and public parks of the world have included fountains. The gardens around the Taj Mahal in Delhi, within the Imperial City of Beijing, and throughout the royal city of Kyoto all feature fountains as a part of their general decor.

Fountain building reached its zenith in the Western world during the period of the Roman Empire. By the end of the Republic, Rome may have had as many as a million citizens. Providing a dependable supply of freshwater for that many people was clearly one of the government's highest priorities. To meet this challenge, the Romans constructed a complex system of aqueducts, tunnels, and pipes to carry water from rivers and lakes many miles outside Rome into the city. The final destination for this water was usually a fountain, placed at the center of a neighborhood where it was easily accessible to all citizens.

In many cases, the fountains out of which this water flowed consisted of a series of basins, one above the other. Water flowed out of pipes into the highest basin, where it was used for drinking and cooking. Excess water then flowed into the next lower basin, where it was used as a source of water for animals. Finally, the overflow from this middle basin was collected in a third basin, where water was used for washing. As water flowed through this system, from top to bottom, it become more polluted and, hence, was suitable only for each of the purposes designated.

The success of the Roman engineers in constructing this water system was quite remarkable. At its height, the city boasted between 1,200 and 1,300 public fountains, which delivered to citizens much of the 38 million gallons of water carried to the city each day. The fountains, were never turned on or off, but continued to supply safe, fresh water day after day, year after year.

The era of great aqueducts and fountains came to an end in the fifth century A.D., when they were largely destroyed by the Ostrogothic warlord Vitiges. The system lay in ruins for nearly a thousand years before government and church officials once more developed an interest in constructing fountains. During the Renaissance, however, the construction of fountains became at least as much of an artistic and architectural challenge as it was a task of supplying water for the general public. Great artists and architects were hired to design and build ever more elaborate structures for the exhibit of water displays.

In 1543, for example, the writer Claudio Tolomei described some of the new fountains that were being constructed in Rome. "But what pleases me more in these new fountains," he said, "is the variety of ways with which they guide, divide, turn, lead, break, and at one movement cause water to descend and at another time to rise." He goes on to observe that "the ingenious skill newly rediscovered to make fountains, in which mixing art with nature, one can't judge if it [the fountain] is the work of the former or the latter" (Symmes 1998, 1).

Religious and secular leaders were soon competing with each other to sponsor the greatest, most beautiful, most complex, or most spectacular fountain yet to be built. The great Italian sculptor, painter, and architect Gian Lorenzo Bernini (1598–1680), for example, was hired by Pope Urban VIII to produce a great fountain in his honor. That fountain, the Triton Fountain in the Piazza Barberini in Rome, is one of the best known fountains and most beautiful works of art in the city. It shows the Greek god of water, Triton, half-man and half-fish, resting on a huge scallop shell supported by four dolphins. A powerful jet of water issues from the conch shell he holds above his head. The fountain is embellished with the papal tiara as well as bees that were symbolic of the Barberini family to which Urban belonged.

Pope Innocent X commissioned Bernini to design another fountain to honor four great rivers of the world and the peoples associated with them. Bernini's fountain contains a giant natural rock in its center, surmounted by a lion, horse, and palm tree, from which the fountain issues. Surrounding the rock are statues representing Europeans (the Danube River), Americans (the Río de la Plata), Indians (the Ganges), and Africans (the Nile).

The largest and arguably most famous fountain in Rome is the Trevi Fountain, constructed on the site of the terminus of the aqueduct Aqua Virgo in A.D. 19. The site had been given the name *trevia*, because it was located at the intersection of three ("tre") streets ("via"). It was widely acknowledged to have the purest water anywhere in the city.

After the original site was destroyed, a number of efforts were made to construct a new fountain at the location, but none of these efforts met with success until the early eighteenth century. Then, Pope Clement XII chose architect Nicola Salvi to design and build a great fountain. Salvi's work shows the Roman god of the seas, Oceanus, riding on a huge chariot across the back of the fountain. Salvi's use of Oceanus was intended to reflect his own feeling that water is the most important element in nature.

The Italian interest in fountains as works of art and architecture eventually spread throughout Europe. Gardens throughout France, Spain, and England incorporated fountains as part of their overall design. The use of fountains at the palaces of Versailles in France and Blenheim in England are among the most outstanding examples of the period. By the end of the nineteenth century, the use of elaborate fountains in private homes and public gardens had also spread to the United States. One of the first such works was constructed at Thomas Jefferson's estate at Monticello.

Fountains continue to hold great appeal for architects, landscape designers, and ordinary men and women. Not only can fountains be great artistic inspirations, but the sights and sounds of water flowing in a fountain can be a great stimulus to meditation and thought. *See also* Aqueducts

Further Reading

Images Publishing Group. *Water Spaces of the World: A Pictorial Review of Water Spaces*. Mulgrave, Victoria, Australia: Images Publishing Group, 1997.

Jellicoe, Susan. *Water: The Use of Water in Landscape Architecture*. London: A. and C. Black, 1971.

Symmes, Marilyn, ed. *Fountains: Splash and Spectacle. Water and Design from the Renaissance to the Present*. New York: Rizzoli International Publications, 1998.

Frost Giants

See Jotuns

G

Ganga (Ganges)

See Sacred Waters

Geothermal Energy

Geothermal energy is heat energy produced when subsurface water flows over hot rocks lying at some distance beneath the earth's surface. This energy can be tapped by drilling wells, similar to water or oil wells, into zones where this heat is concentrated. The heat energy can then either be used to heat spaces and objects directly or be converted to electricity. In *direct utilization*, heat energy is used to heat homes and other buildings, to power industrial processes, for agricultural purposes, or in other applications. In *electricity generation*, heat energy is used to operate turbines, which drive generators.

Sources of Geothermal Energy

Heat is constantly being produced within the earth's interior as the result of the radioactive decay of unstable elements. This heat is responsible for the very high temperature of the earth's core, which may reach 7,500°C (13,000°F). It also accounts for heat in the layer surrounding the core—the mantle. The mantle is also heated by convection currents that carry heat from the core upward, into the mantle. Temperatures in the mantle range from about 2,200°C (4,000°F) at its boundary with the core, to about 870°C (1,600°F) at the upper boundary of the mantle, where it comes into contact with the earth's crust.

The thickness of the earth's crust varies considerably in various locations on the planet. In some places, the upper layer of the mantle may be no more than a few kilometers from the earth's surface. With modern drilling techniques, it is relatively easy to drill through the crust into the upper layer of the mantle in such regions. If geological conditions are suitable, such wells can be used to tap the mantle's thermal energy.

Forms of Geothermal Energy

Geothermal energy can be collected in areas where groundwater flows onto very hot rocks, producing either hot water or steam, or some combination of the two. Such sources can be classified into one of three categories: (1) low-temperature wells, where water temperatures are less than about 90°C (190°F); (2) moderate-temperature wells, with water temperatures between 90 and 150°C (190–300°F); and (3) high-temperature wells, with water temperatures greater than about 150°C (300°F). These sources differ from each other because of the temperature of the magma (molten rock) onto which water flows. The hotter the magma, the more completely water is heated and the more likely it will be converted to steam.

Low-temperature wells produce hot water only; moderate-temperature wells produce a mixture of hot water and steam, known as *wet steam*; and high-temperature wells produce nearly pure steam, known as *dry steam*, exclusively.

Dry-steam geothermal wells are used for the purpose of generating electricity. Wet-steam wells may also be used for electricity generation, but they are also used for direct heat utilization. Hot-water wells are used exclusively in direct heat applications.

The relationship between water, steam temperature, and use of geothermal energy is expressed in a famous diagram first proposed by the Icelandic engineer Baldur Lindal in 1973. The Lindal diagram reproduced on page 133 shows the range of applications possible from a geothermal source, with water temperatures ranging from about room temperature (20°C) to very hot steam (200°C).

Geothermal Technology

Most plants used to extract geothermal energy from the earth can be classified as one of four types: heat pump, direct steam, binary cycle, or direct flash.

Heat Pump Plants

In a *heat pump* facility, a well is sunk deeply enough to reach a source of geothermally heated water. That water is then piped up into a storage tank, from which it is transferred to homes, buildings, and other places where it can be used as a source of heat. As the heated water gives up its energy, it is cooled and returned to the ground.

In a typical facility of this type at the Oregon Institute of Technology, hot water at a temperature of 89°C (190°F) is withdrawn from three wells at the rate of 62 L/s (liters per second). The water is then used to heat eleven buildings on the institute campus through a system of pipes connected to both natural convection and forced-air convection radiators.

Larger systems can be used to provide heat to many more buildings. Perhaps the most famous example of such a system is the one found in Reykjavik, Iceland, where hot water from geothermal wells beneath the city supplies heat to 145,000 people. In France, forty similar projects provide heat for more than 500,000 people.

Hot-water systems have many other applications. For example, they can be used in agriculture and dairy farming to heat greenhouses, aquaculture systems, mushroom-growing facilities, and animal barns. Geothermal hot water can also be used to drive certain industrial processes. The oldest industrial application is in Larderello, Italy, where geothermal hot water has been used in the production of boric acid and borate compounds since 1790.

Direct Steam Plants

A *direct* (or *dry*) steam plant is conceptually simple. Dry steam is piped directly from an underground source to a power plant, where it is used to drive a turbine. The turbine, in turn, makes a generator spin, which produces electricity. Although simple in design, direct steam plants are somewhat rare because dry steam sources are not very common. In the United States, for example, only two dry steam sources have been found—in Yellowstone National Park and at the Geysers, in northern California. The Yellowstone source is of no commercial value, however, since it is protected from development by its national park status.

The Lindal Geothermal Energy Diagram

°C	
180 –	Evaporation of highly concentrated solutions Refrigeration by ammonia absorption Digestion in paper pulp, kraft
170 –	Heavy water via hydrogen sulphide process Drying of diatomaceous earth
160 –	Drying of fish meal Drying of timber
150 –	Alumina via Bayer's process
140 –	Drying farm products at high rates Canning of food
130 –	Evaporation in sugar refining Extraction of salts by evaporation and crystallization
120 –	Fresh water by distillation Most multiple effect evaporations, concentration of saline solution
110 –	Drying and curing of light aggregate cement slabs
100 –	Drying of organic materials, seaweeds, grass, vegetables, etc. Washing and drying of wool
90 –	Drying of stock fish Intense de-icing operations
80 –	Space heating Greenhouses by space heating
70 –	Refrigeration (lower temperature limit)
60 –	Animal husbandry Greenhouses by combined space and hotbed heating
50 –	Mushroom growing Balneological baths
40 –	Soil warming
30 –	Swimming pools, biodegration, fermentations Warm water for year-round mining in cold climates De-icing
20 –	Hatching of fish, fish farming

Saturated steam (left axis label, upper range)

Water (left axis label, lower range)

Source: Lindal, B., Industrial and other applications of geothermal energy, in Armstead, H. C. H., ed., *Geothermal Energy*, UNESCO, Paris, 1973, pp. 135–148.

Binary Cycle Plants

Binary cycle plants get their name from the fact that such facilities use two steps to convert geothermal heat to electricity. They operate with hot water whose temperature ranges from about 100 to about 180°C (212 to 360°F). Pipes bring this water from underground reservoirs into a heat exchange unit that contains a low-boiling substance (the "working fluid"), such as isobutane, isopentane, or ammonia. Heat from the hot water causes the working fluid to vaporize. The vapor is carried out of the heat exchange unit into a turbine. The vapor causes the turbine to spin, which drives a generator, which produces electricity. The vapor is then condensed and returned to the heat exchange unit while the hot water is returned to the ground.

Direct Flash Plants

A *direct flash plant* is constructed over a wet steam deposit where hot water is stored in permeable rock reservoirs under high pressure. The pressure is high enough in the reservoir to prevent the water from boiling even though its temperature reaches 150°C (300°F) or more. A pipe brings this very hot water to a chamber (the *separator*) in the plant, where the reduced pressure causes some of the water to begin to boil. The steam formed is vented away into a turbine, which runs a generator, while the liquid water remaining in the separator (in the form of brine) is returned to the ground.

A modification of the direct flash plant is the *double-flash system*. In the double-flash system, hot groundwater is fed first into a separator, as in the direct flash plant. The steam formed is sent to a turbine, while the unboiled hot water is sent to a flash unit where, under reduced pressure, more of the hot water begins to boil. Steam formed in the flash unit is also sent to the turbine, while leftover water is returned to the ground. A double-flash system has the advantage of extracting more heat from groundwater than can be obtained from the single-stage direct flash plant.

Economics of Geothermal Energy

In areas where reservoirs of hot water and steam are readily available, geothermal energy can be an attractive alternative for the production of electricity to more conventional energy sources, such as coal and petroleum. Unfortunately, those areas are relatively few around the planet. Existing geothermal plants in the United States currently produce electricity at a cost of 5 to 8 cents per kilowatt hour, roughly comparable to the cost of fossil-fuel-produced electricity (Geothermal Energy Program).

The use of geothermal energy for the operation of heat pumps for direct purposes may be an even more promising use of the resource. By some estimates, nearly half of the United States—including a very large portion of the Western states—has the type of geothermal reservoirs that would allow

the development and installation of heat pumps for use in private homes, commercial buildings, agricultural and dairy farms, and similar applications (http://geoheat.oit.edu/images/usmapl.gif).

The actual cost of geothermal energy depends on a number of factors including temperature and depth of the resources, type of resource (hot water and/or steam), chemistry of the water and/or steam, topography of the site, permeability of the reservoir, climatic conditions, proximity of transmission lines, and environmental constraints.

Environmental Issues

From some standpoints, geothermal energy is an environmentally attractive power source. The technologies used are relatively simple and, at this point, have been well developed. Energy production involves the release of no harmful by-products, such as oxides of sulfur and nitrogen and carbon dioxide. The only waste products, such as unused water from the reservoir, are returned to the earth from which they were removed. In most cases, disturbance of the surrounding land is minimal.

Still, the production of geothermal power is not an entirely risk-free activity. Some examples of problems that may arise include loss of vegetation and soil erosion; disturbance of nearby watersheds; disruption of natural hydrothermal features, with the possibility of increased steaming from the ground and lowering of the water table; and possible biological contamination of effluents returned to the ground.

Current Status of Geothermal Projects

Overall, geothermal energy has proved to be a popular alternative to conventional forms of energy production, such as fossil-fuel and nuclear generating plants in places where reservoirs are available and environmental impacts are minimal. As of 1998, the latest date for which data are available, there were 379 geothermal power plants around the world (no data were reported from Italy and New Zealand), with a combined output of 8,148 MW (megawatts) of electricity. The United States had the largest number of plants (203), followed by the Philippines (64), Mexico (26), Japan (18), and Indonesia (15). Geothermal plants in the United States are located in California, Hawaii, Nevada, and Utah.

The first geothermal power plant was built in Larderello, Italy, in 1904. The largest plant in the world is located at the Geysers, in California. This plant produces about 1,700 MW of electricity—about 7 percent of all the electricity produced by the Pacific Gas & Electric Company in 2000.

China leads the world in the use of direct utilization of geothermal energy, with an installed capacity of 1,914 MW in 1998. The United States ranked second with 1,905 MW of installed capacity, followed by Iceland (1,443 MW), Japan (1,159 MW), Hungary (750 MW), Italy (314 MW), and France (309) MW. Worldwide, the most common use for direct utilization of geothermal power was for space heating (33% of all applications), followed by bathing and swimming (19%), greenhouses (14%), heat pumps for heating and cooling (12%), aquaculture (11%), and industry (10%) (Fridleifsson 1998, 6). *See also* Groundwater

Further Reading

Bowen, Robert. *Geothermal Resources*. London: Elsevier Science Publications, 1989.

Dickson, Mary H., and Mario Fanelli. *Geothermal Energy*. New York: Wiley, 1995.

Fridleifsson, Ingvar B. "Direct Use of Geothermal Energy around the World." *GHC Bulletin* (December 1998): 4–8. Online at http://geoheat.oit.edu/pdf/pdfindex.htm.

Geothermal Energy Program, http://www.eren.doe.gov/geothermal/geopowerplants.html.

Geothermal Resources Council. *Geothermal Energy: Bet on It*. Davis, CA: Geothermal Resources Council, 1985.

Further Information

Geothermal Education Office
664 Hilary Drive
Tiburon, CA 94920
Telephone: (415) 435-4574 or (800) 866-4436
Fax: (415) 435-7737
URL: http://geothermal.marin.org
e-mail: mnemzer@marin.org

Geysers and Hot Springs

Geysers and hot springs consist of streams of water released from the earth after being heated by hot underlying rocks. The only difference between geysers and hot springs is the character of the release. In the case of hot springs, water is released as a stream of very hot water; in the case of geysers, water is released in the form of a violent eruption that carries very hot water, steam, and other gases with it.

Properties of Geysers

The size and timing of geyser eruptions differs widely from place to place. In some cases, a geyser erupts on a relatively regular schedule, such as once every hour or once every day. Old Faithful Geyser in Yellowstone is perhaps the most famous example of regular eruptors. It tends to erupt once every 30 to 90 minutes. The eruption of many other geysers, by contrast, is irregular and largely unpredictable.

The size and duration of eruptions also differ from place to place. The largest known geyser in the world, Steamboat Geyser in Yellowstone, emits a column of hot water and steam more than 90 meters (300 ft) into the air. By contrast, many geysers produce eruptions that extend upwards no more than a meter or so.

Eruption patterns may be affected by a variety of factors, such as rainfall, rock movement, and even barometric pressure.

Formation of a Geyser

Three conditions are needed for the formation of a geyser: water, heat, and a particular type of geological system. The heat needed for a geyser is usually provided by hot magma beneath the earth's surface. In many cases, geysers form in the areas where tectonic plates are in motion relative to each other, such as along the Ring of Fire that circles the Pacific Ocean. In such locations, hot magma is able to escape from the mantle and move upward towards the surface, where it can heat water to form a geyser.

Geysers also form in other types of regions. For example, the Yellowstone geysers are located in the center of a tectonic plate, far from any region of interaction with another plate. In such cases, geysers are located over *hot spots*—areas in the crust where heat from the earth's interior is escaping with unusual intensity. The cause of such hot spots is not entirely understood, but they do provide a source of heat that permits the formation of a geyser.

The water needed for the formation of a geyser comes from groundwater present at various depths beneath the earth's surface. Precipitation that falls on the earth seeps into the soil and then moves slowly through the water table to some outlet, which can be a river, lake, or geyser pool. The transit time for underground water can be very long indeed—many hundreds of years, for example. But at some time in some locations, that water eventually drains into a reservoir that is in direct contact with hot rocks.

At such points, the water may experience various fates. In some cases, it may simply drain onto hot rock, begin to boil, and escape through openings in the crust in the form of steam. Such formations are known as *fumaroles*. In other cases, the flow of water may be sufficiently large enough and the geological formations of the right shape to produce *hot springs*, in which very

hot or boiling water issues forth from beneath the surface. If there is a sufficient amount of loose dirt at the point of release, the hot water may mix with this dirt to form a *mud pot*. If geological conditions are exactly right, a geyser may form.

Geyser Eruptions

The mechanisms by which a geyser eruption takes place are not completely understood by scientists. In fact, it appears that a variety of different geological formations may allow for the creation of a geyser. The one condition that must be met is that water draining into the reservoir does not escape until a pocket of steam can form, which expands and drives the hot water out of an opening and into the atmosphere.

One possible sequence of events might be the following: groundwater flows into a cavelike reservoir at the base of which are very hot rocks. In this environment, the incoming water may actually become heated to temperatures greater than the boiling point of water, such as 1,000°C (over 500°F). Boiling does not occur, however, because of atmospheric and vapor pressures on top of the water. The higher the pressure on the surface of the water, the higher its boiling point.

As cool water flows into the reservoir, it is cycled through the existing water by convection currents. At the same time, very hot water from much deeper in the earth works its way upward through cracks in the rock into the water reservoir. The temperature of this water is often much greater than that of the water in the reservoir.

At some point, this very hot water may be hot enough to start boiling at even the elevated pressures within the reservoir. As it does, the steam that forms creates a bubble that expands very rapidly, pushing the hot water around it out the geyser vent and into the air. The eruption carries steam, very hot water, and dissolved gases into the air.

Scientists have now defined at least six distinct types of formations in which geysers can form, though there are likely to be others.

Geographical Distribution

Geysers are relatively rare phenomena on the earth. No more than about fifty geyser fields are known, and most of those consist of five or fewer individual geysers. The geyser field in Yellowstone National Park is by far the largest of its kind in the world. As a result, it has been studied more extensively than any other geyser field.

Other geyser fields are located at Beowawa, Nevada; Steamboat Springs, Nevada; and Umnak Island, Alaska, in the United States; in the Valley of the Geyser along the Geysernaya River on the Kamchatka Peninsula in Russia; in the El Tatio region of Chile; on the northern island of New Zealand; and in Iceland. The construction of geothermal plants in the area around Beowawa and Steamboat Springs resulted in the destruction of these two geyser fields. Overall, about 700 geysers exist on earth, about two-thirds of them in Yellowstone National Park. The Upper Basin of the Yellowstone River contains nearly 180 geysers, the largest concentration of geysers in the world. The park also contains the tallest geyser in the world, Steamboat Geyser, along with what may be the most famous geyser of all, Old Faithful.

Geysers are widespread on the island of Iceland, from whose language they get their name. In Icelandic, the word *geyser* means "gusher." As many as 200 geysers may once have existed in New Zealand, but geothermal construction has destroyed many of these. Waimangu Geyser was perhaps the largest geyser ever observed. It was said to have reached a height of 1,600 feet when it was still active in the early 1900s.

Hot springs occur more widely than do geysers and can be found in many parts of the United States, Japan, Europe, New Zealand, and other parts of the world. Hot Springs National Park in Arkansas is one of the most highly developed visitor attractions for hot springs enthusiasts.

Hot springs have long been a considerable attraction for tourists and for those with medical problems. As hot water rises upward through the earth to its surface, it dissolves minerals which, according to some people, provide it with medicinal values. There has been a long tradition in Europe and Japan for people to "take the waters" at a spa in an attempt to cure or alleviate a host of physical and medical problems.

One of the best known and most popular hot spring locations in the world is around the town of Beppu, in Japan. There are more than 4,000 hot springs in the area, some of which were described as early as A.D. 867. Today, the hot springs attract more than 12 million tourists a year.

The U.S. Geological Survey lists just over a thousand hot springs in the United States. About half of these are used as resorts or for bathing, irrigation, or water supply. The remaining number have not been developed for any practical application. The largest number of hot springs is located in Idaho (203; 120 not developed), followed by California (184; 42 not developed), Nevada (174; 66 not developed), and Oregon (105; 18 not developed). *See also* Geothermal Energy; Groundwater

Further Reading

Bryan, T. Scott. *The Geysers of Yellowstone.* Niwot, CO: University Press of Colorado, 1991.

Schreier, Carl. *Yellowstone's Geysers, Hot Springs, and Fumaroles.* Moose, WY: Homestead Publishing, 1987.

A photoglossary of geysers and related phenomena can be found on the U.S. Geological Survey's website at http://volcanoes.usgs.gov/Products/Pglossary/pglossary.html.

A complete list of hot springs in the United States can be found on the website for the National Geophysical Data Center at http://www.ngdc.noaa.gov/cgi-bin/seg/m2h?/net/web/seg/data/menus/springs.men+MAIN+MENU.

Gibbs, William Francis (1886–1967)

Gibbs was arguably the most famous ship designer in the United States, if not the world, during the twentieth century. He is best know for designing the SS *America* in 1937 and the SS *United States* in 1952, as well as streamlining the production of naval ships needed during World War II.

Personal History

William Francis Gibbs was born in Philadelphia on 24 August 1886. He earned his bachelor's degree in economics at Harvard in 1910 and then entered the Columbia University School of Law, from which he received an M.A. and an LL.B. in 1913. His first job was with a New York law firm that specialized in real estate law, but he discovered that he had relatively little interest in the field.

The turning point in Gibbs's life came in 1914 when the passenger ship *Empress of Ireland* sank after a mid-ocean collision with another ship. More than a thousand passengers and crew died in the mishap. Gibbs began to speculate about ways in which oceangoing ships could be designed to reduce the risk of such tragic accidents. A year after the *Empress of Ireland* tragedy, Gibbs and his brother Frederick founded the shipbuilding firm of Gibbs Brothers, Inc. The first ship produced by the firm crossed the Atlantic in record time, fourteen hours less than the previous record.

Among the many awards given to Gibbs during his lifetime were the Franklin Medal of the Franklin Institute and membership in the National Academy of Science and the National Academy of Engineering. Gibbs died in New York City on 7 September 1967.

Professional Accomplishments

Gibbs developed a number of improvements for improving the safety and efficiency of oceangoing vessels. For example, he proposed dividing the hull into a number of waterproof compartments so that damage to any one part of the hull would not result in a flooding of the whole ship. In addition, he proposed the use of high-pressure, high-temperature steam turbines that operated more efficiently than previous engines.

One of Gibbs's most impressive achievements came during World War II when the U.S. Navy required a large number of vessels in a short period of time. Gibbs proposed the revolutionary concept of building a ship at various locations rather than at a single shipyard. Each of the separate locations assembled one part of the ship, which was then put together at yet another location. The efficiency of this mass-production method of shipbuilding was such that the time required for building a Liberty ship during the war dropped from four years to less than a week. By the mid-1940s, shipyards had turned out more than 2,000 Liberty ships for the U.S. Navy and other users.

In addition to his construction firm, Gibbs held a number of important government positions, including controller of shipbuilding for the Army, Navy, and Maritime Commission; chairman of the American-British-Canadian Combined Shipbuilding Committee on Standardization of Design; and special assistant to the director of the Office of War Mobilization.

Further Reading

"William F. Gibbs—A Man and His Dream," http://www.cybercomm.net/~lynn/wf_gibbs.html.

Global Applied Research Network

Global Applied Research Network (GARNET) is an organization devoted to the development of methods for exchanging information about water supply and sanitation issues in developing countries, using low-cost, informal networks of researchers, users, and funders of research. Since the mid-1990s, GARNET has focused on the establishment of local network centers in South Asia, Latin America, and West Africa as a way of decentralizing the organization's previous global focus. It has also worked to develop network centers on special interest topics, such as wastewater treatment. Finally, it has improved its system of communication by developing electronic forms of networking and document delivery and by disseminating annual newsletters. One of its most important current activities is collating a database of research projects, fieldwork studies, and other references on water use and sanitation issues.

Further Information

Mr. Darren Saywell
GARNET Secretary
Water, Engineering and Development Centre
Loughborough University
Leicestershire LE11 3TU
United Kingdom
Telephone: +44 1509 222885
Fax: +44 1509 211079
URL: http://info.lut.ac.uk/departments/cv/wede/
    garnet/grntover.html
e-mail: d.l.saywell@lboro.ac.uk

Global Energy and Water Cycle Experiment

The Global Energy and Water Cycle Experiment (GEWEX) was established in 1980 as part of the World Climate Research Pro-

gramme (WCRP). The WCRP operates under the joint sponsorship of the International Council for Science, the World Meteorological Organization, and the United Nations' Intergovernmental Oceanographic Commission.

The overall goal of GEWEX is to develop models that will reproduce and predict variations of the global hydrological system, its impact on atmospheric and surface conditions, and regional variations that may occur in water supply and hydrologic processes as a result of natural and man-made changes in the environment, such as changes in greenhouse gases. Four steps are involved in attaining this goal:

- Measuring the status of and changes in the hydrologic cycle and energy flow in the atmosphere and on the earth's surface;

- Developing models of the global hydrologic cycle and its impact on the hydrosphere, atmosphere, and lithosphere;

- Developing the ability to predict changes in global and regional hydrologic processes as a result of changes in the environment;

- Improving techniques of observation, data management, and assimilation systems so that they can be used for long-range weather forecasting and hydrologic and climate predictions.

GEWEX operates in cooperation with a number of national scientific agencies, including the National Aeronautics and Space Administration (U.S.), National Oceanic and Atmospheric Administration (U.S.), National Space Development Agency (Japan), European Space Agency, Centre National d'Etudes Spatiales (France), and Deutsch Forschunganstalt für Luftund Raumfahrt (Germany).

Further Information

International GEWEX Project Office
1010 Wayne Avenue, Suite 450
Silver Spring, MD 20910
Telephone: (301) 565-8345
Fax: (301) 565-8279
URL: http://www.gewex.com
e-mail: gewex@cais.com

Global Rivers Environmental Education Network

Global Rivers Environmental Education Network (GREEN) was created as a part of the Earth Force program in 1999. Earth Force was established in 1994 with a grant from the Pew Charitable Trusts in recognition of two important national trends: an increasing interest among young people to work on environmental issues and their desire to help their communities through voluntary service. GREEN aims to improve middle and high school students' understanding and concern about the watersheds in their communities, and to use this understanding and concern to find creative solutions for local problems.

The GREEN website provides an online database and watershed action tool that allows individuals to enter data from their own research to produce graphs and tables as well as to share those data with other students across the country and around the world. An interactive guide also helps young people identify watershed problems and find ways of solving those problems.

GREEN provides a catalog that includes information on topics such as the Science Olympiad, electronic monitoring equipment, books and multimedia materials, biological monitoring equipment, and individual parameter monitoring kits.

Further Information

Earth Force
1908 Mount Vernon, Second Floor
Alexandria, VA 22301

Telephone: (703) 299-9400
Fax: (703) 299-9485
URL: http://www.earthforce.org/green
e-mail: earthforce@earthforce.org

Global Water

Global Water was founded in 1987 by Americans John McDonald and Peter Bourne, both of whom had been involved in the United Nations International Drinking Water and Sanitation Decade of the 1980s. Both men were appalled when they discovered how difficult it was for people in many developing countries to obtain adequate supplies of safe drinking water. They decided not to focus their efforts on simply finding more efficient ways to meet people's immediate needs but to attack the problem of water supplies at the source—by drilling new water wells and developing methods for purifying and storing existing sources of water.

Over the years, Global Water has carried out projects in South America, the Middle East, Africa, and Southeast Asia. For example, the organization is working on the construction of centralized, state-of-the-art water purification, storage, and distribution systems powered by photovoltaic solar panels for remote villages in Bolivia and Peru. In Kenya, Global Water has provided a team to help local residents learn how to drill for water in rural villages. In Laos, the organization found ways to locate and purchase existing well-drilling equipment that could be used to drill new water wells in rural villages.

Further Information

Global Water
1819 H Street, NW, Suite 1200
Washington, D.C. 20006-3603
Telephone: (301) 656-2818
Fax: (301) 656-0452
URL: http://www.globalwater.org
e-mail: info@globalwater.org

Gods and Goddesses of Water

Water gods and goddesses are deities associated with some specific manifestation of water, such as rain, rivers, or the ocean, or with the concept of water in general. This entry deals with some general aspects of water gods and goddesses. Refer to the end of this entry for references to specific gods and goddesses of special importance.

Prior to the scientific age, humans lacked any rational explanation for the vast majority of natural phenomena around them, phenomena such as fire, wind, rain, thunder, and lightning. They knew nothing about the physical processes by which clouds form, raindrops are produced, floods occur, and other water-related events take place. Acting on an apparently inherent and instinctive human need to understand such phenomena, most cultures developed mythical explanations as to the origin, causes, and expression of these events. In most cases, they assigned responsibility for such occurrences to specific mythical individuals, such as gods and goddesses, spirits, giants, or other supernatural beings.

It is hardly surprising that water deities are to be found in the mythical system of almost every culture. Water occurs almost everywhere and is essential in many ways to the survival of life. It is needed for drinking, cooking, and washing, and serves as a source for food through fishing and other marine activities.

Cultures with Many Water Gods

Water deities appear to be more important in the mythology of some tribes than others. Nations that are surrounded by water, such as Haiti, Cuba, and the islands of Oceania, might reasonably be expected to have developed a host of deities who control the ocean, waves, rain, storms, and other similar phenomena. Haitian culture pays homage, for example, to Adamisil Wedo, Vierge, and Ageweta, water goddesses; Aida Wedo, a goddess of freshwater and rainbows; Christaline, an evil sea goddess; Coatrischie, a goddess of water, winds, and storms; Damballah, a goddess of sweet waters; Sirène, a sea goddess who is the protectress of bathing children; and Guabonito, a goddess of the sea who also teaches humans about medicine.

Aztec culture also has a very complex panoply of deities relating to water. Some examples of the Aztec gods and goddesses connected to water include Acuecucyotici-huati, god of the oceans; Ahuic, god of rivers, streams, and waves on the beach; Apozanolotl, god of white-capped waves, foam, and suds on the water; Atlacamani, god of ocean storms; Ayauhteotl, god of mist; Xipe Totec, god of nocturnal rain; and Chalchihuitlicue, an important figure in the Aztec Great Flood myth.

Special Water Deities

Water occurs naturally in many forms, such as rivers, oceans, rain, mist, snow, and frost. Many cultures have created gods or goddesses with responsibilities for some particular manifestation of water. Some examples include the Eskimo god Alignka, responsible for the tides; As-ava, a Russian god responsible for freshwater; Gawaunduk, the Ojibwa god responsible for the mist; Ho Po, the Chinese god in charge of the Yellow River and supreme Chinese river god; Kun, the Aymaran god of snow; Laga, the Norse deity responsible for wells and springs; Möûll, Scandinavian god of ice and snow; Ochumare, a Puerto Rican goddess of rainbows; Olokum, a West Indies goddess of the ocean depths; and Shango, a Yoruba god who controls thunder and storms.

Perhaps the one specialized form of water deity most common is the rain god or goddess. In many parts of the world, human survival was (and is) fundamentally dependent on a reliable source of rain. In such places (and elsewhere), cultures have invented rain deities who are thought to be responsible for the appearance of rain and to whom sacrifices can be made, prayers offered, and ceremonies conducted to guarantee that rain will appear.

Many Native Americans, for example, lived in the western half of North America, where rainfall was (and is) scant and often unpredictable. One of the most common ceremonies among such tribes was the rain dance, a ritual in which formalized dances were conducted to convince the rain god or goddess to provide badly needed precipitation. Some examples of rain gods and goddesses from around the world include the following:

Abeguwo (Melanesia)
Aryong Jong (Korea)
Atoja (Peru)
Bunbulama (Australia)
Chac (Mayan)
Ch'ih Sung-tzen (China)
Cocijo (Zapotec)
Deng (Dinka)
Domfe (Kurumba)
Eschetweuarha (Chamacoco)
Katsinas (Acoma)
Mam (Mopan)
Mamacocha (Aymara)
Ilyap'a (Incan)
Sio Humis (Hopi)
Tomituka (Australia)
Tsui'goab (Hottentot)
Wuluwaid (Australia)

Famous Myths and Stories

Water gods and goddesses often play an important role in the greatest epics of many cultures. One story that appears in the mythologies of many cultures is that of the Great Flood. In the Great Flood, a ruling deity expresses his or her discontent with human nature by allowing endless rains to cover the face of the earth. In most cases, all human life and that of other animals is destroyed. A single pair of individuals of each species is allowed to survive the flood, making it possible for life on earth to begin anew once more.

According to Aztec legends, for example, the goddess Chalchihuitlicue became angry with humans when she saw how evil they had become. She allowed the rains to fall until the world was covered with water and almost all living beings had been drowned. Legend has it that this flood brought about the end of the fourth world in human history. As the flood receded, the world became populated again and became the fifth world in which we are now living.

Water deities appear in other epic tales as well. For example, the Tuscarora tribe has a legend somewhat similar to the Tower of Babel story that explains why humans speak different languages. According to this legend, a woman by the name of Godasiyo is chief of a village in which is born the cutest puppy dog that has ever been seen. As an argument develops over ownership of the puppy, Godasiyo declares that the people closest to her should travel to a distant island where they can all live in harmony. She then invents the canoe to provide transportation for the trip. During the trip, Godasiyo's colleagues begin to argue, her canoe tips over, and she and the beautiful puppy drown. She is reborn as a sturgeon, and the puppy is restored as a whitefish. When the others attempt to talk with each other about the miracle they have seen, they find that they are now speaking in different languages and cannot understand each other. *See also* Aegir; Flood Legends; Indra; Jotuns; Mermaids and Mermen; Nymphs; Poseidon; Ymir

Further Reading

Andrews, Tamra. *Legends of the Earth, Sea, and Sky*. Santa Barbara, CA: ABC-CLIO, 1998.

"The Book of Gods, Goddesses, Heroes, and Other Characters of Mythology," http://www.cybercomm.net/~grandpa/gdsindex.html.

Leach, Marjorie. *Guide to the Gods*. Santa Barbara, CA: ABC-CLIO, 1992.

The Great (Biblical) Flood

See Flood Legends

Groundwater

Groundwater is water stored under the surface of the earth. It comes in the vast majority of cases from precipitation that falls on the earth's surface. Some of the precipitation washes away immediately into lakes, rivers, and other water reservoirs, but most of the precipitation sinks into the soil, where it may become groundwater.

Zone of Aeration

The uppermost layer of soil is called the *zone of aeration*, or the *vadose zone*. In this region, the pore spaces within soil and rock contain both water and air. Some of the water trapped in these pores is taken up by plant roots. The rest evaporates to the atmosphere and is lost to the soil. Water trapped in the zone of aeration moves either upward by means of evaporation or downward as a result of the force of gravity. Water moves through the zone of aeration relatively quickly. Water that is not taken up by plants tends to return to the atmosphere within a matter of days or weeks.

Zone of Saturation

Water that sinks more deeply into the earth eventually reaches the zone of saturation. In this region, pore spaces are filled with water. No air is present at this level. Water in this layer of the earth moves not because of evaporation or as a result of gravitational forces, but toward other bodies of water at or beneath the earth's surface. For example, water deep in the zone of saturation might move sideways or upward into a nearby lake. Groundwater at this depth moves much more slowly than water in the zone of aeration. Water that reaches this level may not return to the hydrologic cycle for thousands of years.

The upper surface of the zone of saturation is called the *water table*. The water table tends to follow the contours of the surface of the land. That is, if someone living on top of a hill has to dig down 100 feet to reach the water table, it is likely that a neighbor in a valley nearby will have to dig down almost the same amount.

Aquifers

A portion of the zone of saturation from which water can be obtained with reasonable effort is called an *aquifer*. The word *aquifer* comes from two Latin terms meaning "water holding." To be classified as an aquifer, a region of the zone of saturation must meet two conditions. First, it must contain sufficiently large amounts of water to justify drilling. Second, the water must flow easily enough to supply human, animal, or other needs.

In most cases, it is necessary to use a mechanical pump to obtain water from a well drilled into an aquifer. The pump is necessary to force water stored in the zone of saturation upward and out of the well. An important exception is an *artesian well*. In an artesian well, a layer of permeable rock filled with water is trapped between two impermeable layers. If a hole is drilled into the permeable layer at a point lower than the water table, water pressure alone will force water to flow out of the hole.

Importance of Goundwater

Groundwater is an enormously important water resource for humans and other living things on the planet. By some estimates, more than 97 percent of all the freshwater resources in North America (not including frozen water) is stored in the zone of saturation. The volume of groundwater in the United States alone is more than nine times that found in the Great Lakes.

Humans have long relied on groundwater for many of their everyday needs, such as drinking, cooking, washing, and irrigating the land. In the United States, more than two-thirds of all the water drawn from aquifers is used for irrigation. Industries account for the next largest use of groundwater, with public and rural water supplies making up the other two major users of groundwater.

The Groundwater Foundation reports that 34 percent of all water used for agricultural purposes comes from groundwater, 40 percent of all public water supplies come from groundwater, and 53 percent of all drinking water used by Americans (including 97 percent of the drinking water used by rural residents) comes from groundwater.

Concerns about Groundwater Supplies

There are currently two major concerns about groundwater supplies around the world. The first is the rate at which they are being depleted. Most aquifers have taken hundreds of years to form, but humans are withdrawing water from them at a relatively rapid rate. In fact, water is being taken out much more quickly than it is being replaced by precipitation. As a result, some aquifers are beginning to dry up, endangering the water supplies for some large parts of the United States and other parts of the world.

An example is the Ogallala aquifer that lies beneath the states of Nebraska, Colorado, Kansas, New Mexico, and Texas. The aquifer was formed thousands of years ago and, until recently, was more than 1,000 feet thick in some regions. Humans have been pumping water out of the Ogallala aquifer since the 1930s, however, primarily to irrigate farms in the area. So much water has been removed from the aquifer at this point that some experts believe it may dry up in some parts of the region by the year 2020.

A second concern about groundwater is the risk of pollution. Pesticides, fertilizers, farm wastes, and a host of other materials are now being dumped on the land, to be carried into the zone of saturation with precipitation. Since there is no place for these chemicals to go once they reach the zone of saturation, they become part of the water table from which humans later withdraw water for personal, agricultural, and industrial use. In some parts of the American West, water obtained from municipal wells now contains unacceptably high levels of certain hazardous materials. *See also* Hydrologic Cycle; Water Pollution

Further Reading

Everett, Lorne G., and Igor S. Zektser. *Groundwater and the Environment: Applications for the Global Community*. Boca Raton, FL: CRC Press, 2000.

Moore, John E., Alexander Zaporozec, and James W. Mercer. *Groundwater: A Primer*. Alexandria, VA: American Geological Institute, 1995.

Wilson, James. *Ground Water: A Non-technical Guide*. Philadelphia: Academy of Natural Sciences, 1982.

Further Information

The Groundwater Foundation
P.O. Box 22558
Lincoln, NE 68542-2558
Telephone: (402) 434-2740
Fax: (402) 434-2742
URL: http://www.groundwater.org
e-mail: info@groundwater.org

Gulf Stream

The Gulf Stream is a current of warm water that flows northward out of the Gulf of Mexico and through the Straits of Florida, along the eastern coast of the United States and then northeast into the Atlantic Ocean. It originates at the confluence of two equatorial currents—the North Equatorial Current, which flows westward along the Tropic of Cancer, and the South Equatorial Current, which flows to the northwest from the west coast of Africa into the Caribbean.

The Gulf Stream is separated from the U.S. coast north of Cape Hatteras by a thin finger of the cold Labrador Current, which flows southward off the west coast of Greenland. The intersection of the Gulf Stream and the Labrador Current off the Grand Banks is responsible for one of the most concentrated fog centers found anywhere in the world.

When it reaches the mid-Atlantic, the Gulf Stream splits into two major arms, one of which flows to the northeast over the west coast of the British Isles as the North Atlantic Drift. The second arm flows to the east and southeast along the west coast of Spain as the Canaries Current.

History

The Gulf Stream was first described in 1513 by Ponce de León during a visit to the Florida peninsula. It was first mapped in 1665 by a German Jesuit priest, Athanasius Kircher (1601–1680), using data collected from a number of explorers. Benjamin Franklin (1706–1790) took special interest in the Gulf Stream and appointed a U.S. Navy captain named Folger to construct a map of the current in 1770. Franklin believed that the map would help ships select the fastest route from the United States to Europe and thus improve the speed of mail deliveries. As it turned out, ship captains thought Franklin's map to be useless and paid no attention to it. The map itself, however, is one of the most famous depictions of the current to have been made.

During his trips back and forth across the Atlantic, Franklin took every opportunity to make observations and measurements of the Gulf Stream. On his last voyage, he dropped thermometers to a depth of more than 100 feet to measure temperature changes in both horizontal and vertical directions around the current.

Some of the most productive research on the Gulf Stream was carried out under the direction of Matthew Fontaine Maury (1806–1873), sometimes called the father of oceanography. Maury first coined the phrase, "There is a river in the ocean," which has so often been used to describe the Gulf Stream. Maury reported on much of his research on the Gulf Stream in his epic text, *The Physical Geography of the Sea* (1855).

Until the mid-twentieth century, all observations of the Gulf Stream were made by merchant, military, or research ships traveling across the Atlantic or anchored within the Gulf Stream for the purpose of taking measurements. By the 1960s, however, more advanced technologies began to make possible new types of measurements that expanded and enhanced human knowledge about the Gulf Stream. For example, in 1966, the Naval Oceanographic Office began to carry out airborne surveys of the Gulf Stream using infrared thermometers mounted on specially designed Lockheed NC-121K aircraft. A decade later, the National Oceanic and Atmospheric Administration was using the Advanced Very High Resolution Radiometers (AVHRR) installed on earth-orbiting satellites to carry out daily measurements of sea temperatures in and around the Gulf Stream.

The use of satellites to collect better data on the Gulf Stream has now become routine. A variety of instruments on a number of different satellites are able to measure eddies and smaller currents within the Gulf Stream itself, detect and identify the presence of plant and animal life in the current, measure vertical variations within and at the edges of the current, and carry out other measurements of the Gulf Stream's physical, chemical, and biological properties.

Characteristics

At its origin in the Florida Straits, the Gulf Stream is typically about 80 km (50 mi) wide and about 640 m (2,100 ft) deep. Its average speed is about 5 km/h (3 mi/h) and average surface temperature, about 25°C (77°F). As the current travels northward, it becomes wider (up to 350 km [200 mi]) and deeper (about 1,200 m [4,000 ft]). The current's surface temperature decreases somewhat as it moves northward, as does its average speed.

The Gulf Stream also shows a distinct vertical structure, with both temperature and water speed higher in the upper levels and decreasing at greater depths.

As the Gulf Stream reaches its widest points, it may develop irregular, meandering paths known as *Gulf Stream rings* or *eddies*. These eddies tend to travel in a westerly, southwesterly, or southerly direction at rates of a few kilometers per day. They persist to a depth of 1,500 m (5,000 ft) or more and may last for a few months to several years. Eddies are of special importance for biological and climatic reasons because they have the effect of mixing water with different temperatures, salinity, nutrient composition, and other characteristics.

Climatic Effects

As early as the 1820s, some scientists were hypothesizing that the Gulf Stream had a moderating effect on the climate in western Europe. The constant supply of warm water from the Saragasso Sea, it was thought, kept temperatures in western Europe significantly higher than would be the case without the ocean current. This hypothesis is now generally accepted by most oceanographers and climatologists.

The Gulf Stream has some important effects on weather also. The collision of warm air above the Gulf Stream with the Labrador Current produces fog banks that survive more than half the year in some parts of the northern Atlantic. The flow of cold air from above the Labrador Current over the Gulf Stream also tends to produce large, unstable cold air fronts that may survive more than a week and extend upward several kilometers into the atmosphere. These cold fronts result in heavy winds over the ocean and clear weather over the eastern coast of the United States. *See also* Maury, Matthew Fontaine; Oceans and Seas

Further Reading

Gaskell, T. F. *The Gulf Stream*. New York: John Day Company, 1973.

MacLeish, William H. *The Gulf Stream*. Boston: Houghton Mifflin, 1989.

Stommel, Henry. *The Gulf Stream: A Physical and Dynamical Description*. 2d ed. Berkeley: University of California Press, 1976.

H

Haliae

See Nymphs

Hard Water

The term *hardness* is used to describe water that contains certain types of chemicals that make it difficult (or "hard") to form suds with soap. Water that lacks such chemicals is said to be *soft* water. Soft water lathers easily when soap is added to it.

Causes of Hardness in Water

As water runs over or through the ground, it dissolves a variety of chemicals found in the earth. Prominent among these chemicals are the carbonate, bicarbonate, sulfate, and chloride of calcium and magnesium. Common examples of these compounds include the following:

calcium carbonate ($CaCO_3$)
magnesium bicarbonate ($Mg(HCO_3)_2$)
calcium bicarbonate ($Ca(HCO_3)_2$)
calcium sulfate ($CaSO_4$)
magnesium carbonate ($MgCO_3$)
magnesium sulfate ($MgSO_4$)

In general, the carbonates and sulfates listed above tend to be only slightly soluble in water. The solubility of the calcium carbonate in pure water, for example, is about 9 parts per million.

The extent to which carbonates are soluble depends, however, on the pH (acidity) of the groundwater in which they are dissolved. Carbon dioxide, a component of the atmosphere, dissolves in water to form a weak acid known as *carbonic acid* (H_2CO_3). The solubility of calcium carbonate increases significantly in groundwater that contains carbonic acid. In such a case, the carbonate is converted to the bicarbonate, which is much more soluble than is the carbonate:

$$CaCO_3 + H_2CO_3 \rightarrow Ca\ (HCO_3)_2$$

The solubility of the sulfates is not affected in a similar fashion by the acidity of the groundwater.

An important factor that determines the hardness of water, then, is the source from which it is obtained. If water comes from wells that lie within a limestone strata, for example, it will tend to be quite hard. Limestone consists of calcium carbonate, which will dissolve in the well water, making it hard. If water comes from a river that flows through sandstone, by contrast, the water will tend to be soft. Sandstone consists of silicon dioxide (SiO_2), which is insoluble in water.

Types of Hard Water

Hard water is said to be either *temporary* or *permanent*. The two terms reflect both the ease with which the water can be softened and the cause of the hardness. When hard water is boiled, the carbonates and bicarbonates it contains tend to break down:

$$CaCO_3 \xrightarrow{heat} CaO + CO_2$$

In this case, the calcium oxide (CaO) settles out of the water as a precipitate and the carbon dioxide (CO_2) escapes as a gas. The water left behind after boiling tends to be quite soft. Water that can be softened in this way is said to contain *temporary hardness*.

Heat has no comparable effect on water containing sulfates and certain other compounds; thus, boiling has no effect on the hardness of such water. Hard water of this kind is said to be *permanent* hard water. It must be softened by other methods.

Disadvantages of Hard Water

Hard water cannot be used efficiently for bathing, washing of clothes, or other cleaning purposes. The reason is that the chemicals in hard water react with soaps to produce a solid precipitate that forms a scum on clothes, a "ring" in a bathtub, and other telltale deposits on objects being cleaned. The equation below shows what happens in the reaction between the calcium carbonate (for example) in hard water and a typical soap, sodium stearate ($NaCOOC_{17}H_{35}$).

$$(CaCO_3 + 2NaCOOC_{17}H_{35} \rightarrow Ca(COOC_{17}H_{35})_2 + Na_2CO_3))$$

The calcium stearate ($Ca(COOC_{11}H_{35})_2$) formed in this reaction is insoluble in water and precipitates out in the form of a scum or other deposit.

The formation of this deposit is undesirable, of course, because it is repugnant to look at. Just as important, however, is the waste of soap used when washing with hard water. The first portion of soap added to the wash is prevented from taking part in any cleaning action because it is reacting with the chemicals in the water. Only after all these chemicals have been precipitated out of the water (and it has become soft) will additional soap be available for the actual cleaning process for which it was intended.

Hard water has other serious disadvantages. When used in water heaters, boilers, and other equipment carrying hot water, the reaction shown here may occur:

$$CaCO_3 \xrightarrow{heat} CaO + CO_2$$

Calcium or magnesium carbonate or bicarbonate in the hard water is decomposed by the hot water, producing calcium oxide (CaO) as a deposit on the inside walls of the heater or boiler. This deposit, known as *boiler scale*, greatly reduces the heat conductivity of the device, eventually making it less useful for the purpose for which it was designed.

Softening Hard Water

Although temporary hard water can be softened by boiling, that method is not a very efficient process to use in most instances. The two most common methods for softening water in everyday situations is by adding chemicals to the water or passing the water through a water softener.

The purpose of adding chemicals to hard water is to precipitate out the chemicals present that make the water hard in the first place. For example, lime (calcium hydroxide; $Ca(OH)_2$) is often used as a water softener. When lime is added to hard water, the following chemical reaction occurs:

$$Ca(HCO_3)_2 + Ca(OH)_2 \rightarrow 2CaCO_3 + 2H_2O$$

A water softener is a unit in which a reaction similar to the one above takes place.

The unit contains some substance that will react with and remove the chemicals in hard water. A common substance used in water-softening units is zeolite. *Zeolite* is a generalized term for compounds containing sodium, aluminum, silicon, and oxygen, with the general formula $NaAlSiO_4$. The zeolite is attached permanently to a filter inside the water-softening unit. As water passes downward through the unit, a reaction takes place in which calcium and magnesium in the hard water are exchanged for sodium in the zeolite:

$$CaCO_3 + NaZeolite \rightarrow Na_2CO_3 + CaZeolite$$

The sodium carbonate that remains in the water has no adverse effects on the sudsing action of water, so the water leaves the unit as soft water. *See also* Surface Tension

Further Reading

The topic of hard water is discussed in many elementary, high school, and college textbooks.

Heavy Water

Heavy water is a type of water whose molecules consist of one atom of oxygen and either two atoms of deuterium or two atoms of tritium. Heavy water that contains deuterium is also called *deuterium oxide*. Its chemical formula is D_2O. Heavy water that contains tritium is known as *tritium oxide*. Its chemical formula is T_2O.

Isotopes of Hydrogen

Deuterium is an isotope of hydrogen whose nucleus contains one proton and one neutron. By contrast, ordinary hydrogen (also known as *protium*) contains a single proton only. Tritium is another isotope of hydrogen whose nucleus contains two neu-

trons and one proton. The list below summarizes the three isotopes of hydrogen.

Name	Protium	Deuterium	Tritium
Number of protons	1	1	1
Number of neutrons	0	1	2
Atomic mass	1	2	3
Atomic number	1	1	1
Symbol	$_1^1H$	$_1^2H$	$_1^3H$

Discovery of Heavy Water

The English chemist Frederick Soddy (1877–1956) had proposed the existence of chemical isotopes during the 1910s. Almost from the time he made this suggestion, chemists began to consider the possibility that hydrogen, the lightest and simplest element of all, might have isotopes.

The existence of hydrogen isotopes was unusually important. In the case of nearly all elements, two isotopes differ from each other in mass by a relatively small amount, usually no more than a few percent. But isotopes of hydrogen would differ from each other by very large amounts. For example, deuterium has a mass twice as large as that of protium, and tritium has a mass three times as great. These mass differences make it possible for the isotopes of hydrogen to be used diagnostically in a variety of ways.

The search for heavier isotopes of hydrogen was largely unsuccessful, however, until the 1930s. Then, the American chemist Harold Clayton Urey (1893–1981) was able to produce deuterium experimentally. He accomplished this feat by allowing a large volume of liquid hydrogen to evaporate slowly. The lighter isotope of hydrogen (protium) evaporates slightly more rapidly than does the heavier isotope (deuterium). Over time, then, the evaporation of liquid hydrogen results in a product that is slightly richer in deuterium than normal hydrogen. When this process is repeated over and over again, a product increasingly rich in heavy hydrogen can be prepared. Eventually, Urey was able to prepare a sufficiently pure sample of deuterium to identify and characterize the isotope.

The next step was to prepare water whose molecules consisted of deuterium atoms rather than protium atoms. The American chemist Gilbert Newton Lewis (1875–1946) was able to carry out this synthesis in 1933.

Tritium was prepared for the first time shortly after Lewis's production of heavy water by the Australian physicist Marcus Laurence Elwin Oliphant (1901–2000). Heavy water made with tritium is of less interest than deuterium oxide because tritium is radioactive and less easy to work with than is deuterium oxide. Today, deuterium oxide is easily prepared in large quantities for both research and industrial purposes.

Properties of Heavy Water

Heavy water occurs naturally to the extent of about one part in 6,500. It has a freezing point of 3.8°C, a boiling point of 101.4°C, and a density of 1.1056 grams per cubic centimeter at a temperature of 25°C. It is used as a tracer in many research studies because its mass is so significantly different from that of ordinary water, protium oxide. Its most important commercial application is as a moderator in nuclear reactors. A moderator is a substance that slows down the speed of neutrons that pass through the material.

Holy Water

Holy water is water that has been blessed by a priest. In this regard, holy water differs somewhat from holy waters (found in lakes, rivers, and other bodies of water), which possess some inherent magical or supernatural properties without the need for the blessings of a priest or other holy person. The distinction between these two forms of holy water is not always entirely clear, however. For example, most Christians believe that holy water must be blessed before being used. However, water taken from the River Jordan, where Jesus was baptized, is thought to have very special curative and protective powers.

History of Holy Water

In many religions, holy water is thought to have special powers for protecting people against both physical problems and attacks by evil spirits. In both the Roman Catholic and Eastern Orthodox churches, for example, the sprinkling of holy water is an essential sacrament used for a variety of purposes. This custom has a very ancient ancestry and can be traced to practices among the ancient Hebrew tribes as well as later Muslims. In fact, some authorities believe that a machine described by the Greek historian Hero in 200 B.C. may have been used by the Egyptians to dispense holy water.

One of the most frequently cited exhortations about the use of holy water comes from the autobiography of St. Teresa of Avila, written in 1562. She writes about her use of holy water to frighten away the devils who repeatedly tried to distract her from her devotions. She explains that in one instance, she tried making the sign of the cross over the book from which she was reading, but the devils refused to leave. Only when she had sprinkled some holy water on them did the devils actually flee.

Functions of Holy Water

Many commentators stress that frightening away evil spirits is the most important function of holy water. One priest has written that "The devil hates holy water because of its power over him. He cannot abide in a place or near a person that is often sprinkled with blessed water" ("Holy Water," 10/18/99).

Holy water is, therefore, an essential ingredient in exorcisms, in which the devil or other evil spirits are driven out of human bodies over whom they have taken control. In fact, it is the sight and smell of the holy water, according to some believers, that finally drives the evil spirit out of a person's body.

Protection against spirits can be obtained by sprinkling holy water on a person's body or by having the person actually drink the holy water, according to some faiths. In many cases, it is also recommended that holy water be sprinkled on the walls and floors of a room to protect the people who live there from attacks by the devil. In some religions, the sprinkling and drinking of holy water are accompanied by detailed and specific rituals. In the Orthodox faith, for example, the sprinkling of holy water is carried out by family members who carry candles and icons throughout the living quarters.

Holy water may also have broader and more general applications. Some authorities recommend that it be sprinkled on sores or cuts, be drunk during times of temptation or anger, or be used during any type of bodily illness.

Finally, holy water is also a purifying liquid that cleanses a person's soul, just as water itself can be used to cleanse a person's body and garments. Individuals who enter a church may, for example, make the sign of the cross on their bodies with holy water taken from a font. The holy water wipes out any venial sins the person has brought to church with him or her. *See also* Sacred Waters

Further Reading

"Epiphany and the Blessing of Water" (Byzantine Catholic web page), http://www.byzantines.net/feasts/epiphany.htm.

"Holy Water" (entry in *The Catholic Encyclopedia*), http://www.newadvent.org/cathen/07432a.htm

"Holy Water," http://www.mary-mother-of-unity.com/holyh2o.htm,

Holy Waters

See Sacred Waters

Hookah

A hookah is a water pipe used for smoking. It is also called a *nargile* (also *nargilla;* pronounced *nar gee la*), shisha, or hubbly-bubbly. Some historians trace the invention of the hookah to India. Over time, the object's popularity spread westward, across Persia and into Turkey. Although hookahs can still be found almost everywhere in the Middle East, they are most popular in Turkey. In many communities, hookah smoking is still the center of socializing for both men and women.

Construction of the Hookah

The hookah consists of four parts, the *agizlik*, or mouthpiece; the *lüle*, or upper portion of the hookah; the *marpuç*, or tube; and the *gödve*, or container that holds water. The substance to be smoked is placed into the lüle and covered with hot coals. The selection of material to be smoked is a fine art for most hookah users. Western tradition sometimes claims that hashish is a popular smoking material, although the vast majority of smokers use some form of tobacco. The two most popular forms of tobacco are a strong, dark material imported from Iran and a fruity tobacco made by adding fruit or even wine to pure tobacco.

Hookah making has long been a very sophisticated art. Each of the four parts of the hookah is made by a specific type of craftsperson from distinct regions of the country. For example, lüles have traditionally come from the Tophane region of Turkey, and gödves, from the area of Beykov. The finest hookahs are crafted from brass, silver, olive wood, and other materials and are decorated extensively. The best examples of the art have become collectible objects by both smokers and nonsmokers alike.

Man smoking a water pipe in the Khan el-Khalili bazaar in Cairo, Egypt, in 1994. *Dave Bartruff/Corbis.*

Smoking the Hookah

The hookah is smoked by inhaling through the mouthpiece. Smoke from the smoldering tobacco is drawn through the water, up into the *marpuç,* and out of the mouthpiece. The smoke inhaled by this method is cooler than that from a cigarette, cigar, or pipe. It is thought to contain fewer harmful ingredients, such as nicotine and tar, which are trapped by the water in the gödve. Smokers also believe that the filtered smoke tends to be somewhat erotic and intoxicating.

Traditionally, hookah smoking has been accompanied by certain very specific customs, as with the tea ceremony in Japan. Coals are to be lit and placed on the tobacco only in certain ways, and the pipe is then to be smoked according to very formal rules.

Before the introduction of cigarettes, hookah smoking had long been a popular activity practiced at home or in special cafes, where coffee and tea were also available. Among the men and women in a village or neighborhood, smoking was often the center of social interaction. Although hookah smoking has lost popularity in many parts of the Middle East, hookah cafes still exist in many parts of the area, particularly in Turkey.

Further Reading

"Hookah!" http://www.tierracaliente.com/hookahm.shtml. This website contains three articles on hookahs originally printed in the *Turkish Daily News* (3 March 1997), the *New York Times* (10 June 1997), and the *Salt Lake Tribune* (21 April 1996).

Hovercraft

See Boats and Ships

"Hubbly-Bubbly"

See Hookah

Human Water Needs

See Biological Functions of Water; Desalination

Humidity

Humidity is defined as the amount of water vapor present in the atmosphere. Humidity can be measured in a number of ways, including *absolute humidity, mixing ratio*, and *relative humidity*.

Saturation of Air

In all but the most unusual cases, the air over a region contains at least some water vapor. That water vapor reaches the atmosphere through evaporation from surface moisture, lakes, rivers, oceans, and glaciers; through transpiration from plants; and through other mechanisms. The amount of

Saturated Vapor Pressures

Air Temperature (°C)	Saturation Vapor Pressure (mb)
–15	1.9
0	6.1
5	8.7
10	12.3
15	17.0
20	23.4
25	31.7
30	42.5
40	73.8
50	123.5
60	196.5
70	312.0
80	474.2
90	701.8
100	1,014.2

moisture present in a sample of air depends on various factors, including the amount of surface moisture available, temperature, and atmospheric pressure.

Any given volume of air can hold a certain maximum amount of water vapor for any given temperature and atmospheric pressure. The usual way of expressing this quantity is the *saturation vapor pressure* of the air. To determine the saturation vapor pressure, imagine a closed container to which a small amount of water has been added. The temperature in the container is noted and kept constant; over time, some of the liquid water in the container evaporates and becomes part of the air overlying the water as water vapor. The water vapor formed by this process exerts pressure on the surface of the liquid water, just as does the air itself. For any given temperature, only some given amount of water vapor can be added to the air above the liquid. This maximum amount of water vapor is the saturation vapor pressure. At

room temperature of about 20°C, the saturated vapor pressure is about 23 mb (millibars).

If the temperature of the container is now increased, a greater amount of water will evaporate and the water vapor formed will exert a greater pressure. For example, if the temperature of the container is increased to 30°C, the saturated vapor pressure becomes about 42 mb. The saturation vapor pressure in millibars (mb) for water at various temperatures is given in the table above.

Absolute Humidity and Mixing Ratio

Absolute humidity and mixing ratio are two methods for expressing the actual amount of water vapor present in a sample of air. Absolute humidity is expressed as the mass of water vapor in a given volume of air. The usual units for expressing absolute humidity are grams per cubic meter. For example, a given sample of air might have

Saturated Mixing Ratios

Temperature (°C)	Saturation Mixing Ratio (g/kg)
0	3.8
10	7.8
20	14.8
30	27.7
40	49.8
50	88.1

an absolute humidity of 5 g/m³ (grams per cubic meter).

Absolute humidity has one important disadvantage as a measure of the amount of water vapor in the atmosphere. A given parcel of air may rise or sink in the atmosphere, changing its volume in the process. Since the volume of the parcel changes, so does its absolute humidity. Yet, the amount of water vapor in the parcel may not have changed at all.

To deal with this drawback, a measure known as *mixing ratio* is sometimes used to express the amount of water vapor in air. Mixing ratio is defined as the mass of water vapor in a given mass (not volume) of air. The usual units used in measuring mixing ratio are grams of water vapor per kilogram of dry air. For example, a parcel of air might be said to have a mixing ratio of 5 g/kg (grams per kilogram). If this parcel of air rises or sinks, its volume may change, but its mass does not. Therefore, its mixing ratio remains constant even though its absolute humidity may change.

As with saturation vapor pressure, the mixing ratio of air at its saturation point changes with temperature and pressure. The saturation mixing ratio for various temperatures at sea level are given in the table above.

Relative Humidity

The most commonly used and most useful way of expressing humidity is by means of *relative humidity*. Relative humidity refers to the amount of water vapor contained in a sample of air compared to the maximum amount of water vapor that *could* be held at that temperature and is expressed as a percentage. Relative humidity can be determined by dividing the actual vapor pressure by the saturated vapor pressure or by dividing the actual mixing ratio by the saturated mixing ratio at the given temperature, or:

$$RH = p_{(H20)a} \div p_{(H20)m} \times 100\%, \text{ or}$$

$$RH = [H_2O(g)_a] \div [H_2O(g)_m] \times 100\%,$$

where $p_{(H20)a}$ and $p_{(H20)m}$ represent the actual and saturated vapor pressures and $[H_2O(g)_a]$ and $[H_2O(g)_m]$ represent the actual and saturated mixing ratios of the sample.

For example, suppose a sample of air has an actual vapor pressure of 16.2 mb at a temperature of 30°C. Then the relative humidity of that air is 16.2 mb ÷ 42.5 mb (the saturated vapor pressure, from Table 1 above), or 38.1 percent. Suppose a second sample of air has an actual mixing ratio of 4.9 g/kg. Then the relative humidity of that air is 4.9 g/kg ÷ 27.7 g/kg (the saturated mixing ratio, from the table above), or 17.7 percent.

Relative humidity, like absolute humidity and mixing ratio, is affected by two factors: availability of moisture and temperature. Suppose the relative humidity in a region is 40 percent. If a storm center moves through the area and brings rain, water vapor will be added to the atmosphere and the relative humidity will increase. If a cold front moves through an area, the temperature will drop, some water vapor will condense from the air, and the relative humidity will fall.

Dew Point

Weather forecasters often refer to the *dew point* in their reports. The dew point is the temperature to which a sample of air must be cooled for water vapor to condense. Consider the second example in the paragraph above. Suppose that the mixing ratio in the atmosphere prior to the approach of the cold front were 50.0 g/kg when the temperature was 50°C. From Table 2 above, the saturated mixing ratio at 50°C is 88.1 g/kg, so the relative humidity would be 50.0 g/kg ÷ 88.1 g/kg, or 56.7 percent.

But suppose the air temperature dropped to 40°C as the cold front passed through. At a temperature of 40°C, the saturated mixing ratio (from Table 2) is 49.8 g/kg. At just below 40°C, then, water vapor would begin to condense out of the air. The dew point under these conditions would be just less than 40°C. As is obvious from this example, the dew point is the temperature at which the relative humidity is 100 percent.

Measurement of Humidity

Both absolute humidity and mixing ratio can be determined experimentally though not very easily. For example, one could use some method for extracting all of the water vapor in a given sample of air, weigh that water, and determine the number of grams present per kilogram of air sample to find the mixing ratio. Determining the relative humidity is much simpler. A number of instruments, known as *hygrometers*, have been invented for this purpose. Once the rel-

ative humidity of an air sample is known, the absolute humidity and mixing ratio can be determined.

For example, recall the formula for the relative humidity, based on absolute humidity:

$$RH = [H_2O(g)_a] \div [H_2O(g)_m] \times 100\%$$

In this formula, the relative humidity (RH) can be determined by means of a hygrometer, and the saturated mixing ratio ($[H_2O(g)_m]$) can be found by consulting a table like Table 2. Rearranging the above equation, then, allows one to determine the absolute humidity ($[H_2O(g)_a]$) from the other two measurements:

$$[H_2O(g)_a] = RH \times [H_2O(g)_m]$$

See also Clouds; Precipitation

Further Reading

Textbooks on meteorology always contain a section on humidity and related topics. Two popular college-level books are the following:

Ahrens, C. Donald. *Meteorology Today: An Introduction to Weather, Climate, and the Environment*. 6 ed. Pacific Grove, CA: Brooks/Cole, 2000.

Lutgens, Frederick K., and Edward J. Tarbuck. *The Atmosphere: An Introduction to Meteorology*. 8 ed. Upper Saddle River, NJ: Prentice Hall, 2001.

See also the following:

Konvicka, Tom. *Teacher's Weather Sourcebook*. Englewood, CO: Teacher Ideas Press, 1999.

Hydrate

A hydrate is a chemical compound that contains one or more molecules of water in a definite ratio to the other components of the compound. For example, copper sulfate ($CuSO_4$) frequently occurs in the form of *hydrated* copper sulfate, in which five molecules of water are associated with a single copper sulfate unit: $CuSO_4 \cdot 5H_2O$. The name of the hydrated form of copper sulfate is *copper sulfate pentahydrate*. Compounds may form more than one kind of hydrate. For example, sodium sulfate forms both a decahydrate ($Na_2SO_4 \cdot 10H_2O$) and a pentahydrate ($Na_2SO_4 \cdot 5H_2O$). The former is also known as *Glauber's salt*, used as a cathartic, and the latter is better known as *photographer's hypo*, used in developing photographs.

Hydrates are commonly formed when a salt is formed during the evaporation of its water solution. In such cases, molecules of water may be trapped within the crystal structure of the salt being formed. This water is called *water of hydration*.

Hydrates have a greater or lesser tendency to give up their water of hydration. For example, sodium tetraborate decahydrate ($Na_2B_4O_7 \cdot 10H_2O$) begins to lose its water at about 60°C (140°F), while calcium sulfate dihydrate ($CaSO_4 \cdot 2H_2O$; gypsum) loses its water of hydration only at temperatures of above 163°C (325°F). The parent compound of a hydrate that remains after it has lost its water of hydrate is said to be *anhydrous* ("without water").

In some compounds, water of hydration is so loosely held that it tends to give up its water at room temperature. Such is the case, for example, with sodium carbonate decahydrate ($Na_2CO_3 \cdot 10H_2O$; washing soda). Compounds of this type are said to be *efflorescent*. They may give up all of their water of hydration and remain partly hydrated, or they may lose all of their water hydration to become a dry powder.

Some compounds have a tendency to take on water from the atmosphere. Such compounds are said to be *hygroscopic*. In some cases, hygroscopic compounds may take up so much water that they actually turn into a liquid. The formation of the liquid occurs for one of two reasons: first, the compound may take up so much water that it actually dissolves, second, the addition of the water to the compound may lower the melting point of the substance, turning it into a liquid. Examples of deliquescent compounds include calcium chloride ($CaCl_2$), lithium sulfide (Li_2S), and magnesium iodide (MgI_2).

Hydrates have a number of practical applications. For example, the naturally occurring mineral gypsum (calcium sulfate dihydrate; $CaSO_4 \cdot 2H_2O$) can be heated to produce plaster of paris (calcium sulfate monohydrate; $CaSO_4 \cdot H_2O$), which is then used to form casts and molds. Anhydrous calcium chloride has long been used as a *desiccant*, a substance that removes water from the air. In this role, it is widely used on dirt roads. When calcium chloride is spread on such roads, it takes moisture out of the air and helps to settle the dust on the roads. Another common desiccant is alumina (aluminum oxide; Al_2O_3), used as a additive in many prepared foods. The purpose of the additive is to absorb moisture that may form in a package and keep the main ingredients dry.

A particular kind of hydrate of some commercial interest is methane hydrate. Methane (CH_4; the primary component of natural gas) is usually found in nature as a gas. In some circumstances, however, it may occur as a hydrate in which methane molecules are trapped inside ice formations. Methane hydrates usually occur in solid layers under the ocean floor and under permafrost. They are of considerable commercial interest since some experts believe that methane hydrate deposits are very extensive and may constitute an important energy source for the world.

Hydraulic Device

A hydraulic device is one that operates because of the power provided by the motion of water or some other liquid.

The Hydraulic Press and Hydraulic Jack

Perhaps the best known of all hydraulic devices is the *hydraulic*, or *hydrostatic, press*, invented by Joseph Bramah (1748–1814). The hydraulic press is based on Pascal's Law—namely, that a pressure exerted to a confined liquid is transmitted equally to every point in the liquid. A hydraulic press consists of two cylinders, each fitted with a piston and filled with a liquid. The two cylinders are of different diameters and are joined to each other by a pipe, which is also filled with the liquid. The hydraulic press invented by Bramah used water as the working liquid, although modern presses may use other fluids, such as an oil or a gas.

When pressure is applied to the piston in the smaller cylinder, that pressure is exerted equally in all directions within the liquid, including to the larger piston. For example, suppose that the two pistons in a hydraulic press have surface areas of 10 square inches and 100 square inches. And suppose that a force of 100 pounds is applied to the smaller piston. The pressure on that piston, then, is 100 lb/10 in.^{2}, or 10 lb/in.^{2}. When that pressure is applied to the larger piston, the force exerted is 10 lb/in.^{2} times 100 in.^{2}, or 1,000 pounds.

A simple modification of the hydraulic press is the *hydraulic jack*, used to lift heavy loads or exert large forces. A hydraulic jack also consists of two cylinders of different sizes joined to each other by means of a pipe. The system is filled with a fluid, such as water or oil. The smaller cylinder is also fitted with a pump, which can be used to force liquid into the larger cylinder. By exerting a relatively small force on the smaller piston, a much greater force can be exerted through the larger piston.

Other Hydraulic Devices

The *hydraulic ram* was a device invented in the late eighteenth century for moving water uphill. The general principle of the ram was invented by John Whitehurst in England in 1772, although the device was later perfected by the Frenchman Joseph-Michel Montgolfier, best known for his research on hot-air balloons, in 1796. The device soon became widely popular because it was simple to build and operate, and was highly reliable. Hydraulic rams are said to work efficiently for up to a hundred years without the need for major repairs.

In a hydraulic ram, water from some natural source, such as a stream or artesian well, is fed into one end of a long *drive pipe*. At some distance along the drive pipe is located a specially designed vertical *poppet* valve through which water can escape until a certain minimum pressure is reached within the valve. At that point, the valve closes and water is forced to continue along the drive pipe until it reaches a second valve—called a *check valve*—set within the drive pipe. Located just beyond the check valve is a vertical compression chamber filled with air.

As water flows through the check valve, it compresses air in the chamber. The more air is compressed within the chamber, the greater the pressure it exerts on water in the drive pipe. Eventually, this pressure becomes great enough to push water back down and out of the chamber, closing the check valve and forcing water out the end of the drive pipe with a force many times greater than that of the water entering the drive pipe.

As the check valve closes, the poppet valve opens once more, and the whole cycle is repeated. In modern hydraulic rams, the cycles occur about once every second.

Hydraulic rams fell out of favor in many locations when electrical power became generally available to rural areas. Electric motors often provided a simpler and more economical method for pumping water then did the ram. However, in recent years, rams have once more become popular as a device for moving water in areas where electricity has become more expensive or landowners prefer alternative methods.

Other devices still carry the term *hydraulic* in their name although they do not—and in some cases, never did—involve the use of water. An example is the *hydraulic accumulator*, a device that acts as either a source of fluid or a shock absorber. Hydraulic accumulators come in a number of forms, most commonly a bladder, diaphragm, piston, or weighted type. All forms consist of a pressurized gas (usually nitrogen), a spring, or a weighted plunger that is caused to push on fluid (often some type of oil) stored in a container, releasing the fluid through an outlet pipe or absorbing some type of shock.

Another so-called hydraulic device is the *hydraulic actuator*. A hydraulic actuator is a device for using the energy of a moving fluid to push a piston back and forth in a cylinder, to turn a rotor, or to produce some other form of useful mechanical energy. Depending on their design, hydraulic actuators can produce forces of up to several hundred tons.

Further Reading

Ewbank, Thomas. *A Descriptive and Historical Account of Hydraulic and Other Machines for Raising Water*. New York: Arno Press, 1972.

Pippenger, John J. *Fluid Power: The Hidden Giant*. Jenks, OK: Amalgam Publishing, 1992.

Schetz, Joseph A., and Allen E. Fuhs, eds. *Handbook of Fluid Dynamics and Fluid Machinery*. New York: Wiley, 1996.

Watt, S. B. *A Manual on the Hydraulic Ram for Pumping Water*. London: Intermediate Technology Publications, 1981.

Hydraulic Press

See Bramah, Joseph

Hydraulics, Hydrostatics, and Hydrodynamics

The terms *hydraulics*, *hydrostatics*, and *hydrodynamics* refer to three different disciplines, but they are closely related to each other. Developments in one field often bring about or are caused by developments in one or both of the other fields. While scientists and engineers may specialize in any one of the three fields, they are familiar with the general principles and developments in the other fields.

The term *hydraulics* refers to all the ways in which water at rest or in motion can be put to use in practical applications, such as the construction and operation of dams, canals, and water supply systems. *Hydrostatics* is the study of the properties of fluids at rest, especially with regard to the pressure exerted by a fluid on objects immersed within them. *Hydrodynamics* deals with the properties of fluids in motion, with special attention to the forces exerted on bodies immersed within the fluid and the interaction between such bodies and the fluid.

As their names suggest, all three of these terms were first used in connection with the properties and behavior of water (*hydro-* = "water"), although these properties and behaviors have long since been generalized to other types of fluids, such as oils and air.

Early History

Hydraulics is by far the oldest of the three fields. Humans developed methods for using the energy of running water in the earliest stages of civilization. For example, engineers from both the Egyptian and Mesopotamian civilizations discovered methods for building dams, irrigation systems, and wells with which to supply communities with the water they needed for farming and domestic purposes.

Their contemporaries in Asia had developed similar technologies. The Harappan civilization, for example—centered on the Indus River—developed a complex system of waterways used for irrigation and transportation. Homes excavated from one of the civilization's great cities, Mohenjo Daro, had sanitation systems not unlike those found today, with indoor bathrooms, covered sewer lines, and cesspools.

In China, a major challenge facing hydraulic engineers was to find a way of taming the Hwang Ho (Yellow) River. Regular, periodic flooding made (and still makes) much of the river unusable for navigation, irrigation, and other purposes. By 2500 B.C., however, Chinese engineers had developed systems of dikes and canals that brought the river under at least some degree of control for much of the year.

From Technology to Science

The accomplishments described above are to be classified as works of art rather than scientific achievements. Early inventions were the result of trial-and-error research over long periods of time rather than derivations from basic scientific principles. In fact, the first scientific law dealing with the properties of water was not discovered until the third century B.C. It was during that period that the Greek natural philosopher Archimedes (287–212 B.C.) discovered the law of buoyancy—namely, that the loss of weight of an object immersed in water (or any fluid) is equal to the weight of the water (or fluid) displaced by the body.

Archimedes' principle was essentially a dead end in the field of hydrostatics. His discovery did not lead to further developments in the field, nor was it applied to any great extent in the invention of practical devices. In fact, there was, with one great exception, relatively little progress in the fields of hydraulics, hydrostatics, and hydrodynam-

ics until the Renaissance. The one exception was the work of Roman engineers in the development of one of the most advanced and complex systems for using water to be created. When this system was destroyed during the fall of the Roman Empire, the art and science of hydrology was essentially lost for more than a thousand years.

The Rise of Hydrostatics

With the revival of interest in scholarship during the Renaissance, scholars began to investigate the properties and behavior of liquids, especially water. In 1586, the Belgian-Dutch mathematician and physicist Simon Stevinus (also *Stevin*) published the first modern book on hydrostatics, *De Beghinselen des Waterwichts*. In this book, he announced his explanation of the *hydrostatic paradox*, the fact that the force exerted by a liquid on the base of its container depends only on its vertical height, not on its volume or the shape of the container.

The next great contributor to the study of hydrostatics was the French mathematician and physicist Blaise Pascal (1623–1662). Pascal found that the pressure exerted by a fluid in a closed vessel is transmitted undiminished throughout the fluid and in all directions. Pascal also studied the behavior of liquids confined to closed glass tubes (barometers) and showed that it was the force of air pressure acting on the liquids that maintained them at certain levels.

The Italian physicist and astronomer Galileo Galilei (1564–1642) also studied the subject of hydrostatics. In fact, his first published work in 1612 was an extension of Archimedes' studies on the properties of floating bodies. It was this work that first brought Galileo to the attention of the scientific community.

Quite remarkably, the work of Archimedes, Stevinus, and Pascal largely established the field of hydrostatics, even as it is studied and practiced today. As the *Dictionary of the History of Science* reports, "[l]ittle more has been added to the subject except mathematical refinement and the theory of the spontaneous rise of liquids in narrow tubes [capillarity]" (1981, 193).

The Rise of Hydrodynamics

Humans have been confronted with problems involving the flow of water across objects and over surfaces for thousands of years. The construction of an aqueduct, reservoir, or sailing ship required that builders know something about the manner and rate at which water would flow through an opening or across a surface. But there was virtually no effort to apply theory or mathematics to the solution of such problems until the seventeenth century. Prior to that time, whatever knowledge was available about hydrodynamics would come from trial-and-error experiences.

Partly because of his interest in the motion of the planets through space, Isaac Newton (1642–1727) was one of the first scientists to attempt a theoretical study of the way fluids travel. In his *Principia*, he reports on one discovery in this field—namely, that the minimum cross section of a stream issuing from the hole in a vessel occurs just outside the hole. Newton's work on hydrodynamics came to relatively little, however, as he made a number of incorrect assumptions about the way fluids flow.

Streamline

Credit for the first true analysis in hydrodynamics goes to Johann Bernoulli (1667–1748) and his son Daniel (1700–1782). The Bernoullis focused on the simplest possible form of flow: *streamline*, or *laminar*. Streamline flow occurs when every particle that makes up a fluid travels in a well-defined continuous path from one end of a container to the opposite end. In streamline flow, the velocity with which any one particle

passes a point in the container is exactly the same as the velocity of every other particle that passes the same point.

The mathematical analysis of streamline flow is a useful first step in unraveling the problems of hydrodynamics. However, this kind of flow is seldom found in the real world. Instead, the path followed by a liquid in traveling through a pipe, a canal, or along some other passage is likely to be erratic as bubbles of gas collect in the liquid, the liquid rubs against imperfections in the side of the passage, or some other event occurs to disrupt the smooth flow of the liquid.

The flow of water that does not occur in a streamline pattern is said to be *erratic*, or *turbulent*. The mathematical analysis of turbulent flow is much more complex than that of streamline flow.

A mathematical description of hydrodynamical phenomena can be derived from two simple concepts. The first concept is Newton's second law, namely that the force exerted by an object is proportional to the product of its mass and its acceleration, or F = ma. The second concept is now known as the *equation of continuity*. This equation is a mathematical expression of the fact that the mass of a fluid entering one end of a tube must be equal to the mass of the fluid leaving the other end of the tube. It can be written as:

$$\rho_1 v_1 A_1 = \rho_2 v_2 A_2$$

The two Bernoullis are often given credit for deriving the fundamental equations of hydrodynamics. The son published his book on the subject, *Hydrodynamica*, in 1738, and the father, his *Hydraulica*, some time later (although he predated it by ten years because of a disagreement about priority of discoveries reported in the two books).

In fact, credit for the most complete description of the flow of fluids should probably go to the Swiss mathematician and physicist Leonhard Euler (1707–1783). In 1775, he published a set of equations, based on but more carefully derived than those of the Bernoullis, that accurately portray the flow of water through a pipe. The equation of continuity was one element of the explanation provided by Euler.

At about the same time, one of the most surprising discoveries in the young field of hydrodynamics was announced by the French mathematician and physicist Jean Le Rond d'Alembert (1717–1783). D'Alembert discovered that an ellipsoidal-shaped object immersed in water exerted no resistance to the flow of water across its surface. The discovery was later termed the *d'Alembert paradox* because of its unexpected nature.

Viscous and Turbulent Flow

The next step in understanding the nature of fluid flow—that which occurs in viscous liquids—came in the early nineteenth century with the work of the French engineer C.L.M.H. Navier (1785–1836). Navier took into consideration the internal attractive forces among particles (the viscosity) of a liquid flowing through a pipe. The equations he derived more closely describe the motion of fluids in the real world than those proposed by the Bernoullis and Euler. Quite remarkably, Navier made some incorrect assumptions at an early point in his analysis of viscous flow, yet managed to come out with a correct answer nonetheless.

Navier's work was later extended by the Irish mathematician and physicist G.G. Stokes (1819–1903). In contrast to Navier's approach, Stokes's original assumptions about the flow of liquids were correct, although his results were essentially the same as those of Navier. The equations developed by these two men for the description of turbulent flow are now known as *Navier-Stokes equations*.

A final fundamental accomplishment in the analysis of fluid flow came with the work of the German physicist Hermann von Helmhotz (1821–1894). Helmholtz attacked the problem of the turbulent flow of liquids and developed a theory that could explain the flow of eddies, vortices, and other irregular patterns of movement. With the Helmholtz additions, hydrodynamical theory had reached a point that it could describe with at least some degree of accuracy the way water (and other fluids) flows through pipes, canals, and other tubes.

Research in hydrodynamics during the twentieth century has focused on the solution of more detailed problems. For example, one of the questions that puzzled researchers during the nineteenth century was the so-called *boundary problem*. The question was how, if at all, the thin layer of liquid coating the inside wall of a pipe, canal, or tube affected the flow of the major body of the fluid. Did that film flow along with the rest of the fluid, or did it stick to the inside of the pipe, as would a viscous liquid?

This problem was solved in the first two decades of the twentieth century by the German physicists Ludwig Prandtl (1875–1953) and Theodor von Kármán (1881–1963). Prandtl and von Kármán showed that the thin film acted as a viscous liquid and acted as a relatively permanent coating on the inside wall of a pipe. Thus, the equations of flow that had been derived for a nonviscous liquid in the previous century had to be revised.

Much of the work in hydrodynamics through the end of the nineteenth century was, indeed, focused on the flow of water itself. In the early twentieth century, however, the results obtained for the study of water were being applied to other types of fluids, such as the flow of oil through pipes and the movement of air over the ground, airplane wings, and other objects. The various applications of hydrodynamics to other fields are known as aerodynamics (for the flow of air and other gases), fluid dynamics (for water and other kinds of fluids), gas dynamics (the flow of non-air gases), and magnetohydrodynamics (the flow of ionized gases and plasmas).

The field of hydraulics has come full circle during the twentieth century. Once practiced as an art with virtually no scientific basis, hydraulics is now as firmly based in research and mathematics models as is hydrostatics or hydrodynamics. A search of current literature in the field demonstrates how today's hydraulic engineers and researchers in the field have adapted discoveries in hydrostatics and hydrodynamics for the solution of everyday problems involving the flow of water and other fluids. Titles of these reports—covering various fields—include "Role of Bed Discordance at Asymmetrical River Confluences," "Riprap Protection at Bridge Piers," "3D Numerical Simulation of Deposition Patterns Due to Sand Disposal in Flowing Water," "Risk of Sediment Erosion and Suspension in Turbulent Flows," and "Impact of Turbidity Currents on Reservoir Sedimentation." *See also* Bernoulli, Daniel

Further Reading

Brater, Ernest F., et al., eds. *Handbook of Hydraulics for the Solution of Hydraulic Engineering Problems*. New York: McGraw-Hill, 1996.

Bynum, W. F., E. J. Brown, and, Roy Porter, eds. *Dictionary of the History of Science*. Princeton, NJ: Princeton University Press, 1981.

Hwang, Ned H. C., and Robert J. Houghtalen. *Fundamentals of Hydraulic Engineering Systems*. Upper Saddle River, NJ: Prentice Hall, 1996.

Polevoy, Savely. *Water Science and Engineering*. London: Blackie Academic & Professional, 1996.

Spellman, Frank R., and Joanne Drinan. *Water Hydraulics*. Lancaster, PA: Technomic Publishing, 2001.

Hydroelectric Power

See Energy from Water

Hydrofoil

See Boats and Ships

Hydrogen Bonding

Hydrogen bonding is a force of attraction between two molecules caused by the electrostatic attraction between a hydrogen atom in one molecule and a strongly negative charge on another atom in the second molecule. Although hydrogen bonding may occur in many kinds of molecules, only its applications to water will be discussed in this entry.

Polarity of the Water Molecule

The water molecule consists of two hydrogen atoms bonded to a single oxygen atom. The nucleus of the oxygen atom exerts a strong attraction not only on its own electrons but also on those of the hydrogen atoms. By comparison, the attraction exerted by the nucleus of the hydrogen atoms is weak. As a result, the electrons in the water molecule tend to be associated more strongly with the oxygen atom than with the hydrogen atoms. The water molecule is, therefore, a polar molecule in which the oxygen atom tends to be more strongly negative and the hydrogen atoms, more positive.

A common way of representing this condition is shown below. In this diagram, the Greek letter delta (δ) represents a partial charge—a partial positive charge ($\delta+$) on the hydrogen atoms and a partial negative charge ($\delta-$) on the oxygen atom.

Hydrogen Bonding in Water

Hydrogen bonding occurs in water because of the attractive force between two molecules. A hydrogen atom carrying a partial positive charge in one molecule is attracted to an oxygen atom carrying a partial negative charge in a second molecule. As a consequence, the water molecules in a sample of water are not distributed randomly throughout the liquid, but take on a somewhat orderly arrangement. The diagram below shows how these molecules might be arranged because of the influence of hydrogen bonding.

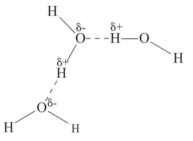

The effectiveness of hydrogen bonding in ordering water molecules is a function of temperature. As the temperature rises, water molecules move more rapidly, and hydrogen bonding is less successful in maintaining an ordered structure among molecules. As the temperature falls, water molecules move more slowly and an ordered structure can occur more readily.

When the temperature falls sufficiently, water freezes. In the solid state, water molecules move slowly enough so that hydrogen bonding can bring about the formation of crystals with definite geometric structures whose shape is determined by the forces between adjacent molecules.

Effects of Hydrogen Bonding

Hydrogen bonding is responsible for many of the unusual physical properties of water. For example, the process of boiling takes place when sufficient heat is added to a liquid. The heat is used to make the mole-

cules of the liquid move more rapidly. When those molecules are moving rapidly enough, they enter the vapor state: the liquid boils.

In the case of water, an additional amount of heat must be added to produce boiling. The heat is needed not only to make molecules move more rapidly, but also to overcome the force of attraction between molecules caused by hydrogen bonding.

Evidence of this fact can be found by comparing the boiling points of water and other liquids with comparable molecular masses. Other factors being equal, two substances with similar molecular masses should have similar boiling points. They should require similar amounts of heat to produce the molecular motion needed to convert a liquid to a vapor.

For example, water (H_2O) and methane (CH_4) have similar molecular masses—18 for water compared to 16 for methane. But the boiling point of water (100°C) is much greater than the boiling point of methane (−161°C). The difference in these two values reflects the extra energy needed to break the hydrogen bonds in water, which do not exist in methane.

Another physical property of water that depends on hydrogen bonding is surface tension. All liquids display some amount of surface tension—the tendency for the liquid to behave as if it were covered by a thin skin. In most liquids, the surface tension is quite small; in water, however, it is substantial. In fact, the strong surface tension displayed by water explains the ability of certain types of water animals to walk on the surface of water.

Water's surface tension is explained by hydrogen bonding between adjacent water molecules. The molecules tend to pull on each other, forming a skinlike effect that requires energy to break. Light objects, such as tiny animals and needles, can actually float on the water because they do not exert sufficient force to break the hydrogen bonding between molecules. *See also* Properties of Water; Surface Tension

Further Reading

Most introductory high school and college chemistry textbooks contain some discussion of hydrogen bonding.

Hydrological Sciences Branch (NASA)

The Hydrological Sciences Branch (HSB) is a subdivision of the National Aeronautics and Space Administration (NASA) devoted to the understanding, quantification, and analysis of different components of the hydrological cycle, with particular emphasis on the interaction between those processes and the atmosphere. In its attempt to better understand the hydrological cycle, HSB is developing, testing, and using remote-sensing methods for measuring soil moisture content, snow mass, precipitation, evapotranspiration, vegetation density, and other hydrological variables. At the same time, hydrological and atmospheric models are being developed and tested for making better use of data obtained from its studies and for producing better predictions about atmospheric behavior.

Special areas of interest at HSB include Hydrological Modeling; Remote Sensing of Snow; Microwave Measurement of Oceanic Precipitation; Land Surface Reflectance and Emission Modeling; Remote Measurements of Soil Moisture; and Global Hydrologic Processes and Climate.

Further Information

Hydrological Sciences Branch, Code 974
NASA Goddard Space Flight Center
Greenbelt, MD 20771
Telephone: (301) 614-5734
Fax: (301) 614-5808

URL: http://hsb.gsfc.nasa.gov
e-mail: Paul.R.Houser.1@gsfc.nasa.gov

Hydrologic Cycle

The hydrologic cycle consists of a series of processes by which water moves through various parts of the planet. It includes such mechanisms as precipitation, evaporation, sublimation, and condensation.

Earth's Water Resources

Earth's water resources are distributed among three primary zones: the lithosphere (crust, mantle, core, and other solid components), atmosphere (the gaseous envelope surrounding Earth), and hydrosphere (all liquid components of the planet). Water constantly moves from one zone to another in time periods that range from a few days to many years.

By far the largest water reservoir on the planet is the world's oceans, where about 97 percent of all water resources are found. The next largest water reservoir consists of ice caps and glaciers, which hold just over 2 percent of the planet's water. The third most important reservoir is groundwater, accounting for about 0.6 percent of all water resources.

Water in the atmosphere accounts for about 0.001 percent of all water resources. Finally, the water found in rivers, lakes, streams, ponds, wetlands, and other sources makes up the remaining water on the planet, a total of less than 0.01 percent of all water found on Earth.

Water is constantly being recycled through each of these reservoirs but at very different rates. On average, water in the atmosphere is replaced at the rate of about 9 to 12 days. River water, similarly, is recycled at nearly the same rate. By contrast, water in ice caps and glaciers is recycled about every 16,000 years, and water in the oceans, about every 3,100 years for surface water and every 37,000 years for deep water.

Elements of the Hydrologic Cycle

Some of the stages of the hydrologic cycle that are easiest to observe are those involving the hydrosphere and the atmosphere. Water is constantly evaporating from oceans, lakes, rivers, and other bodies of water. Some water also evaporates from moisture found in the soil. The water vapor thus formed rises into the air, where it remains in a gaseous state or condenses to form water droplets or very small ice crystals. These water droplets and ice crystals may remain suspended in air, invisible to the human eye, or may condense to form clouds.

When condensation occurs and clouds form, precipitation is likely to follow. Precipitation may occur as rain, snow, sleet, or hail. When this precipitation reaches the earth's surface, water has passed through a second stage of the hydrologic cycle, *deposition*.

Evaporation is not the only mechanism by which water passes into the atmosphere. For example, green plants give off water from their leaves by the process known as *transpiration*. And some water may pass directly from the frozen state (for example, ice in glaciers) into the atmosphere by means of *sublimation*.

Water that falls on the oceans, lakes, rivers, and other bodies of water in the form of precipitation is available immediately for evaporation. It may evaporate and pass through the water cycle very quickly (in the case of rivers) or more slowly (in the case of lakes and oceans).

Water that falls on the land during precipitation may follow one of two pathways. First, it may simply drain off into a lake, river, or other body of water. It then becomes part of the hydrospheric reservoir, available for evaporation and recycling. Second, it may soak into the ground and join the water reservoir contained in groundwater.

Groundwater consists of water that is contained in the open spaces between solid particles of rocks and minerals. Some supplies of groundwater are relatively shallow, while others may extend to a depth of more than 2,000 meters (about 3 mi). Water that makes up this reservoir is not static but moves slowly underground, usually toward a surface reservoir, such as a river or lake. Groundwater has a recycling time much longer than that of surface water, of the order of three hundred years (for shallow groundwater) to more than four thousand years (for deeper groundwater). *See also* Clouds; Evaporation; Evapotranspiration; Groundwater; Oceans and Seas; Precipitation; Rivers and Streams; Transpiration

Further Reading

Berner, Elizabeth Kay, and Robert A. Berner. *Global Environment: Water, Air, and Geochemical Cycles*. Upper Saddle River, NJ: Prentice-Hall, 1996.

Speidel, David H., Lon C. Ruedisili, and Allen F. Agnew. *Perspectives on Water: Uses and Abuses*. New York: Oxford University Press, 1988.

Hydrology

Hydrology is the science that deals with the physical and chemical properties of water; its distribution and circulation in Earth's atmosphere, lithosphere, and hydrosphere; and the interaction of water with other parts of the biotic and abiotic environment.

History

Water plays such an essential and pervasive role in human society that primitive forms of hydrological studies must have been conducted just after the dawn of civilization. For example, archaeologists have discovered the existence of excavated tunnels—known as *qanats* or *kanats*—in the Middle East, dating to the seventh century B.C. These tunnels were apparently dug into the side of hills to tap sources of underground water. The most ambitious of these structures were sunk to a depth of about 150 meters (400 ft) and extended a distance of more than 30 kilometers (20 mi).

There are also a number of biblical references that describe efforts by humans to locate water and extract it for their use. In Genesis 26:17–19, for example, we are told:

And Isaac digged again the wells of water which they had digged in the days of Abraham his father; for the Philistines had stopped them after the death of Abraham; and he called their names after the names by which his father had called them. And Isaac's servants digged in the valley, and found there a well of springing water.

The key issue in any understanding of hydrology is the hydrological cycle, the sequence of reactions by which water passes through the atmosphere, lithosphere, and hydrosphere by means of evaporation, transpiration, and precipitation. Why is it, early humans asked, that the oceans did not continue to rise when they were constantly being fed by river water? And where did water come from to fill the rivers? Was there enough rain and snow to account for all existing river water, or were there other sources, such as underground caverns, from which rivers were fed?

During ancient times, there was neither the theoretical understanding nor the technology that would allow humans to answer such questions. Speculations about the hydrological cycle were common, however. Records indicate that the Chinese may have understood the fundamental elements of this process as early as the fifth century B.C. At about the same time, the ancient Greeks were hypothesizing the relationship among rivers, lakes, oceans, precipitation, and groundwater. Anaxagoras of Clazomenae (500–428 B.C.) suggested, for example, that "Rivers depend for their existence on the rains and on water within the earth, as the earth is hollow and has water in its cavities."

Explanations for the movement of water throughout the earth advanced little over the next two thousand years. Leonardo da Vinci (1452–1519), so advanced in his thinking in other areas, still believed that water cycled through the earth through underground channels, moving from the oceans to eventually become the source of rivers and streams.

The first person to explain correctly the role of rivers and streams in the hydrological cycle was French scholar Bernard Palissy (1510–1589). In his book, *Discours admirables*, Palissy wrote that rainwaters "falling on these mountains through the ground and cracks, always descend and do not stop until they find some region blocked by stones or rock very close set and condensed. And they rest on such a bottom and having found some channel or other opening, they flow out as fountains or brooks or rivers according to the size of the opening and receptacles" (Fetter 2001).

The Birth of Modern Hydrology

Palissy's misfortune was that he wrote in French rather than the Latin preferred by most scholars of the day. As a result, his book was largely ignored for nearly two hundred years. Then, in the mid-seventeenth century, a turning point in the history of hydrology was reached when two French scientists applied experimental methods to resolve the question of water flow in rivers.

The two researchers were Pierre Perrault (1608–1680) and Edme Mariotte (1620–1684). In 1674, Perrault measured the amount of precipitation falling on the watershed (the source) of the Seine River and the amount of water flowing from the river into the ocean. He found that the amount of precipitation was six times as great as the amount of water flowing from the river into the ocean. It was clear, he said, that rainfall alone was more than sufficient to account for all the water found in the river. This conclusion ran contrary to the traditional belief that some source other than precipitation, such as underground reservoirs, must be needed to supply the water in rivers and streams.

Mariotte later repeated and confirmed Perrault's experiments. He also discovered that some precipitation soaked into the ground and did not flow directly into rivers and streams. This water penetrated into the soil, as Palissy had said, until it reached an impermeable layer. Natural and artificial openings in such layers, Perrault and Mariotte said, explained the existence of streams, artesian wells, and artificial wells.

The final step in explaining the hydrological cycle was accomplished by the famous English astronomer Edmond Halley (1656–1742). Halley devised a method for calculating the rate at which water evaporates on a warm summer day. He concluded that evaporation by sunlight alone was sufficient to transfer water on Earth's surface to the atmosphere, from where it would then fall back to Earth in the form of precipitation at a rate that would supply all of the rivers on the planet. The cycle—from clouds to rivers to oceans to clouds—had been completed.

Scope of the Field

Hydrology is a very large field that consists of a number of subdisciplines. For example, *hydrogeology* involves the study of groundwater; *glaciology*, the study of glaciers; *hydrometeorology*, the study of water in the atmosphere; *geochemistry*, the study of groundwater quality; and *dendrohydrology*, the study of ancient water patterns based on the study of tree rings.

Some of the roles played by hydrologists include the following:

- Locating water sources for cities and farms

- Developing methods for controlling river flooding and soil erosion
- Analyzing water quality from a variety of sources
- Recommending methods for treating sources of water pollution
- Conducting research on the properties of water and water reservoirs
- Studying the role of water in ancient civilizations
- Designing, testing, and using mathematical and physical models for various types of water supplies and uses
- Constructing and using topographic maps of water sources and supplies
- Serving in governmental and nongovernmental agencies responsible for the development and implementation of water-use policies

See also Hydraulics, Hydrostatics, and Hydrodynamics

Further Reading

Fetter, C. W., Jr. "Historical Knowledge of Ground Water," http://www.appliedhydrogeology.com/history.htm.

Gupta, Ram S. *Hydrology and Hydraulic Systems*. 2d ed. Prospect Heights, IL: Waveland Press, 2001.

Landa, Edward R., and Simon Ince. *The History of Hydrology*. Washington, D.C.: American Geophysical Union, 1987.

Maidment, David R., ed. *Handbook of Hydrology*. New York: McGraw-Hill, 1993.

Further Information

American Institute of Hydrology
2499 Rice Street, Suite 135
St. Paul, MN 55113-3724
Telephone: (651) 484-8169
Fax: (651) 484-8357
URL: http://www.aihydro.org
e-mail: aihydro@aol.com

Hydrology and Water Resources Programme

The Hydrology and Water Resources Programme (HWRP) is an activity of the World Meteorological Organization. The activities of HWRP fall into four major categories:

- measurement of data about hydrological phenomena from networks of meteorological and hydrological observation stations;
- collection, processing, storing, retrieval, and publication of these hydrological data, including data on the quantity and quality of surface water and groundwater;
- provision of such data and related information to consumers involved in the planning and operation of water resource facilities;
- installation and operation of hydrological forecasting systems.

HWRP has five major components: Basic Systems in Hydrology; Forecasting and Applications in Hydrology; Sustainable Development of Water Resources; Capacity Building in Hydrology and Water Resources; and Water-Related Issues. The Hydrological Operational Multipurpose System (HOMS) is an example of the kind of work being done in these programs. The purpose of HOMS is to provide a mechanism by which the technology used in operational hydrology can be transferred among users. Information about this technology is collected by HWRP from users who have already tried out and validated certain equipment and procedures. The information is then made generally available to other water users with similar interests and needs.

Another example of the work being done by HWRP is the Hydrological Information Referral Service, INFOHYDRO. This service collects information from and provides information to both governmental and nongovernmental national and international agencies involved with hydrological issues. The information includes data banks on principle international river and lake basins of the world, networks of hydrological observing stations, and national and international hydrological data banks.

Further Information

Hydrology and Water Resources Programme
World Meteorological Organization
P.O. Box 2300
1211 Geneva 2
Switzerland
URL: http://www.wmo.ch/web/homs/hwrphome.
    html
e-mail: dhwr@gateway.wmo.ch

Hydrophone

A hydrophone is a kind of underwater microphone. It is designed to pick up sounds and other types of vibrations such as those created by marine animals, water movement, and shifting of the ocean floor. Most hydrophones consist of a piezoelectric element—a device that converts mechanical motion into electric current. In its simplest form, a hydrophone can operate as either a transmitter or a receiver. The addition of a preamplifier to the device converts it to an exclusive receiver, the function for which it is most commonly used.

Hydrophones are used for a variety of purposes. Some marine aquaria, for example, have forms that are designed specifically to receive and transmit the sounds produced by whales, dolphins, and other sea animals. Such devices are also available for sale to the general public. Researchers use hydrophones to detect seismic signals that may indicate a forthcoming earthquake or volcanic event. Hydrophones have also been used to study the flow of water and hazardous wastes underground, and to see how rock structures affect such flow.

Hydrophones are designed for each of these specific functions by building them to detect the specific wavelengths (or frequencies) expected in the given application.

Hydroponics

Hydroponics is the technique of growing plants in a nutrient solution, with or without the use of a supporting medium. Hydroponics has been practiced as a method for growing food and decorative plants for thousands of years. Interest in the use of hydroponics for feeding large numbers of people grew in the mid–twentieth century, although efforts to achieve this potential often met with discouraging results.

History

Since the dawn of human civilization, some cultures have been confronted with the challenge of growing crops in areas not well suited for the survival of plants, such as deserts and small islands. It is not surprising, therefore, to find references to the use of hydroponic methods in very early writings.

For example, some historians now believe that some, if not all, of the famous Hanging Gardens of Babylon were sustained by hydroponics. These authorities cite the writings of early historians, such as Strabo and Philo of Byzantium who reported that

"The Hanging Garden has plants cultivated above ground level, and the roots of the trees are embedded in an upper terrace rather than in the earth. . . . Streams of water emerging from elevated sources flow down sloping channels. These waters irrigate the whole garden saturating the roots of plants and keeping the whole area moist." (quoted in "The Seven Wonders," http://ce.eng.usf.edu/pharos/wonders/gardens.html)

Among the best known examples of hydroponic gardening in the ancient world were the floating gardens developed by the Aztecs, who dominated Mexico until the early 1500s. Apparently, the Aztecs first developed this form of gardening when they were driven by their enemies out of areas in which traditional soil-based agriculture could be used. They were forced to develop methods of growing plants on reed rafts in the middle of Lake Tenochtitlán, in the region where Mexico City is now located.

Early uses of hydroponic gardening such as these were largely developed by trial-and-error methods. Humans knew little or nothing about the scientific basis for plant growth. That situation began to change in the seventeenth century when the first experiments on plant physiology were conducted. Among the most important of these for the development of hydroponics was the work of the English scientist John Woodward. In 1699, Woodward grew plants in water to which small amounts of soil had been added. He found that plants grew better when larger amounts of soil were used. Some authorities have called Woodward's solutions "the first man-made hydroponic nutrient solution[s]" (History of Hydroponics, 2).

By the 1860s, sufficient progress in the study of plant physiology had been made to allow scientists to prepare aqueous solutions that contained all of the necessary nutrients needed by growing plants and, hence, the basic element needed in the development of hydroponics. The German botanist Julius von Sachs (1832–1897) developed the first standard formula for a nutrient solution in 1860 and thereby created the field of study now known as *nutriculture*.

Until about 1930, most of the research done on hydroponics was designed for research purposes, to learn more about the nature of plant growth. By the 1920s, however, a few researchers had begun to consider the possibility of using hydroponics as a method for growing crops on a large-scale basis. Some of the fundamental work in this area was carried out by William F. Gericke of the University of California. Gericke eventually achieved quite remarkable success and was featured in news stories of the time, often accompanied by photographs of his 25-foot-tall tomato plants grown by hydroponic techniques. Gericke also invented the name for this process, *hydroponics*, based on two Greek words meaning "water-working."

Commercial Applications

One of the earliest commercial successes of Gericke's research was a hydroponic farm built by Pan American Airways in the 1930s on Wake Island. Pan Am used the farm to grow fresh vegetables for passengers on its transpacific flights.

A vital factor in the growing interest in commercial applications of hydroponics was the beginning of World War II. Providing fresh vegetables to fighting units in the Pacific, the Near East, and other remote and relatively infertile areas posed a serious logistical problem for the U.S. and British militaries. One solution for this problem was the creation of hydroponic farms on Pacific islands, such as Wake, Okinawa, and Iwo Jima; on other island bases, such as the U.S. Air Force base on Ascension Island in the South Atlantic and the British air base on Bahrein in the Persian Gulf; and on desert bases, such as the British air base at Habbaniya in Iraq.

The success of hydroponic farming during World War II turned out to be a short-lived triumph. Modest efforts were made after the war to encourage the development of large-scale commercial hydroponic farms, expecially by scientists at Purdue University. But these efforts were foiled by the relatively large cost of constructing the equipment needed for such facilities, espe-cially in view of the huge success of most traditional agricultural operations at the time. Only in areas with specialized agricultural problems, such as the desert areas of California and Arizona, or in desert nations, such as Saudi Arabia, Abu Dhabi, and Iran, did hydroponic farming experience any substantial success.

Today, hydroponic farming, except in specialized conditions such as those described above, tends to be largely a hobby or small-scale approach to commerical plant growing. The science has many enthusiastic adherents, however, and research efforts are still being conducted to improve the efficiency and reduce the cost of hydroponic farming as a large-scale commercial method for food production.

Methods

There are primarily two general approaches to commercial hydroponic gardening. The first method is one that goes back to the earliest days of hydroponics. It makes use of a rectangular, watertight container filled with a nutrient-rich aqueous solution. In early systems, the container was made of concrete or some other ceramic material. As synthetic polymers became available during the mid–twentieth century, however, plastic-lined boxes became more popular for such systems, and continue to be so today.

Plants are grown in a thin layer of some medium—such as rock wool, peat, rice hulls, or wood fiber—embedded in a wire framework suspended a few centimeters above the aqueous solution. Plant roots extend below the medium into the solution, from which they obtain the nutrients they need for growth.

In the second system, plants are anchored in an inert medium, such as sand, gravel, perlite, or vermiculite. The nutrient solution is then poured into the medium and allowed to drain out the bottom. In most systems, the addition of nutrient solution is a cycling process. Solution drained from the bottom of the container is fed once more through pipes back into the container. Recycling can continue as long as the solution contains nutrients in high enough concentration to permit normal plant growth.

One challenge with commercial hydroponic gardening is obtaining exactly the right conditions to obtain maximum growth. Important factors include proper temperature, amount of sunlight, type and concentration of nutrients, and protection from pests. In many locations, constructing and maintaining a facility that meets all of the necessary growing conditions may be too costly to compete with traditional agricultural crops. In locations with specialized conditions, such as an abundance of sunlight, commercial hydroponic operations can compete with traditional agricultural systems economically for at least some crops. With additional research, the scope of potential hydroponic applications may well increase substantially.

Further Reading

History of Hydroponics, http://archimedes.galilei.com/raiar/histhydr.html.

Mason, John. *Commercial Hydroponics*. New York: Simon & Schuster, 2000.

Resh, Howard M. *Hydroponic Food Production: A Definitive Guidebook for Soilless Food-Growing Methods*. 5th ed. Santa Barbara, CA: Woodbridge Press Publishing, 1995.

Roberto, Keith F. *How-to Hydroponics*. Farmingdale, NY: FutureGarden, 2000.

"The Seven Wonders," http://ce.eng.usf.edu/pharos/wonders/gardens.html.

A good general source of information is the magazine *Practical Hydroponics and Greenhouses* (Web site: http://www.hydroponics.com.au).

Hydrothermal Vents

Hydrothermal vents are openings in the ocean floor that usually occur along the Mid-Ocean Ridge, a system of underwater mountains that circles the earth's surface. They occur in places where seawater is able to seep downward into the ocean floor and come into contact with hot magma. These locations are usually found in zones a few kilometers long and a few hundred meters wide. Water temperatures around such vents may be as high as 350°C (660°F). In comparison, adjacent seawater is close to freezing, with a temperature of about 1–2°C (about 35°F).

Hydrothermal vents were first discovered in 1977 in 2.5-kilometer (1.5 mile) deep waters off the western coast of South America. They have since been thoroughly studied by using sophisticated research submersible devices, especially the underwater ship known as *Alvin*.

Characteristics

When seawater comes into contact with magma, both water and liquid rock undergo physical and chemical changes. The water often begins to boil quite suddenly, with the subsequent release of huge clouds of steam that rise upward through the ocean with velocities of up to five meters per second (11 miles per hour). These *hydrothermal plumes* continue to rise until they have cooled to a temperature equal to that of the surrounding water. They then spread out horizontally under the influence of ocean currents, producing the characteristic appearance from which they get their name.

Hydrothermal plumes are generally white or black in appearance, depending on their temperature and the type of solid materials they carry. The former are often called white-smokers, while the latter are known as black-smokers.

Hydrothermal vents come in a great variety of sizes, the smallest being no more than a few centimeters in diameter and the largest (the Transatlantic Geotraverse) covering an area of nearly 500 meters.

Formation

Hydrothermal vents form along the ocean floor in regions where two crustal plates are pulling away from each other. As a result of this action, magma from Earth's interior is pushed upward, where it can come into contact with seawater that is seeping downward into the crust. The interaction of magma and seawater can occur at almost any depth up to a distance of about 5 kilometers (about 3 mi).

Hydrothermal vent activity is also related to volcanic action on the ocean floor. This relationship was first discovered in 1986 during the eruption of a "megaplume" along the Strait of Juan de Fuca off the coast of the northwestern United States. The megaplume covered about a million cubic meters, the largest hydrothermal plume ever observed. Shortly after the plume had been detected, fresh lava flows on the ocean floor near the vent were discovered. Scientists now believe that hydrothermal vent activity may be an indication of possible future volcanic activity in an area.

Marine Life

The regions surrounding hydrothermal vents are home to a unique ecosystem not found elsewhere on Earth. Organisms found in this ecosystem include giant tube worms, giant clams, bristle worms, yellow mussels, pink sea urchins, tripod (three-legged) fish, and limpets. The existence of any life within such regions was a great surprise to scientists. Both temperatures and pressures are far greater than anything on Earth's surface and far greater than almost any known organism can survive.

An Atlantic Ocean black smoker at a mid-ocean hydrothermal vent. *P. Rona, OAR/National Undersea Research Program (NURP), NOAA.*

In addition, sunlight does not penetrate to the depths at which vents are found, so any organism that depends on the process of photosynthesis could not survive here. Researchers have found, however, that certain types of chemosynthetic bacteria thrive in the region around vents. Chemosynthetic bacteria are organisms that obtain the energy they need from chemical reactions—those involving hydrogen sulfide in the case of bacteria living near vents. These bacteria provide food for higher forms of organisms that also live near vents. *See also* Oceanography; Submarines and Submersibles

Further Reading

Cone, Joseph. *Fire under the Sea: The Discovery of the Most Extraordinary Environment on Earth.* New York: Morrow, 1991.

Parson, L. M., C. L. Walker, and D. R. Dixon, eds. *Hydrothermal Vents and Processes*. London: Geological Society, 1995.

Hygrometer

A hygrometer is an instrument used to measure the relative humidity of air. The first hygrometer was probably invented by the Italian artist and scientist Leonardo da Vinci (1452–1519). Da Vinci's hygrometer was based on the fact that many materials change in one or more physical properties—such as their mass, length, or color—when they absorb moisture. An early form of the hygrometer, for example, contained a strip of paper weighted in the middle with a pointer attached to one side of the paper. A scale showing the relative humidity was attached next to the strip of paper. An increase in the amount of moisture in the air caused the paper to become longer, causing the pointer to move downward against the scale. The more moist the air, the farther the pointer moved on the scale.

One of the most familiar simple hygrometers—the hair hygrometer—was invented in 1687 by the French physicist Guillaume Amontons (1663–1705). The hair hygrometer works on the same principle as the paper-strip hygrometer described above, except that it uses a single human hair, or a bundle of hairs. As the air becomes more moist, human hair lengthens. The amount by which a single hair or bundle of hairs stretches, then, can be used to determine the change in humidity of air. Hair hygrometers are simple enough to make that schoolchildren can construct such devices and obtain approximate measurements of the relative humidity of air. They are not accurate enough, however, for precise scientific measurements, such as those needed in meteorology.

The most common type of hygrometer in use today is the psychrometer ("cold measurer"), invented in the late 1700s by the Scottish geologist James Hutton (1726–1797). The form of the instrument most widely used today was developed by John Leslie (1766–1832) in 1800. This form of the psychrometer consists of two thermometers mounted next to each other on a board. A thin piece of absorbent cloth is wrapped around the bulb of one thermometer, known as the *wet-bulb thermometer*. The second thermometer has nothing wrapped around its bulb and is known as the *dry-bulb thermometer*.

To use the psychrometer, the cloth on the wet-bulb thermometer is first dampened with water. The air around the cloth is then set in motion, either by passing air over the bulb or by swinging the combination of two thermometers in the air. Because of this method for using the instrument, it is sometimes called a *sling psychrometer*.

As air passes over the dampened cloth, water evaporates from it. Evaporation removes heat from the thermometer bulb and causes the temperature in the web-bulb thermometer to fall. The less moisture in the air, the more water will evaporate from the damp cloth and the more the temperature will fall. The more moisture in the air, the less water will evaporate and the less the temperature will fall. The action of passing air over the two thermometers or swinging them has no effect, however, on the temperature recorded by the dry-bulb thermometer. The difference in the temperatures shown on the two thermometers, then, is the measure of the relative humidity of the air.

Tables now exist that allow one to determine the relative humidity based on the readings of the dry-bulb thermometer and the difference between the wet- and dry-bulb readings. A simplifed form of such a table is shown below. The table shows the percent of humidity for the conditions given.

Using this table, one can find the relative humidity for when the wet-bulb ther-

Relative Humidity

Dry-Bulb Reading (°C)	Decrease in Wet-Bulb Reading over Dry-Bulb Reading						
	1	*2*	*3*	*4*	*5*	*6*	*7*
0	81	63	45	28	11		
4	85	70	56	42	27	14	
8	87	74	62	51	39	28	17
12	88	78	67	57	48	38	28
16	90	80	71	62	54	45	37
20	91	82	74	66	58	51	44
24	92	84	76	69	62	55	49

mometer reads 14°C and the dry-bulb thermometer reads 20°C. The difference between the two readings—20°C − 14°C—is 6°C; therefore using, the table, the relative humidity would be 51 percent.

Other types of hygrometers have also been developed for specialized purposes. For example, an electric hygrometer operates on the fact that the electrical conductivity of a moist material varies depending on the amount of moisture present in the material. For example, a perfectly dry piece of paper will not conduct any electric current. If the paper is moistened, however, a weak current may flow through the paper. The greater the amount of moisture in the paper, the greater the flow of electric current.

An electric hygrometer contains a plate coated with carbon. The electric conductivity of the carbon changes as it absorbs more or less moisture from the air. These changes in electrical conductivity can easily be read and recorded as changes in current flow.

Electric hygrometers have two important advantages over traditional hygrometers. First, they measure the relative humidity of air more precisely than do hair hygrometers or traditional sling psychrometers. Second, they can be used more easily in determining the humidity in remote locations. Weather balloons that carry instruments high into the atmosphere, for example, are now equipped with electric hygrometers because the readings from such instruments can be relayed as an electric signal to observers on Earth.

Yet another form of hygrometer is the infrared hygrometer. In an infrared hygrometer, two beams of light are passed through a sample of air. One beam travels with a wavelength that is absorbed by water molecules, while the second beam has a wavelength that is not absorbed. In one commercial infrared hygrometer, the two beams have wavelengths of 2.6μ and 2.5μ, the first of which is absorbed and the second, not absorbed. The differences in the amount of absorption between the two beams provides a precise measure of the relative humidity of the air being sampled. Infrared hygrometers provide very accurate measures of relative humidity for samples ranging in dimensions from a few centimeters to a few kilometers. *See also* Humidity

Further Reading

Wiederhold, Pieter R. *Water Vapor Measurement*. New York: Marcel Dekker, 1997.

I

Ice

Ice is the name given to water in its solid state. The term is also used to describe other gases and liquids that may occur in the solid state, such as "methane ice"—methane gas in its solid state. This entry refers to ice only as frozen water.

Physical Properties

Ice can occur in many crystalline forms—more forms than any other known material. The most common form of ice, found in everyday life, is known as *ice I*. Ice I, in turn, occurs in two forms, known as ice Ih (*hexagonal ice I*) and ice Ic (*cubic ice I*). Normal ice is ice Ih, while ice Ic can be produced by evaporating water on a very cold surface (less than 140 K). Other forms of ice, up to ice XIV, can be produced by varying the pressure and other conditions under which it forms.

Some physical properties of ice are similar to those of water, while others are quite different, as shown in the list below.

Ice Masses in Nature

Large masses of ice exists in nature in many forms, including ice sheets, ice shelves, valley glaciers, icebergs, and sea ice. The term *glacier* is used as a comprehensive term to describe any large mass of ice that usually includes air, water, and rocky material—at least part of which is located on land. Glaciers may exist in the form of ice caps, ice sheets, ice streams, ice shelves, ice tongues, valley glaciers, and continental glaciers.

About 10 percent of Earth's land area is covered with one form or another of glacier. These glaciers hold about three-quarters of the world's freshwater. The largest glaciers in the United States are found in Alaska, where about 75,000 square kilometers (about 30,000 sq mi) of land is covered by glaciers. Glacier sources in Washington state provide about 470 billion gallons of water each year to local residents.

Physical Properties of Ice

Property (at 0°C)	Ice	Water
density	0.917 kg/m3	0.999 842 kg/m3
specific heat	2.01 kJ/kg · K	4.180 kJ/kg · K
thermal conductivity	2.2 W/m · K	0.60 W/m · K
vapor pressure	0.61173 kPa	0.6107 kPa
dielectric constant	96.5	87.7
refractive index	1.31	1.333
acoustic velocity	1 951 m/s	1 49 m/s

Formation of Glaciers

Glaciers form primarily in one of two regions: near Earth's poles or at high elevations. In such places, any snow that falls never has a chance to melt completely. Whatever melting may occur is followed by freezing, during which process air is forced out of the snow. The longer snow resides in the glacier, the less air it contains and the more compact it becomes. Each successive snowfall covers layers of snow and ice left by earlier storms. Every addition of snow increases the pressure on the glacier's lowest layers. Eventually, this pressure becomes great enough to convert any snow in the bottom layer to ice.

Layers of snow and ice continue to build up year after year until the mass of the accumulated mixture is large enough to start squeezing ice from the bottom of the glacier. The ice moves outward at very slow rates, ranging from a few tens of feet to a few thousands of feet per year. The fastest moving glacier ever measured was the Kutiah Glacier in Pakistan. In 1953, its outward flow was measured to be about 112 meters (367 feet) per day. Glaciers that move this fast are extremely rare, however.

Ice Caps and Ice Sheets

An ice cap is a large mass of ice, usually in the form of a dome, that covers an area of more than 50,000 square kilometers (about 20,000 sq mi). The dome is large enough for deformation and flow of the ice to occur. In most cases, the ice flows outward from the center of the ice cap at nearly equal rates in all directions.

Ice sheets are similar to but much larger than ice caps. They are the largest concentrations of ice on the planet today. The Antarctic Ice Sheet consists of two parts, the East Antarctic Ice Sheet and the West Antarctic Ice Sheet. The former covers an area of about 10 million square kilometers (about 3.8 million sq mi), while the latter covers an area of about 2 million square kilometers (about 770,000 sq mi). The East Antarctic Ice Sheet is the thickest of existing ice sheets, with a depth of more than 4 kilometers (about 2.5 mi) in some areas.

Two types of ice sheets occur—those with bottoms primarily above sea level, and those with bottoms primarily below sea level. The East Antarctic Ice Sheet is an example of the former, with well over 90 percent of its body above sea level. Most of the West Antarctic Ice Sheet, by comparison, lies below sea level, at a depth of more than 2 kilometers (about 1.25 mi) in some places. Scientists believe that the weight of the glacial ice in the West Antarctic Ice Sheet exerts enough force to depress the land beneath it to a depth of about 2.5 kilometers (1 mi) below sea level.

Outlet Glaciers and Ice Streams

Glaciers flow, much as do rivers and streams. The major difference is the rate at which they move, a rate measured in meters or kilometers per year (for glaciers) rather than meters per second (for rivers and streams). The mass of moving ice that leaves a glacier is known as either an *outlet glacier* or an *ice stream*. The two formations are much the same except that the ice flows that make up an outlet glacier are separated from each other by mountain ranges, hills, or other topographic features, while the sides of ice streams are in contact with each other. The ice in an outlet glacier or ice stream moves considerably more rapidly than the ice within a glacier itself, often moving at the rate of a kilometer or so a year.

Ice Shelves and Icebergs

Rivers of ice that flow out of a glacier often flow into the ocean. When they remain frozen, they may develop into large masses of ice known as *ice shelves*. An ice shelf is attached to a land-based glacier but floats on the water. The largest ice shelf in the world

An April 2002 photo of the collapsed Larsen ice shelf in Antarctica. *NASA/GSFC/ LaRC/JPL, MISR Team.*

is the Ross Ice Shelf in the Antarctic. It is the size of the state of Texas.

Ice shelves are dynamic structures, constantly changing because of the addition of new snow and ice from an ice sheet or ice dome and the loss of ice to the ocean. This loss occurs when large chunks of ice break off and fall into the water, a process known as *calving*. The chunks of ice formed during calving eventually float away and are then known as *icebergs*.

The largest iceberg ever seen was probably one that broke off the Ross Ice Shelf in March 2000. The iceberg, named B15 by scientists, was about 300 kilometers (180 mi) long and 37 kilometers (23 mi) wide. Its total surface area was about equal to that of the state of Connecticut. The tallest iceberg ever seen calved from an ice sheet in Melville Bay, Greenland, in March 1957. It was about 180 meters (550 ft) high, about the height of the Washington Monument.

Icebergs are a serious risk for marine travel. Only about 10 percent of an iceberg is above water, and this part of the structure tells an observer little or nothing about the shape of the portion underwater. Collisions with icebergs have been the cause of many shipwrecks—most prominently that of the *Titanic*, in 1912.

Valley Glaciers and Ice Tongues

The ice from glaciers that form in mountainous areas usually flows outward and down the mountain as *valley glaciers*. Valley glaciers behave in a manner somewhat similar to that of rivers and streams, except that they flow much more slowly. Like rivers and streams, valley glaciers may flow into each other, forming complex tributary systems with dendritic patterns. They also scour the land over which they travel, eroding both bottom and sides of their beds. Because they travel at slower speeds, how-

An Antarctic iceberg. *Paul A. Souders/Corbis.*

ever, their pattern of erosion is different from that of rivers and streams. Erosion caused by rivers and streams produces sharply angled, V-shaped valleys in the mountains, while erosion caused by ice produces more rounded, U-shaped valleys.

Valley glaciers sometimes terminate in a mountain lake, where ice flows outward into the lake in a long narrow structure known as an *ice tongue*. Ice tongues are somewhat similar to ice shelves, except that they are much smaller and narrower in shape. The outflow from polar glaciers may also form ice tongues in the ocean. The largest known ice tongue is the Drygalski Ice Tongue, in the Ross Sea off Antarctica. Drygalski is 70 kilometers (43 mi) long and 20 kilometers (12 mi) wide. It has been increasing in length by as much as a kilometer a year over the past few decades.

Sea Ice

Sea ice is ice that forms over relatively large parts of the ocean in the winter, but then melts in the summer. It may melt completely or almost completely, as it does in the Antarctic, or it may only melt partially, as it does in the Arctic. Sea ice covers an area on Earth's surface of about 25 million square kilometers (about 8 million sq mi) in the winter and about 3 million square kilometers (about 1 million sq mi) in the summer. Sea ice is typically about a meter in thickness, although it may grow to a thickness of up to 10 meters.

Sea ice is also known as *pack ice* and *ice pack*, and may exist in a variety of forms. During an early stage of sea-ice formation, ice crystals develop randomly at temperatures of about $-1.8°C$ (28.8°F), forming *frazil ice*. Those ice crystals may then begin to organize into thin plates known as *grease ice*. As grease ice thickens, it forms larger plates that look like pancakes, giving it its name of *pancake ice*. Pancake ice often has raised edges where adjacent plates have collided with each other. As pancake ice continues to grow and thicken, it develops into its mature form. In areas where sea ice remains permanently or semipermanently attached to the land, it is also known as *fast ice*.

Sea ice is an important part of Earth's climate budget. It is an excellent insulator, preventing ocean waters beneath it from being warmed by solar energy. It is also highly reflective, returning nearly all of the sunlight that strikes it to the atmosphere. Scientists have estimated that ocean water now covered by sea ice would raise the temperature of air above it by as much as 40°C (100°F) if that ice were removed. Without ice, water would absorb solar energy and itself become warmer.

Evidence has been accumulating for many years that significant changes in sea ice may be taking place. Studies have shown a decrease of about 3 percent per decade in the total volume of sea ice measured. In one area, the thickness of an ice sheet decreased by 40 percent over normal in a single year (1998). Because of its climatic importance, sea ice has become the subject of some important research programs in the last two decades.

Ice Volcanoes

Ice volcanoes are somewhat unusual phenomena that occur during the winter when strong onshore waves batter the front edge of an ice shelf on a lake. A variety of conditions are necessary for the formation of an ice volcano. These conditions occur at temperatures well below freezing when waves crash into the ice shelf. Wave action may cause the formation of conelike structures within the ice sheet. These ice cones are generally open at first but may then close over.

As waves continue to wash over and under the ice shelf, they may force their way upward through cracks and openings within the cones. Wave action may then force water upward through the cone, much as magma is released from volcanoes on land. The water may spout upward in the form of very cold water, slush, or ice particles.

Ice Ages

The ice ages were periods in Earth's history during which significant portions of the planet's surface were covered by extensive glaciers. Scientists sometimes use the term *ice era* to describe seven major periods of severe cooling that have occurred over the past 4.6 billion years. Little is known about any of these ice eras except the last, which began about 65 million years ago. During this period, the planet experienced alternate periods of warming and cooling, known as *ice epochs*. During the 2.4 million years of the latest ice epoch, there have been about two dozen *ice ages*.

The last major ice age reached its peak about 18,000 years ago, at which point it covered all of Canada and the northern United States; all of Scandinavia and most of Great Britain; mountainous areas, such as the Alps and Andes; and both the Arctic and Antarctic. Scientists estimate that about 50 million cubic kilometers (17 million cu mi) of ice were trapped in these glaciers, causing a drop in sea level of about 150 meters (400 ft).

Controversy about the precise cause of an ice age continues, although three major factors seem almost certainly to be involved. One is the planet's angular tilt—the angle at which its axis is oriented to the plane of its orbit around the Sun. This tilt slowly changes over time, causing Earth to receive more radiation at some times and less at others. Ice ages may be related to periods during which the tilt results in a decrease in the amount of solar radiation reaching the planet's surface.

A second factor is the change that takes place in the shape of Earth's orbit around the Sun over time. At times, the orbit is nearly circular; at other times, it is more elliptic (flattened). During the latter periods, Earth spends more time at greater distances from the Sun, resulting in a decrease in the average annual temperature on Earth's surface.

A third factor is axial precession, the "wobble" in the orientation of the Earth's axis to its orbit around the Sun. This wobble is responsible for changes in the amount of sunlight striking Earth's surface and, therefore, changes in its annual average temperature over time.

Finally, certain terrestrial factors may be involved in the development of an ice age. Many scientists believe that the planet's annual average temperature is determined to at least some extent by the amount of carbon dioxide and other "heat-capturing" gases in Earth's atmosphere. Decreases in the concentration of such gases may reduce the planet's annual average temperature, resulting in an ice age, while increases may raise temperatures, producing an overall warming effect.

Ice on the Moon and Mars

The presence of ice on the Moon was first detected in November 1996 by the spacecraft *Clementine*, a project of the Strategic Defense Initiative Organization and the National Aeronautics and Space Administration (NASA). This discovery was confirmed two years later, on 5 March 1998, by the NASA spacecraft *Lunar Prospector*.

The discovery of ice on the Moon was somewhat of a surprise to scientists. The Moon has no atmosphere, so any water present on the satellite's surface would quickly be evaporated by sunlight and escape into space. The two spacecraft found, however, that some parts of the Moon are always in shadow, protected from solar energy. In addition, ice was found to be mixed with soil at depths up to about 40 centimeters (18 in.). This ice was protected sufficiently from sunlight to allow it to remain permanently on the Moon's surface.

The ice was concentrated at the Moon's two poles, covering an area of about 1,850 square kilometers (650 sq mi) at each pole. The total volume of ice present has been estimated at about six trillion kilograms (6.6 billion tons). This estimate is based on computer models developed from the *Lunar Prospector* data and may be fairly inaccurate. Meteorites that have collided with the Moon throughout its history are thought to be the source of the ice discovered by the two satellites.

The presence of ice on the Moon may be of some importance for future exploratory visits. It could be used not only as a source of water but also as a raw material for the manufacture of hydrogen and oxygen needed for future research stations on the Moon.

Like the Moon, Mars has a very thin atmosphere, about 1 percent the density of Earth's atmosphere. As a result, most of the water that may have once been present on the planet has long since evaporated and escaped into space. Some water remains in the form of ice on the planet's surface, primarily at the two poles. Both poles are covered with what appears to be a relatively thick ice cap containing frozen water mixed with dust. Both ice caps are also covered with another layer of frozen carbon dioxide. The carbon dioxide layer at the north pole appears to evaporate during the Martian summer, exposing the water ice cap, while the frozen carbon dioxide cap at the south pole appears to persist throughout the year.

Scientists also believe that water and ice were once abundant on Mars and were responsible for the formation of many topographic features now visible on the planet. For example, well-defined canyons were discovered in 1999 in the planet's mid-latitudes with a similar shape to those on Earth carved out by glaciers. *See also* Extraterrestrial Water; Icebergs

Further Reading

Bennett, Matthew. *Glacial Geology: Ice Sheets and Landforms*. Chichester: Wiley, 1996.

Knight, Peter G. *Glaciers*. Cheltenham, UK: Stanley Thornes, 1999.

Petrenko, Victor F., and Robert W. Whitworth. *Physics of Ice*. Oxford: Clarendon Press, 1999.

Post, Austin, and Edward R. LaChapelle. *Glacier Ice*. Seattle: University of Washington Press, 2000.

Further Information

National Ice Center
Federal Building #4
4251 Suitland Road
Washington, DC 20395
Telephone: (301) 457-5303
URL: http://www.natice.noaa.gov
e-mail: webmaster@natice.noaa.gov

National Snow and Ice Data Center
449 UCB, University of Colorado
Boulder, CO 80309-0449
URL: http://nsidc.org
e-mail: nsidc@kyros.colorado.edu

Icebergs

An iceberg is a large mass of ice that has broken loose from a glacier. The vast majority of icebergs originate in Greenland or the Antarctic, although a small number are formed from Alaskan glaciers or glaciers in the Arctic islands off the Eastern shore of Canada. Icebergs are to be distinguished from *pack ice*, which is a large mass of ice formed by the freezing of seawater.

Formation of an Iceberg

The breeding grounds for icebergs are the massive ice fields that cover Greenland, the Antarctic, and other regions near the poles. In most cases, those ice fields have existed for hundreds of thousands of years or more. Over time, the addition of fresh snowfall on top of an ice field produces pressures great enough to cause underlying ice to begin moving. When that happens, rivers of flowing ice—glaciers—are formed.

Glaciers differ widely in terms of their size and rate of movement. In general, most glaciers move at a rate between 100 and 1,000 meters (330–3,300 ft) per year, with an average speed of about 360 meters (1,200 ft) per year.

Life Cycle of an Iceberg

The ice at the foot of a glacier is unstable; it is constantly subject to winds, tides, and wave action. These forces eventually cause pieces of the glacier to break off, drop into the sea, and form icebergs. The process by which icebergs are formed from a glacier is known as *calving*.

Once they begin to float, icebergs travel along routes determined by prevailing winds and ocean currents. For example, icebergs formed in Greenland tend to be pushed northward along the west coast of the island in the West Greenland Current. At some point, they intersect the south-flowing Labrador Current, which takes them along the eastern coast of Canada.

On average, only about 20 percent of the icebergs formed in Greenland ever reach Canada. The rest break apart, melt, dissolve, or follow some other route. The trip across the northern Atlantic may take up to two years. Of those that do survive the trip, most have lost up to 90 percent of their mass through melting, evaporation, and sublimation.

Icebergs that travel the North Atlantic route are of great concern to ships that traverse this part of the ocean. Even relatively small icebergs can cause severe damage to a ship. Most ships now carry radar, sonar, or other detection systems to warn them of icebergs.

By contrast, icebergs formed from the Antarctic ice shelf tend to be less of a hazard to shipping simply because there is less ship travel in that part of the world. Nonetheless, Antarctic icebergs tend to be far larger and far more numerous than those found in the North Atlantic.

Iceberg Descriptions

Scientists classify icebergs not only according to their size and rate of movement but also according to their shapes. The six major terms used to describe icebergs are as follows:

Tabular: Long, flat, tablelike shape
Blocky non-tabular: Flat top with steep sides
Drydock non-tabular: U-shaped with prominent peaks at each side
Dome non-tabular: Smooth, rounded upper surface
Pinnacle non-tabular: Craggy upper surface
Wedge non-tabular: Flat top with steep sides, much higher at one end than the other

Icebergs can be virtually any size, from that of a refrigerator to a ten-story building. The largest Arctic iceberg ever found was discovered off Baffin Island in 1882. It was about 13 kilometers (8 mi) long and 6 kilometers (3.5 mi) wide, with a height of about 20 meters (65 ft) above the surface of the water. The total mass of the iceberg was estimated to be about 9 billion metric tons.

By comparison, one of the largest icebergs ever seen in the Antarctic was found in 1987. It had a total area of 6,350 square kilometers (2,450 sq mi) and an estimated mass of 1.4 trillion metric tons.

The everyday phrase "The tip of the iceberg" reflects one characteristic of all such objects: about 87 percent of the berg is actually below water level. Since ice is slightly less dense than pure water (about 0.9 grams per cubic centimeter compared to 1.0 grams per cubic centimeter), only a small portion of the iceberg is actually lifted out of the water. The fact that so much of a berg is actually hidden from view only adds to its risk for ships.

Towing of Icebergs

Icebergs represent a potentially enormous source of freshwater for parts of the world where water is scarce. In theory, one might find a way of attaching an iceberg to a fleet of tugs and then pulling the berg to some desired location. Some research has been done on the possibility of using this method to bring freshwater in the form of icebergs to Saudi Arabia and other nations in the Mideast where water is scarce.

The feasibility of this concept was first demonstrated in an actual test in 1971. So far, however, the method has not been used to deliver icebergs to water-scarce regions. It has, however, been employed to divert icebergs from pathways that could lead to problems. For example, icebergs can now be towed out of their natural path when they are headed for off-shore drilling rigs off the coast of North America. *See also* Ice; Oceans and Seas

Further Reading

Iceberg Information, http://www.wordplay.com/tourism/icebergs/index.html, 10/30/99.
Wadhams, Peter. *Ice in the Ocean*. Sydney, Australia: Gordon and Breach, 2000.
Current information on iceberg formation and movement is available from the National Ice Center at ⟨http://www.natice.noaa.gov/main-one.htm⟩.

Ice Hockey

See Aquatic Sports, Ice Hockey

Indra riding an elephant. *Historical Picture Archive/Corbis.*

Indra

Indra is one of the most important gods of the Hindu religion. He is described in various parts of the Vedas as the god of rain, of storms, of thunder and lightning, of war, and of the sky. He is often pictured as having four arms that hold a thunderbolt, a spear, a quiver of arrows, and a net and hook with which he captures his foes.

At one point, Indra was the most important of all Hindu gods and goddesses. Late in the Vedic period, however, he loses his priority over other deities. More than 250 hymns in the Vedas are dedicated to Indra—more than to any other Hindu deity. He is sometimes compared with Hercules, Zeus, or Jupiter from Greek and Roman mythology.

Indra's reputation was originally based on his creation of the world as we now know it. According to mythical history, the world was once controlled by the snake god Vratra and covered with darkness. The land was totally dry and infertile. No life could survive in this environment.

Indra went to battle with Vratra, however, and killed the snake god by cutting open his belly with a blow of his mace. With that act, water stored in the snake's belly was released to form the oceans, rivers, and lakes that exist today. As a result of Indra's act, life was able to bloom on Earth.

After the battle with Vratra, Indra continued to have dominion over many natural phenomena. He was generally credited with producing rain by striking the earth with his thunderbolts, accounting for the lightning and thunder that often accompany storms. He is also given credit for keeping earth and sky separate from each other.

Many stories are told of Indra's life and accomplishments in the *Ramayana* epic. As it progresses, the stories become less flattering, such as the one recounting the time he was reputed to have seduced the wife of the holy man Gautama. As a result of this act, he was punished by being given a thousand eyes (or, in some versions, a thousand vaginas) all over his body.

Indra also appears in the mythology of other cultures. For example, he is honored in Indonesian myths by the same name and for many of the same feats for which he is recognized in India. *See also* Gods and Goddesses of Water

Further Reading

Choudhuri, Usha. *Indra and Varuna in Indian Mythology.* Delhi: Nag Publishers, 1981.
Gonda, Jan. *The Indra Hymns of the Rgveda.* Leiden: Brill, 1989.

Inter-American Water Resources Network

The Inter-American Water Resources Network (IWRN) is an association of over a hundred governmental and nongovernmental; public and private; and state, national, and international organizations concerned with water issues in the Western Hemisphere, as they relate to sustainable development and environmental protection. Some examples of the organizations that make up IWRN are the Antigua Public Utilities Authority, American Water Works Association, Brazilian Ministry of Environment, Canadian Wildlife Service, Colorado State University Engineering Research Center, Florida Center for Environmental Studies, Instituto Nacional de Recursos Naturales of Panama, MTP Group, Inc. (USA), National Hydrographic Directorate of Uruguay, United Nations Food and Agriculture Organization, the University of New Mexico, and the World Bank.

The mission of IWRN is to build and strengthen water-resources partnerships that transcend national, organizational, political, language, and disciplinary boundaries; to promote education and the exchange of technical expertise; and to enhance communication, cooperation, collaboration, and financial commitment to the improvement of integrated land and water resources within the Americas.

A major focus of the network has been the sponsorship of a series of Inter-American Dialogues designed to carry out the organization's goals and objectives. The fourth of the dialogues was held in Foz do Iguazú, Brazil, April 22–26, 2001. Other activities carried out during the biennium 2000–2001 included designing a Water Resources Practitioners database; updating directories of water agencies and water networks in the Americas; publishing the Water Resources Practitioners database, the Business Directory, and the Water Agencies and Network Directory in electronic form; continuing support for the Water Policy Round Tables held on a regular basis at the Organization of American States (OAS) in Washington, D.C.; assisting OAS states in the establishment of regional and subregional forums; and promoting and facilitating the use of the IWRN website forum for specific discussion groups.

Further Information

Inter-American Water Resources Network
Unit for Sustainable Development and Environment
Organization of American States
1889 F Street, N.W.
Washington, DC 20006
Telephone: (202) 458-3745
Fax: (202) 458-3560
URL: http://www.iwrn.net

International Association for Hydrologic Research and Engineering

The primary objectives of the International Association for Hydrologic Research and Engineering (IAHR) are to promote applied research, scientific exchange, technology transfer, and research management in the field of hydrologic engineering. The association consists of both individuals and corporate members. IAHR deals with only one phase of water resources issues (hydraulic research and engineering), but cooperates with other associations and organizations interested in other aspects of such issues.

IAHR is divided into four regional divisions (Asia and Pacific, Latin America, Africa, and Europe), three national chapters (Hong Kong, United Kingdom, and Russia), two student chapters (Universities of Iowa and Stuttgart), and three technical sections. The three technical sections are Methods in Hydraulics (including topics such as fluid mechanics, hydroinformatics, experimental hydraulics, and probabilistic methods); Applied Hydraulics (hydraulic machinery and systems, fluid phenomena in energy exchanges, water resources management, and hydraulic structures); and Geophysical Hydraulics (fluvial hydraulics, maritime hydraulics, groundwater hydraulics, and ice research and engineering).

The association produces a number of publications, including *Journal of Hydraulic Research, Journal of HydroInformatics, Journal of Water and Maritime Engineering, IAHR Newsletter,* and *IAHR NewsFlash*, as well as section newsletters, monographs, and proceedings of biennial congresses and specialty conferences.

Further Information

International Association for Hydrologic
    Research and Engineering
Rotterdamsweg 185
2629 HD Delft
The Netherlands
Telephone: +31 15 285 8819
Fax: +31 15 285 8417
URL: http://www.iahr.nl
e-mail: iahr@iahr.org

International Association of Hydrogeologists

The International Association of Hydrogeologists (IAH) was founded in 1956 for the purpose of promoting the study and knowledge of hydrogeological science and its applications for the common good throughout the world. Hydrogeology is the science that deals with the way in which water reacts with geological systems. An important field of research, for example, focuses on groundwater reserves and the ways in which groundwater flows. IAH currently has about 4,000 members and associated members in 135 countries.

The goal of IAH is to promote international cooperation among hydrogeologists and others who are interested in problems of groundwater management. To that end, the association sponsors a number of international conferences and an annual congress at which hydrogeologists present papers and discuss developments in the field. IAH also supports the work of hydrogeologists in developing countries through the Burdon Fund.

The association publishes a scholarly journal, *Hydrogeology Journal*; a triannual newsletter, *News and Information*; and a book series, *International Contributions to Hydrogeology*.

Further Information

IAH Secretariat
P.O. Box 9
Kenilworth CV8 1JG
United Kingdom
Telephone: +44 1926 450677
Fax: +44 1926 856561
URL: http://www.iah.org
e-mail: iah@iah.org

International Association of Theoretical and Applied Limnology

The International Association of Theoretical and Applied Limnology (Societas Internationalis Limnologiae, SIL) was founded in 1922 to promote the study and understanding of all aspects of limnology—the science of inland aquatic ecosystems. One of the most important activities of the association is a triennial conference dealing with current research on topics such as fisheries, pollution, and water-supply issues. Any member of the association is allowed to present a paper at one of these conferences.

Much of the work of SIL is done through working groups, whose objective it is to improve research and communication among workers in their interest areas. Current working groups are concerned with plankton ecology, deep coring operations and paleolimnological work in relict lakes, aquatic microbial ecology, periphyton of freshwater ecosystems, aquatic primary productivity, African great lakes, rquatic badioecology, biomonitoring, conservation and management of running waters, saline inland waters, aquatic birds, biodiversity, physical limnology, wetlands, and aquatic invasive species. The association publishes *SIL News* three times a year, as well as other special publications, such as *Limnology in Developing Countries* and *A Guide to Limnetic Species of Cladocera of Africa Inland Waters*.

Further Information

International Association of Theoretical and
    Applied Limnology
c/o Dr. Robert G. Wetzel
Department of Biological Sciences
University of Alabama
Tuscaloosa, AL 35487-0206
URL: http://www.limnology.org
e-mail: rwetzel@biology.as.ua.edu

International Ground Water Modeling Center

The International Ground Water Modeling Center (IGWMC) was established in 1978 at the Holcomb Research Institute at Butler University in Indianapolis, Indiana. It was relocated in 1991 to the Colorado School of Mines in Golden, Colorado, where it is currently a part of the university's Department of Geology and Geological Engineering.

IGWMC has three primary objectives:

- to stimulate the development and correct use of simulation models and other computer-based technology for the management and protection of groundwater resources;
- to train groundwater professionals in the use of these materials;
- to facilitate communication and interaction among groundwater professionals on modeling issues.

The center carries out a number of activities in attempting to reach these goals. For example, it is involved in the development, testing, distribution, and technical support of groundwater management software; basic research on methods of modeling; development of educational courses and materials; organization of conferences; and dissemination of information via the Internet, mail, and workshops. IGWMC also provides a telephone advisory service through which information is provided both on modeling and software questions in general and on specific questions in the same areas.

Further Information

International Ground Water Modeling Center
Colorado School of Mines
Golden, CO 80401-1887
Telephone: (303) 273-3103
Fax: (303) 384-2037
URL: http://www.mines.ed/igwmc
e-mail: igwmc@mines.ed

International Hydrological Programme

The International Hydrological Programme (IHP) is a cooperative scientific program that operates under the sponsorship of the United Nations Educational, Scientific, and Cultural Organization (UNESCO). It was originally established as the International Hydrological Decade by UNESCO in 1965. Its purpose is to provide a mechanism by which nations around the world can improve their understanding of the hydrologic cycle and, thereby, enhance their ability to manage and develop their water resources.

Much of IHP's work is carried out through the International Hydrological Programmes—five-year programs focusing on methodologies for studies and training and education in the water sciences. The fourth phase of IHP—"Hydrology and Water Resources Sustainable Development in a

Changing Environment"—ran from 1990 to 1995, while the fifth phase—"Hydrology and Water Resources Development in a Vulnerable Environment"—ran from 1996 to 2001. Some of the projects completed as part of that program were Global Hydrological and Geochemical Processes, Groundwater Resources at Risk, Integrated Water Resources Management in Arid and Semi-Arid Zones, Humid Tropics Hydrology and Water Management, and Integrated Urban Water Management.

IHP supports a very active publishing program that includes four regular series of publications: Studies and Reports in Hydrology, Technical Documents in Hydrology, International Hydrology series, and IHP Humid Tropics Programme series. In addition, the organization publishes a number of nonserial works, such as *Decision Time for Cloud Forests, Ecohydrology: Advanced Study Course*, and *Environmental Isotopes in the Hydrological Cycle*. Finally, IHP co-publishes a number of other technical works with the International Association of Hydrogeologists and the International Association of Hydrological Scientists.

Further Information

International Hydrological Programme
UNESCO
Division of Water Sciences
1, rue Miollis
75732 Paris Cedex 15
France
Telephone: +33 (0) 1 45 68 40 02
Fax: +33 (0) 1 45 68 58 11
URL: http://www.unesco.org/water/ihp
e-mail: ihp@unesco.org

International Maritime Organization

The International Maritime Organization (IMO) was established as the result of an international conference sponsored by the United Nations in 1948. Originally called the Inter-Governmental Maritime Consultative Organization (IMCO), the organization actually came into force in 1958, when a sufficient number of members ratified the original treaty. The organization changed its name to its present form in 1982.

The purpose of the organization, according to the original treaty, is "to provide machinery for cooperation among Governments in the field of governmental regulation and practices relating to technical matter of all kinds affecting shipping engaged in international trade; to encourage and facilitate the general adoption of the highest practical standards in matters concerning maritime safety, efficiency of navigation and prevention and control of marine pollution from ships."

The first task to which IMO turned its attention was the development of maritime safety. In 1960, it produced a revised and improved version of an older document, the International Convention for the Safety of Life at Seas (SOLAS), which is still in force. The organization next attacked a variety of problems, such as the facilitation of international maritime traffic, the transport and handling of dangerous goods, and the improvement of the system by which the tonnage of ships is measured.

The *Torrey Canyon* disaster of 1967, in which 120,000 metric tons of oil were spilled off the coast of southern England and northern France, posed a new challenge to the IMO: pollution of ocean waters. In the ensuing years, the organization took on the challenge of developing rules and regulations designed to prevent tanker accidents and minimize their environmental consequences, and to deal with routine tanker operations, such as cleaning of oil cargo tanks and disposing of engine-room wastes. The major result of this work was the International Convention for the Prevention of Pollution from Ships, adopted in 1973 and later modified in 1978.

Over the last twenty-five years, the organization has continued to work on issues involving maritime safety and ocean pollution. In 1999, the IMO assembly adopted a resolution outlining the organization's mission for the 2000s. Included among the goals were developing a safety culture and environmental consciousness, ensuring an effective uniform implementation of existing IMO standards and regulations, avoiding excessive regulation, and strengthening the organization's technical cooperation programs.

Further Information

International Maritime Organization
4 Albert Embankment
London SE1 7SR
United Kingdom
Telephone: +44 (0) 20 7735 7611
Fax: +44 (0) 20 7587 3210
URL: http://www.imo.org
email: mharvey@imo.org

International Oceanographic Commission

The International Oceanographic Commission (IOC) was founded as a program of the United Nations in 1960. The commission currently consists of 128 member states who meet in general assembly once every two years. Day-to-day operations are supervised by an Executive Council and the IOC Secretariat. IOC's primary objective has been to promote marine science research and related ocean services so that humans can learn more about the general nature and resources of the planet's oceans. At the present time, IOC activities are focused on four major themes:

- developing, promoting, and facilitating international oceanographic research;
- promoting effective planning, establishment, and coordination of a global ocean-

observing system to provide useful data for oceanic and atmospheric forecasting, for ocean and coastal zone management by coastal nations, and for global climate change research;

- providing international leadership for education and training programs and for technical assistance in oceanic and coastal zone research;
- ensuring the efficient handling and distribution of data and information on the oceans obtained by research and observation.

Some examples of the types of activities currently being sponsored by IOC include Ocean Science in Relation to Living Resources, Ocean Science in Relation to Non-Living Resources, Ocean Mapping, Marine Pollution Research and Monitoring, International Oceanographic Data and Information Exchange, International Tsunami Warning System in the Pacific, Global Ocean Observing System, Integrated Global Ocean Services System, and Training, Education, Mutual Assistance and Capacity Building.

Further Information

Intergovernmental Oceanographic Commission
    of UNESCO
1, rue Miollis
75732 Paris Cedex 15
France
Fax: +33 (1) 45 68 58 12
URL: http://ioc.unesco.org/iocweb
e-mail: n/a

International Office for Water

The International Office for Water (IOW) was established in 1991 for the purpose of bringing together bilateral and multilateral organizations, ministries, water agencies, local communities, universities, engineering schools, research centers, land development companies, water suppliers, industrialists, professional associations, and nongovernmental organizations for the purpose of establishing a network for information exchange and training. There are currently 149 member organizations.

The work of IOW is divided into three major categories: CNFME, the National Training Centre for Water Professions; AQUA-COOPE, the International Institute for Water Administration; and SNIDE, the National Centre for Water-Related Information and Documentation.

CNFME offers a series of 135 courses for engineers, executives, technicians, and other workers in the field of water. The courses cover all fields of specialization, including design, implementation, operation, maintenance, and management. The center is currently training about four thousand men and women per year.

AQUACOOPE provides members with expertise and advice in areas such as forming national or local strategies for water resources management, organizing the management of specialized administrations and institutions, preparing legal and regulatory texts, setting up partnerships, and undertaking economic simulations and the planning and study of master plans. This division of IOW now trains more than four hundred individuals a year.

SNIDE calls itself "Your international memory for water." It contains more than 170,000 references on water with nearly 400 international periodicals and 15,00 books indexed each year. The center provides document retrieval, scanning, and other services for members. The division also manages the Permanent Inventory of Water Studies and Research being developed by national and regional water-related organizations. IOW also acts as the water center for France's GRISELI water-research program. The office has also been directed by the French government to establish the National Data Reference Centre for Water, which prepares a list of available data on inland, surface, and groundwater; defines protocols, coding, geographical references, and standards necessary for an efficient data network; and updates the catalog on a regular basis with data producers and database managers.

Further Information

International Office for Water
21 rue de Madrid
75008 Paris
France
Telephone: 01.44.90.88.60
Fax: 01.40.08.01.45
URL: http://www.oieau.fr
e-mail: dg@oieau.fr

International Programme for Technology and Research in Irrigation and Drainage

The International Programme for Technology and Research in Irrigation and Drainage (IPTRID) was founded in 1991 and was originally hosted by the World Bank. The programme was transferred in 1998 to the United Nations Food and Agriculture Organization (FAO). IPTRID's mission is to improve the standard of irrigation and drainage research and development in developing countries, with the goal of improving production of food and other agricultural commodities in order to help eliminate poverty.

IPTRID carries out its operation through two kinds of networks: the IPTRID Central Regional Network and IPTRID Country/ Regional Networks. The Central Network is made up of seven members, consisting of the U.S. Bureau of Reclamation, Cemagref (France), HR Wallingford (UK), ICID Central

Office (India), ILRA (Netherlands), FAO, and the IPTRID Secretariat. Some of the services provided by the Central Network include a database of research projects in irrigation and drainage; a bibliographic research service; a text-delivery service; technical advice; and publication of the organization's biannual newsletter *GRID*, issue papers, country reports, and technical reports.

Fourteen Country/Regional Networks currently exist. Their primary function is to provide a conduit between the Central Network and individual members. A goal of IPTRID is for Country/Regional Networks to begin providing all or many of the services now being offered by the Central Network.

Further Information

International Programme for Technology and Research in Irrigation and Drainage
Land and Water Development Division
Food and Agriculture Organization of the United Nations
Viale delle Terme di Carcalla
00100 Rome
Italy
Telephone: +39 06 570 54 033
Fax: +39 06 570 56 275
URL: http://www.hrwallingford.co.uk/projects/IPTRID/iptrid.htm
e-mail: iptrid@fao.org

International Water Association

The vision for the International Water Association (IWA) is that it become the leading international membership association for the improvement of "all aspects of water supply and treatment, wastewater collection, treatment and disposal and overall management of water quality and quantity, including environmental and public health issues." IWA hopes to attain that vision by collecting and distributing information on the latest skills, techniques, and knowledge concern-

ing all aspects of water management. The organization intends to use all means possible, such as meetings, publications, expert networks, and the electronic media, to reach this goal.

Among the association's achievements thus far are

- conferences, including the biennial IWA world water conference as well as regional conferences in Southeast Asia, the Middle East, and southeastern Europe;
- publications, including scientific and technical journals, the magazine *Water21*, and a variety of technical reports, specialized newsletters, and books on water management;
- specialist groups covering about fifty major topics of interest in the field of water management;
- the IWA Foundation, established for the exchange of expertise to and among developing countries.

Among the publications available from IWA are the *Journal of Hydroinformatics; Journal of Water Supply: Research and Technology; Water Research*; and *Water Science. Water21* is published six times a year and carries regular news features as well as editorials, information on products and services, updates on research and development projects, forthcoming events, and interviews with important individuals in the field.

Further Information

International Water Association
Alliance House, 12 Caxton Street
London SW1H 0QS
United Kingdom
Telephone: +44 (0) 20 7654 5500
Fax: +44 (0) 20 7654 5555
URL: http://www.iwahq.org.uk
e-mail: water@iwahq.org.uk

International Water Management Institute

The International Water Management Institute (IWMI) is a scientific research organization concerned with water issues in developing nations in the Southern Hemisphere. The institute is staffed by scientists from ten nations, who carry out projects centered on four core programs: Irrigation and Water Resources; Policies, Institutes, and Management; Health and Environment; and Applied Information and Modeling Systems. The institute also sponsors a Poverty, Gender, and Water Project. IWMI researchers come from a variety of disciplines, including hydrology, agronomy, economics, social science, environmental health, statistics, and information technology.

An important part of IWMI's work is the development and validation of software tools on topics such as monitoring and measuring of water resources in a river basin, global information on water and moisture availability, and planning tools for policy makers in developing countries. Some of these tools are available for downloading from the institute's website at no cost to users. They are designed for use by governmental and nongovernmental organizations working on poverty, agriculture, and water issues, but they are not intended for use by commercial groups.

IWMI publishes books, workshop proceedings, newsletters, and research reports, many of which are also available on the organization's website. The institute has also been designated the convening center for the Systemwide Initiate on Water Management (SWIM), whose reports and papers can be accessed through IWMI's website.

Further Information
International Water Management Institute
P.O. Box 2075
Colombo, Sir Lanka
Telephone: (94-1) 867404, 869080/1, 872178, 872181
URL: http://www.cgiar.org/iwmi
e-mail: k.nonis@cgiar.org

International Water Resources Association

The International Water Resources Association (IWRA) was established in 1972 for the purpose of promoting the sustainable management of water resources throughout the world. The major foci of the association's activities are (1) the advancement of research on water resources and related environmental issues; (2) the promotion of water resources education; (3) improvement in methods of exchanging information and expertise; (4) networking with other organizations interested in water resources issues; and (5) providing an international forum for water resource issues.

IWRA publishes a quarterly journal, *Water International Journal*; a quarterly newsletter, *Update Newsletter*; an *International Expert Directory*; and a listserver, *IWRAnews*. The association sponsors regional conferences at various locations around the world each year, as well as the World Water Conference every three years.

Further Information
International Water Resources Association
4535 Faner Hall
Southern Illinois University
Carbondale, IL 62901-4516
Telephone: (618) 453-5138
Fax: (618) 453-2671
URL: http://www.iwra.siu.edu
e-mail: iwra@siu.edu

Irrigation

Irrigation is the process of providing water to land by artificial means. Irrigation has been used by humans for thousands of years to supply water to areas of land as extensive as thousands of acres and as limited as small lawns and gardens.

History

Archaeologists have found evidence that irrigation systems were being used in Egypt as far back as about 5000 B.C. to carry water from the Nile River during flooding to cultivated fields. By 2100 B.C., the Egyptians had constructed complex systems of canals to bring water from the Nile to Lake Moeris, about 20 km (12 mi) from the river. Somewhat earlier, the Sumerians had built irrigation canals to make use of water taken from the Tigris and Euphrates Rivers. Irrigation systems were also being widely used by the Chinese as early as 2200 B.C. and by the Incas and Native Americans before the beginning of the Christian era. One of the most extensive of these systems covered about 100,000 hectares (250,000 acres) of land in the Salt River valley of Arizona.

A number of mechanical devices were constructed early in history to move water from rivers and streams to agricultural land at higher elevations. One of these devices was the *Archimedes' screw*, a tube bent in a spiral around a central axis operated by a crank. When the crank is turned, water is carried up through the spiral from the river or stream into the irrigation ditch. Another lifting device is the *Persian wheel*, a waterwheel-like device in which buckets are attached to the outer perimeter of the wheel. When the wheel is turned by draft animals, the buckets lift water from a river or stream into an irrigation ditch. The Persian wheel is still widely used in countries around the world with relatively primitive irrigation systems.

Dams were also built early on to collect the water needed for irrigation systems. They not only accumulated dependable supplies of water that could be used when needed but also raised the water to a height sufficient that it could be transferred to irrigation systems by gravity alone.

Irrigation Use

The extent to which irrigation is used for agricultural purposes depends to a large extent on the climate of an area. In regions where rainfall is limited, farming can take place only when systems of irrigation are developed. In Egypt, for example, 100 percent of all farming depends on irrigation systems that draw water from the Nile or large oases. In other countries of the Middle East with low annual rainfall, such as Iraq, Iran, Saudi Arabia, Israel, and Jordan, agriculture is also very dependent on irrigation.

Certain regions of a country may also require irrigation for farming to be possible. The huge agricultural output of California, and other western states of the United States, for example, is largely dependent on extensive irrigation systems that may bring water from many miles away to areas that would otherwise be deserts.

Factors other than climate may influence the use of irrigation systems. Especially in nations with large populations, irrigation can increase the extent and efficiency of agriculture. It comes as no surprise, then, to learn that the nation with the second largest percentage of farmed land maintained by irrigation is Pakistan (80%) and that the two nations with the largest amount of irrigated farmland are China (52,943,200 hectares; [130,825,000 acres]; 52% of all farmed land) and India (50,101,000 hectares [123,802,000 acres]; 28% of all farmed land) ("Land under Irrigation").

Worldwide, about 15 percent of all farmland is maintained by irrigation. In the United States, that number is about 11 percent. As of 1995, the latest year for which data are available, about 134 bgw/d (billion gallons of water per day) were being used in the United States for irrigation. That number is somewhat less than the highest use rate ever recorded in this country, 150 bgw/d in 1980, but about average for the first half of the 1990s.

The largest state user of irrigation water in the United States is California, which used about 28.9 bgw/d in 1994 for farming purposes. Other major users of irrigation water are Idaho (13.0 bgw/d), Colorado (12.7 bgw/d), Texas (9.45 bgw/d), Montana (8.55 bgw/d), and Nebraska (7.55 bgw/d). Of the fifty states, only West Virginia used no water for agricultural irrigation (U.S. Bureau of the Census 2000, tables 387 and 388).

Methods

Irrigation systems can be classified into two major categories: surface and subsurface systems. Surface systems, in turn, can be subdivided into flood, furrow, sprinkler, and drip systems.

Flood irrigation is one of the oldest irrigation systems and is used for crops that grow close together, such as rice. In some cases, as with rice, a field is allowed to remain flooded with water while the crop grows. In other cases, as with certain types of fruit trees, an area is flooded with water and then allowed to dry up before being reflooded. An alternative version of flood irrigation is *basin irrigation*, in which bowl-shaped depressions are made around the base of trees and then flooded, allowed to dry up, and reflooded.

In *furrow irrigation*, crops are planted in long rows separated by ditches through which water can flow. Furrow irrigation is most widely used with vegetable crops.

Sprinkler irrigation systems most closely resemble natural rainfall. They usually consist of large devices similar to lawn and garden sprinklers found in many yards. They may be either anchored to the ground or connected to each other and movable by large wheels that travel across a field. Another type of sprinkler system consists of

large guns that fire huge sprays of water across a field, much like the hoses used by firefighters. Sprinkler irrigation is most commonly used with low-lying crops, such as grass, hay, and alfalfa.

Drip irrigation systems make use of long pipes or hoses containing many tiny holes. Water is released through the holes at a small but steady rate. One important advantage of drip irrigation is that relatively small amounts of water are lost through evaporation. Drip irrigation systems have become very popular for use in lawns and home gardens. Commercially, they are most commonly found in Australia, Israel, and the United States.

Sub-surface irrigation systems may be used when there is not enough room to lay down a surface system or the soil is too productive to give up to irrigation. Such systems are similar to drip irrigation systems except that the pipes or hoses they use are buried underground rather than resting on top of the ground.

Environmental Issues

The use of irrigation in agriculture can pose two kinds of environmental problems. First, the sheer volume of water needed for irrigation is enormous. By one estimate, nearly 90 percent of all freshwater used by humans around the world goes to irrigation systems (Miller 1985, 156). The removal of this water from natural sources can cause serious problems. In the United States, for example, the removal of water from the Ogallala aquifer in the western United States has proceeded to the point that within the next two decades, the aquifer may no longer be able to provide water for any form of human consumption.

Another major problem arising from irrigation is *salinization*. All water contains some level of dissolved salts. When water is used in irrigation systems, most of it eventually evaporates, leaving behind those salts.

Over time, the quantity of salts can accumulate on and in the soil to a level where they become toxic to plants growing in the soil.

History records a number of instances in which salinization became so serious that entire regions were no longer capable of supporting plant growth. Perhaps the most dramatic of these examples is modern-day Iraq. Before the second millennium B.C., the land between the Tigris and Euphrates Rivers was very fertile. After irrigation systems were built by the Sumerians, however, salinization began to occur and the land became less productive. Within four hundred years, harvests dropped by 75 percent. This loss in productivity is generally regarded as an important factor in the decline of the Sumerian civilization that occurred at about this time.

Similar problems have arisen in modern times. For example, salinization began to appear at significant levels in Pakistan's large Indus valley in the early 1960s. About one-fifth of the land on which irrigation was being used had begun to experience a loss in productivity ("Irrigation," Microscoft Encarta Online Encyclopedia). Salinization can be prevented and reversed with flushing programs, but such programs are very expensive to build and use.

A related environmental problem associated with irrigation systems is waterlogging, in which some of the irrigation water sinks so far into the soil that it does not evaporate. Over time, the amount of water reaching the water table increases, and the water table itself begins to rise. Eventually, it may become so high that plant roots are permanently surrounded by saturated soil, and the plants begin to die. *See also* Canals; Dams

Further Reading

"Land under Irrigation," http://www.fao.org/ag/agl/aglw/aquastat/dbase/dbase2.

Michael, A. M. *Irrigation: Theory and Practice.* Columbia, MO: South Asia Books, 1999.

Miller, G. Tyler, Jr., *Living in the Environment.* Belmont, CA: Wadsworth Publishing, 1985.

On-Farm Irrigation Committee of the Environmental and Water Resources Institute. *Selection of Irrigation Methods for Agriculture.* Reston, VA: American Society of Civil Engineers, 2000.

Postel, Sandra. *Pillar of Sand: Can the Irrigation Miracle Last?* New York: W. W. Norton, 1999.

Stewart, B. A., and D. R. Nielsen, eds. *Irrigation of Agricultural Crops.* Madison, WI: American Society of Agronomy, 1990.

U.S. Bureau of the Census. *Statistical Abstract of the United States.* Washington, DC: Government Printing Office, 2000.

Further Information

The Irrigation Association
6540 Arlington Blvd.
Falls Church, VA 22042-6638
Telephone: (703) 536-7080
Fax: (703) 536-7019
URL: http://www.irrigation.org
Irrigation Journal (online), at http://www.greenindustry.com

J

Johnstown Flood of 1899

The Johnstown Flood of 1899 is generally thought to be the worst flood disaster ever to strike the United States. It occurred in Johnstown, Pennsylvania, on 31 May 1899.

Johnstown was founded in 1794 at the confluence of the Little Conemaugh and Stony Creek Rivers. It was built to a large extent on the floodplain formed by these two waterways. The town grew rapidly after the construction of the Pennsylvania Marine Canal in 1834 and the arrival of the Pennsylvania Railroad and the Cambria Iron Company in the 1850s.

The prosperity brought by these developments involved an environmental price: a dam, the South Fork Dam, had to be built 14 miles upstream of Johnstown on the Little Conemaugh. The purpose of the dam was to capture and supply water for the canal. It held back the artificial Lake Conemaugh, 3 miles long and 450 feet higher than Johnstown itself.

The arrival of the railroad, at just about the time the dam was being completed, made the dam unnecessary, and it soon fell into disrepair. Only the interest of a local fishing and hunting club provided any reason to maintain the dam at all. The club closed up the relief gates and began to raise the height of the dam. Engineers began to worry about the stability and safety of the structure, and residents of Johnstown were used to hearing reports that the dam had failed or was about to fail.

Thus, it is easy to understand the incredulity with which they received reports on the morning of 31 May 1899 that the dam was beginning to give way. Those reports followed a heavy storm that released an estimated trillion gallons of water in the Little Conemaugh's drainage basin within a matter of days. Volunteers desperately tried to open sluiceways to allow release of the flooded Lake Conemaugh, but their efforts were in vain. Water was pouring into the lake far faster than it could be released through the sluiceways. In late afternoon, the dam collapsed. Twenty million tons of water were released into the Little Conemaugh. The resulting wave rushed into Johnstown at 4:07 P.M. with a height of up to 60 feet and speeds of up to 40 miles per hour.

The flood swept through the city in no more than ten minutes. At the end of that time, 2,209 people had been killed and 99 families totally lost. About 80 of those people had survived the flood, but were killed in fires that swept the area when debris carried by the floodwaters began to burn. Today, the terrible disaster is remembered by the Johns-

The aftermath of the Johnstown, Pennsylvania, Flood; ruins of the Cambria Iron Works from Willis Fletcher Johnson's *History of the Johnstown Flood,* **1889.** *Archival photograph by Steve Nicklas, NOS, NGS.*

town Flood National Memorial, established on 31 August 1964. *See also* Floods

Further Reading

"The Johnstown Flood of 1889" (original newspaper article about the flood), http://www.voicenet.com~ginette/flood.htm.

McCullough, David G. *The Johnstown Flood.* New York: Simon & Schuster, 1968.

Muson, Howard. *The Triumph of the American Spirit: Johnstown, Pennsylvania.* Lanham, MD: Johnstown Flood Museum and American Association for State and Local History Library, 1989.

O'Connor, Richard. *Johnstown: The Day the Dam Broke.* Philadelphia: Lippincott, 1957.

Jordan River

See Sacred Waters

Jotuns

The Jotuns are figures from Norse mythology, said to be Earth's oldest living creatures. According to legend, they inhabited the world even before gods or humans.

The oldest of all Jotuns was Ymir, who was killed by Odin, the supreme god in the Norse pantheon, and his two brothers, Vili and Vei. As Ymir died, he released so much blood that all other Jotuns except Bergelmir and his wife were drowned. The two remaining Jotuns then mated and reproduced a new race of giants, who continued to carry on wars with the gods and with humans.

The Jotuns are generally divided into four groups: mountain giants, fire giants, sea giants, and frost giants. The physical attributes, personalities, and characters of these four groups differed considerably from each other. In most cases, all Jotuns were very ugly, but were very wise and much larger and stronger than either men or gods. For example, the Jotun Mimir was considered to be the wisest of all creatures. In fact, Odin sacrificed one of his eyes to drink from the well of wisdom guarded by Mimir.

The sea giants were the least numerous of the Jotuns. They lived at the bottom of the sea and were wonderfully skilled in the use of magic. They were able to make sea creatures do their bidding and serve them in their underwater homes. The most famous of the sea giants was Aegir.

The frost giants originally came from Niflheim, the land of the dead. Niflheim was created from the flesh of Ymir after he was killed. The frost giants are the oldest and most powerful of all Jotuns. A number of Norse tales tell of the conflict between the frost giants and the gods, especially Thor. Many of these stories are based on the battle between Thor and the other gods and goddesses and the frost giants over Thor's magic hammer. The magic hammer was thought to be the only means by which the gods could protect themselves against the frost giants.

In one tale, the frost giant Thrym had found a way to steal Thor's magic hammer from him. He agreed to return the hammer, however, in exchange for the hand of Freyja, the goddess of love, fertility, and beauty. Thor agreed to this demand, but tried to deceive Thrym by appearing himself at Thrym's castle, disguised as Freyja. Thor's friend, Loki, was able to convince Thrym to give up the magic hammer before the frost giant learned of Thor's true identity. As soon as Thor got his hammer back, he killed Thrym and all of the frost giants in Thrym's castle. *See also* Aegir; Gods and Goddesses of Water; Ymir

Just, Ernest Everett (1883–1941)

Just was an African American scientist who spent most of his research career at the Marine Biological Laboratory at Woods Hole, Massachusetts (now the Woods Hole Oceanographic Institute). He was interested in a variety of biological problems, especially the embryological development of marine organisms. He made important contributions to the field of parthenogenesis, the process by which the eggs of frogs and sea urchins can be induced to begin growth and development without being fertilized.

Personal History

Ernest Everett Just was born on 14 August 1883 in Charleston, South Carolina. After studying at the Colored Normal, Industrial, Agricultural, and Mechanical College of South Carolina, Just transferred to Kimball Union Academy in Meriden, New Hampshire. He finished the four-year program at Kimball in three years, and then entered Dartmouth College. He earned his B.A. in biology from Dartmouth in 1907.

Professional Accomplishments

With employment opportunities limited because of his race, Just accepted a position in the English Department at Howard University. He later became professor of zoology and physiology at Howard's School of Medicine. In 1909, he was accepted as a summer research assistant at Woods Hole, where he soon became interested in the biology of marine organisms. In 1911, he was appointed research assistant to the director at Woods Hole, Frank Rattray Lillie. Lillie also arranged for Just to enter a doctoral program at the University of Chicago, where he was awarded his Ph.D. in 1916.

Just was active in other aspects of the biological community as well. He was associate editor of Woods Hole's *Biological Bulletin*, the *Journal of Morphology, Physiological Zoology*, and *Protoplasma*. In 1939, he published *The Biology of the Cell Surface*, summarizing much of his research work at Woods Hole during the 1920s.

For all his accomplishments, Just was disappointed that many doors remained closed to him because of his race. For example, he was never elected to the National Academy of Sciences although he served on nominating committees for the academy. In 1929, he accepted an invitation to work at the Kaiser Wilhelm Institute for Biology in Berlin. He later served on the staff at the Sorbonne in Paris and the Naples Zoological Station. He had begun to think of Europe as his academic home, and returned to the United States only for short visits. During one of these visits, on 27 October 1941, he died from cancer.

Further Reading

Haber, Louis. *Black Pioneers of Science and Invention*. New York: Harcourt Brace, 1970.

Manning, Kenneth R. *Black Apollo of Science: The Life of Ernest Everett Just*. New York: Oxford University Press, 1983.

K–L

Kayaking

See Aquatic Sports, Kayaking

Lake Dwellers

See Water Dwellers

Law of the Sea

See United Nations Convention on the Law of the Sea

Lemuria

See Atlantis

Leopold, Luna (1915–   )

Luna Leopold comes from a distinguished family of environmental advocates. His father was Aldo Leopold, one of the founders of the modern environmental movement in the United States. Leopold is widely recognized as one of the most influential hydrologists of his time, as well as being an important spokesman for environmental issues. He is perhaps best known for his theory of the way rivers form and evolve over time.

Personal History

Leopold was born in Albuquerque, New Mexico, on 8 October 1915. He earned his bachelor's degree in civil engineering at the University of California at Berkeley in 1936. He then continued his studies at the University of Wisconsin and Harvard University, where he took courses in geology, botany, ecology, and climatology. He also worked for the U.S. Soil Conservation Service in Albuquerque. During World War II, Leopold served in the U.S. Corps of Engineers, where he eventually reached the rank of second lieutenant. After the war, Leopold continued his studies at Harvard, from which he received his doctoral degree in geology in 1950.

Professional Accomplishments

In that same year, Leopold accepted a position as senior engineer with the U.S. Geological Survey. He was later promoted to chief hydraulic engineer (1956), chief hydrologist (1960), and senior research hydrologist (1966). It was during this period that Leopold developed some of his most significant ideas about the ways rivers and streams develop. Leopold argued that rivers and streams follow a well-known scientific principle, forming and following a path that will result in an expenditure of the least possible amount of energy. That is, water flowing over a piece of land will follow a course that encounters the least resistance and, hence, produces the smallest amount of erosion. This principle is now widely accepted in the field of natural hydrology.

In 1973, Leopold retired from the Geological Survey to accept a position as professor of geology and landscape architecture at the University of California at Berkeley. He is now professor emeritus at Berkeley.

Leopold has long been interested in environmental issues, especially the conservation of Earth's natural resources. He has been a member of the Sierra Club and has served on its board of directors. Given his own academic background, it is not surprising that one of Leopold's special interests is in the use—and misuse—of water resources. He has long argued that nations must develop new forms of agriculture that make more efficient use of our water resources.

Leopold has received many honors and awards during his lifetime. He was awarded the Kirk Bryan Award of the Geological Society of America in 1958, the Veth Medal of the Royal Netherlands Geographical Society in 1963, the Cullen Medal of the American Geographical Society in 1965, and the Rockefeller Public Service Award in 1966. He was the first hydrologist to be elected to the National Academy of Sciences, and served as president of the American Geological Society in 1972.

Limnatides

See Nymphs

Limnology

The word *limnology* comes from two Greek words meaning "pool" or "marsh" (*limne*) and "the study of" (*ology*). In modern terms, limnology refers to the study of all types of inland waters, including lakes, streams, springs, ponds, wetlands, and other bodies of water completely enclosed by land. By such a definition, even bodies of salt water, such as inland seas, are included in the science of limnology. Limnology is often considered to be the sister science of oceanography—the study of the world's open waters.

Limnology is an interdisciplinary science that deals with all facets of inland waters, including their geological and physical structure and their chemical composition and biotic composition.

The term *limnology* was coined by the Swiss physician François Alphonse Forel (1841–1912) who spent much of his life studying Lake Leman (also known as Lake Geneva) near his home. In 1892, he suggested the word *limnology* for "the oceanography of lakes." Forel eventually wrote a four-volume monograph summarizing his studies of the lake, earning him the title "Father of Limnology." Today, the F.-A. Forel Institute in Versoix, Switzerland, carries on his tradition by sponsoring research in limnology, environmental geochemistry, ecotoxicology, and quaternary geology.

History

Limnological research was conducted informally even before Forel's studies of Lake Léman. Among the earliest of such studies was that conducted by the Swiss scientist Horace-Bénédict de Saussuré (1740–1799), who was the first scientist to measure the temperature at various depths in lakes. In 1850, the Swiss-American naturalist Louis Agassiz (1807–1873) published *Lake Superior: Its Physical Character, Vegetation and Animals, Compared with Those of Other and Similar Regions*, a monograph describing his travels along the north shore of the lake. In it, he described the geology, flora, fauna, and fossils found in and around the lake.

By the late nineteenth century, research in limnology was becoming more popular. In 1870, for example, the Swiss biologist P. E. Muller used a plankton net originally developed by Johannes Muller in 1843 to discover and study plankton in Swiss lakes. Forel also continued his limnological studies, conducting the first analysis of dissolved oxygen in a lake in 1885. In 1901, he wrote the first textbook on limnology, *Handbook of Limnology*.

One of the seminal works in limnology in the United States was *The Lake as a Microcosm*, published in 1887 by Stephen Forbes of the Illinois Natural History Survey. Forbes and his colleagues studied the interaction between science, the fisheries industry, and political pressures in regional lakes. In 1908, the first course in limnology was taught at the University of Wisconsin, which continues to host one of the preeminent programs in limnology in the United States.

Scope of Limnology

Limnology is an interdisciplinary subject that draws from a large variety of other scientific fields. Students who major in the subject or who choose to earn a Ph.D. in limnology are expected to have background information on a wide variety of subjects, such as general ecology, algae, physiological plant ecology, hydrology, water-resources engineering, water motions in lakes, air-sea interactions, hydrodynamics, environmental law, environmental toxicology, water-chemistry, ecology of fishes, plankton ecology, environmental microbiology, aquatic insects, water-rights law, river-basin planning, and water-use policy and planning. *See also* Aquatic Plants and Animals; Estuaries; Groundwater; Rivers and Streams

Further Reading

Brönmark, Christer, and Lars-Anders Hasson. *The Biology of Lakes and Ponds*. Oxford: Oxford University Press, 1998.

Kalff, Jacob. *Limnology*. Upper Saddle River, NJ: Prentice Hall, 2001.

Monson, Bruce A. *A Primer on Limnology*. 2d ed. St. Paul, MN: Water Resources Research Center, 1992.

Wetzel, Robert G. *Limnology: Lake and River Ecosystems*. San Diego: Academic Press, 2001.

Further Information

American Society of Limnology and Oceanography
(see website first)
c/o Dr. Denise Breitburg
The Academy of Natural Sciences
Estuarine Research Center
10545 Mackall Road
St. Leonard, MD 20685
Telephone: (410) 586-9711
Fax: (410) 586-9705
URL: http://www.aslo.org
e-mail: secretary@aslo.org

Literary References to Water

Water, ice, rain, oceans, rivers, and other aspects of our aquatic environment are common themes in books, poems, and other literary works. Many of those works deal with scientific and technical aspects of the subject, written for both the very youngest readers as well as specialists in a water-related field, and everyone in between. The bibliography at the conclusion of this book lists many examples of this type of work. The present entry focuses on references to water-related topics in fictional literature.

Novels and Short Stories

The adventures of men and women and boys and girls on rivers, lakes, and oceans have been the subject of countless numbers of novels and short stories in countries around the world for many years. Some examples of these tales are the following:

Moby Dick (1851), by Herman Melville (1819–1891), is a tale told by the narrator Ishmael about the search by Captain Ahab for a great white whale nicknamed Moby Dick, who has taken his leg in an earlier encounter. When Ahab does encounter the whale again, a three-day battle ensues in which the captain is killed by the whale and all of his crew are drowned when their boat capsizes and sinks. Only Ishmael himself survives the disaster. Melville wrote a number of other adventure stories about the sea, including *Typee* in 1846, *Omoo* in 1847, *Mardi* in 1849, and *Billy Budd*, published posthumously in 1924.

20,000 Leagues under the Sea (1869–1870), by Jules Verne (1828–1905), tells of the search by three men for a giant sea monster that has been attacking ocean-going shipping vessels. They discover that the "sea monster" is actually a submarine operated by one Captain Nemo who has developed a hatred of the dry land and everyone who lives on it. The book tells of the men and Nemo during the long year they spend together on the submarine *Nautilus*.

Typhoon (1903) was written by Joseph Conrad (1857–1924), who was himself a sailor in the British navy for seventeen years before devoting all his time to writing. *Typhoon* is an action story about the fate of the merchant ship *Nan Shan* as it is caught in the middle of a typhoon. The crisis is complicated by a riot that breaks out in the hold of the ship among Chinese laborers who are returning home. Among Conrad's other sea stories are *Children of the Sea* (reprinted as *The Nigger of the 'Narcissus'* [1897]), which tells of Conrad's experiences on a trip from Bombay to England; "Heart of Darkness" (1902), which tells of one of his voyages to Africa; and *Mirror of the Sea* (1906), which details his experiences after signing up as a sailor at the age of seventeen; and *The Rover* (1923), which is a story of a French sailor who plots to foil Admiral Nelson's attack against France.

The Sea Wolf (1904), by Jack London (1876–1916), chronicles the adventures of one Humphrey Van Weyden after his ship is involved in a collision in San Francisco Bay and is picked up by the sailing schooner *Ghost*. Van Weyden meets a wealthy industri-

alist, Wolf Larson, on the schooner, and they compete for the attention of poet Maude Brewster. After being shipwrecked and abandoned on a desert island, the three major characters are once again saved by a nearly destroyed *Ghost*, and two of the three escape from the island with their lives.

Mutiny on the Bounty (1932), by Charles Bernard Nordhoff (1887–1947) and James Norman Hall (1887–1951), is an account of one of the most famous mutinies in history, when Fletcher Christian and members of H.M.S. *Bounty* set Captain Bligh and his faithful crew members adrift in the South Pacific. Nordhoff and Hall followed up this hugely successful book with two books telling about the 3,600-mile voyage of Bligh and his eighteen loyal followers (*Men against the Sea* [1934]) and about the life of the mutineers who escaped from their ship and lived "happily ever after" (*Pitcairn's Island* [1934]).

Captains Courageous (1897), by Rudyard Kipling (1865–1936), tells the story of one Harvey Cheyne, the spoiled son of a millionaire, who falls overboard from an ocean liner off the coast of Newfoundland in the 1890s and is rescued by the crew of a fishing boat. Cheyne spends the next year as a member of the crew of the fishing boat before returning home, a wiser and more self-sufficient young man.

The Happy Return (1937), by Cecil Scott Forester (1899–1966), was the first in a series of books about one of the most famous naval heroes of all time, Horatio Hornblower. Forester wrote a total of fifty-six books, ten of which dealt with the adventures of the mythical naval officer who lived during the Napoleonic Wars. The books followed Hornblower's rise from a midshipman to admiral of the fleet in volumes such as *Captain Hornblower* (1939), *Lord Hornblower* (1946), *Mr. Midshipman Hornblower* (1950), *Lieutenant Hornblower* (1952), and *Hornblower and the Atropos* (1953). Forester's most famous book was probably *The African Queen* (1935), later made into a successful motion picture starring Katherine Hepburn and Humphrey Bogart.

The Old Man and the Sea (1952), by Ernest Hemingway (1899–1961), is a short novel about an aged Cuban fisherman, down on his luck, who snags a giant marlin while out fishing one day. After a long battle with the fish, it is devoured by a school of sharks, leaving the fisherman nothing but a skeleton to bring home as a reminder of his heroic struggle.

Beyond the Sea (1999), by contemporary author Mary Kingsley, is the tale of a young woman who travels to the arctic on a sailing ship and falls in love with the ship's captain.

Loch Ness Monster

See Sea Monsters

Lock

See Canals

Lourdes

See Sacred Waters

Luge

See Aquatic Sports, Luge

M

Magellan, Ferdinand (ca. 1480–1521)

Magellan was a Portuguese explorer who led the first expedition to circumnavigate the globe. Before he set out on this historic journey in 1519, Europeans were largely unaware of the planet's vast ocean resources and knew of the Pacific Ocean only through Vasco Balboa's (1475–1519) brief description of it from the shores of Panama.

Personal History

Magellan was born in Sabrosa, Trás-os-Montes, Portugal, in about 1480. In 1505, he made his first sea voyage, a trip to India to supply the Portuguese garrison there. In 1512, he took part in another expedition, this time to Morocco, where he was wounded and permanently crippled. He was denied a pension and was accused of treason, leading to his dismissal from the military in 1517.

Professional Accomplishments

Two years later, Magellan was able to convince Emperor Charles V of Spain to fund an expedition that would sail westward as far as possible to explore the parts of the globe awarded to Spain by an edict of Pope Alexander VI. Magellan left Spain on 20 September 1519 with five ships. He traveled across the Atlantic Ocean, down the eastern coast of South America, around the southern tip of the continent (now called the Strait of Magellan), and into the Pacific Ocean.

While traversing the Pacific, Magellan attempted to measure the depth of the ocean with a piece of rope more than 2,000 feet long. He was unable to reach the ocean bottom and concluded that he had found the deepest point in Earth's oceans.

Magellan's voyage later took him to Guam and the Philippine Islands, where he was killed by the natives. Eventually, four of his five ships were lost, but one managed to return across the Indian Ocean, around the southern tip of Africa, and back to Spain. *See also* Oceanography

Further Reading

"Ferdinand Magellan." *Catholic Encyclopedia.* Vol. 9. New York: Robert Appleton Company, 1910. Online at http://www.newadvent.org/cathen/09526b.htm.

Guillemard, F. H. H. *The Life of Ferdinand Magellan and the First Circumnavigation of the Globe, 1480–1521.* New York: AMS Press, 1971.

Hawthorne, Daniel. *Ferdinand Magellan.* Garden City, NY: Doubleday, 1964.

Parr, Charles McKew. *Ferdinand Magellan: Circumnavigator.* 2d ed. New York: Crowell, 1964.

Mar

See Atlantis

Mariculture

See Aquaculture and Mariculture

Marine Plants and Animals

See Aquatic Plants and Animals

Maury, Matthew Fontaine (1806–1873)

Maury has been called by one of his biographers the Father of Naval Oceanography and Meteorology. One of his most important discoveries was the presence of the Gulf Stream—"a river in the ocean," as he then described it. This discovery meant that ships could plot their course across the Atlantic in such a way as to make use of the Gulf Stream's energy rather than fighting against it.

Matthew Fontaine Maury, the American hydrographer and naval officer, ca. 1865. *Hulton-Deutsch Collection/Corbis.*

Personal History

Maury was born near Fredericksburg, Virginia, on 14 January 1806. He entered the U.S. Navy in 1825 as a midshipman and served on a ship that circumnavigated the globe between 1826 and 1830. In 1839, by then a lieutenant in the navy, Maury was injured in a stagecoach accident that made it impossible for him to continue on active duty. Instead, he was made superintendent of the navy's Depot of Charts and Information (later to become the U.S. Naval Observatory).

Professional Accomplishments

For some men, the post to which Maury was appointed would have become a sinecure to occupy one's time until retirement, a job with no important tasks or responsibilities to fulfill. Maury took his assignment very seriously, however. He decided to use his new position to carry out extended studies of the world's ocean winds and currents. He carried out these studies in part by providing sea captains with logbooks in which they were to record relevant water and weather data on their sailings. Maury used these data to prepare maps and charts showing ship captains the shortest and fastest ways to travel across the seas.

Maury also conducted studies of the ocean bottoms and was the first person to recognize the presence of a mountain range running along the floor of the Atlantic Ocean. That mountain range is now known as the Mid-Atlantic Ridge, and is known to be part of an extended underwater mountain range that girdles the globe. Maury eventually produced detailed seafloor maps of vast areas of the Atlantic, Pacific, and Indian Oceans. Based on his research, he was able to report the feasibility of laying a transatlantic telegraph cable, a project that was completed in 1858.

Maury realized that research on Earth's oceans was an international problem, not that of any one nation. In the mid-1850s, he became involved in planning the first marine international conference, held in Brussels in 1853. At about the same time, Maury also published the first textbook on oceanography, *The Physical Geography of the Sea* (1855).

With the outbreak of the Civil War in 1861, Maury left his post in Washington to take charge of coast, harbor, and river defenses for the Confederate Navy. At the conclusion of the war, he traveled to Mexico, where Emperor Maximilian encouraged him to establish a colony for war refugees from Virginia. When that plan came to naught, Maury returned to the United States and accepted a position as professor of physics at Virginia Military Institute. He held that post until he died on 1 February 1873 in Lexington, Virginia. *See also* Oceanography

Further Reading

Iselin, Columbus O'Donnell. *Matthew Fontaine Maury, 1806–1873, Pathfinder of the Seas; The Development of Oceanography*. New York: Newcomen Society in North America, 1957.

Lewis, Charles Lee. *Matthew Fontaine Maury*. 1927. Reprint, New York: Arno Press, 1980.

Williams, Frances Leigh. *Matthew Fontaine Maury: Scientist of the Sea*. New Brunswick, NJ: Rutgers University Press, 1963.

Mermaids and Mermen

Mermaids and mermen are creatures from mythology whose upper bodies resemble humans and whose lower bodies resemble large fish.

Non-Western Creatures

Although mermaids and mermen tend to be associated most commonly with the cul-

tures of Western civilization, comparable creatures can be found in the folklore of nations throughout the world. For example, the bynyip was a legendary creature in the folk history of many aboriginal tribes in Australia. It was a very large animal reputed to have lived in the swamps, lakes, and rivers of the Outback. It usually had no contact with humans, but was known to take one if very hungry.

A creature sometimes mentioned in Japanese mythology is the ningyo, which had the head of a human and the body of a fish. It was known to cry tears of pearl rather than water. Any woman who was able to capture a ningyo was guaranteed a life of eternal youth and beauty.

Another creature from Japanese folklore is the kappa, a water spirit with a special "liking" for children. It was said to have grabbed unsuspecting children walking near the water, taken them into the depths, and drowned them. The kappa was thought to have survived exclusively on cucumbers and the blood of children.

Western Creatures

By far the majority of mermaid and mermen tales are associated with the European continent. The earliest mention of such creatures occurs in folktales from the early Babylonians, although they were somewhat similar to gods and goddesses of the sea worshipped by that culture. Among the Syrians of a later date, a creature called Atargatis was worshipped as a moon goddess but depicted as a mermaid.

The first true mermaids were probably the Sirens, creatures from Greek mythology. They were described as beautiful women who lived on a rocky island and sang enchanting songs to sailors passing by. Hypnotized by the music, the sailors headed for the island and were usually shipwrecked. In the *Odyssey*, Homer describes how Odysseus ordered his men to pack their ears with beeswax to avoid hearing the song of the Sirens, thus saving their ship.

The most extensive collection of mermaid and mermen tales appears to come from the British Isles and Scandinavia. Some Norse legends, for example, tell about the water spirits known as nixes. Female nixes have the traditional appearance of mermaids, with beautiful faces and upper bodies and the tail of a fish. Male nixes, by contrast, can take on virtually any shape they like. The nixes are said to lure humans to the seas in which they live, though they don't necessarily intend to harm them.

In Russian mythology, mermaids and mermen are known, respectively, as the rusalka and the vodyany. The rusalka were known to tempt swimmers deeper into the water, where they were drowned, while the vodyany were more likely to follow fishermen and sailors, attempting to cause them harm.

A popular figure in Scandinavian folklore is the näcken. The name comes from the Swedish word *näck*, which means "nude." Näcken may take any form, including that of a horse, a bull, a cat, a dog, or a human. It often appears in the shape of a very attractive naked man playing a violin. The music he plays is so beautiful that unsuspecting humans are lured into the water and drowned.

Some version of mermaids and mermen has been reported in nearly every part of the British Isles bordering the ocean. Among the Irish, for example, they were called *merrows*, or *muirruhgach*—creatures who lived on dry land at the bottom of the oceans. They were thought to have magic caps that allowed them to travel from their homes in the sea to dry land, where they often worked mischief among humans. The female creatures were described as beautiful and enchanting, while the males were usually thought to be quite ugly, with green hair and a special liking for brandy.

A mermaid sculpture in the harbor of Copenhagen, Denmark. *Paul Almasy/Corbis.*

Along the northwest coast of Scotland existed a race of mermen-like creatures known as the Blue Men of Minch. These creatures were said to live in caves deep underwater and to be responsible for the many shipwrecks that occurred in the passage between Long Island and Shiant Island just off the coast. Legend says that the only way to avoid an attack by the Blue Men was for the captain of a passing ship to speak to the creatures in rhyme and always have the last word in the conversation.

The Mermaid in Modern Times

Mermaid stories have survived much longer than legends about other mythical features. Some authorities believe that the early Christian church was willing to acknowledge the possible existence of such creatures because they reinforced the church's position that women were the source of original sin. It used the comb and mirror usually associated with the mermaid as proof of a woman's inherent conceit and vanity.

In more recent times, scholars have pointed out that some sea creatures have just enough resemblance to humans to permit the continuation of legends about mermaids and mermen. Among these creatures are the manatee, dugong, porpoise, and seal.

Some unscrupulous individuals have taken advantage of the fascination that has existed among humans with mermaids and mermen. For many years, for example, all Japanese circuses and sideshows included a young woman fitted with an artificial tail to simulate culture images of the kappa, ningyo, or some other form of the mermaid. In the United States, showman P. T. Barnum for many years exhibited his version of the character in the form of the Feegee (for "Fiji") Mermaid.

In any case, stories about mermaids continue to be popular to the present day, although mermen seem largely to have disappeared from any form of modern literature. The story by Hans Christian Andersen (1805–1875), "The Little Mermaid," is now a classic of children's literature. Recent films incorporating a mermaid theme include *Mad about Men, Splash*, and Disney's *The Little Mermaid. See also* Gods and Goddesses of Water

Further Reading

The Historical Mermaid, http://rubens.anu .edu.au/student.projects/mermaids/folk-lores.html.

Mermaid Net . . . Where the Sirens Scream, http://www.mermaid.net/cgi-bin/ survey/earth.htm.

"Mermaids," http://www.vshost.net/fantasy/ LLFW/mermaids.html.

Motion Pictures about Water

See Water in Motion Pictures

Mu

See Atlantis

Munk, Walter (1917–)

Walter Munk is an oceanographer and geophysicist who has been associated with the Scripps Institute of Oceanography (SIO) for nearly half a century. His research has focused on tidal patterns, wave propagation, ocean currents, and global climate change, among other topics.

Personal History

Munk was born in Vienna, Austria, on 19 October 1917. He moved with his family to the United States in 1932, where he briefly worked in a bank owned by his grandfather. Munk earned his bachelor's degree in physics and his master's degree in geophysics at the California Institute of Technology in 1939 and 1941, respectively. He then attended the Scripps Institute of Oceanography at the University of California at San Diego, from which he received his Ph.D. in oceanography in 1947.

During World War II, Munk served in the military as a meteorologist for the Army Air Force and as an oceanographer at the University of California's Division of War Research. He was also a member of the U.S. Army's Ski Battalion. One of his most important contributions to the war effort was his work on tidal patterns in the English Channel prior to the invasion of Normandy in 1944. After the war, Munk was involved in the study of disruptions in ocean properties caused by nuclear bomb testing in the South Pacific.

Professional Accomplishments

At the completion of his graduate studies in 1947, Munk accepted a position as assistant professor of geophysics at SIO. He later became associate professor (1949) and professor of geophysics (1954) at the institute. In the latter year, he was also named to the staff of the University of California's Institute of Geophysics in Los Angeles. Six years later, he was chosen to create a new branch of the Institute in La Jolla, California—a branch whose mission it was to carry out research in the fields of atmospheric science, oceanography, and geology. Munk served for many years as the director of the new institute.

During the 1960s, Munk's research focused on the properties of ocean waves and tidal patterns. He developed methods for analyzing changes in wave patterns as they traveled over thousands of kilometers of ocean. He and his colleagues also invented methods using acoustical patterns for studying the movement of water along the ocean's floor. This research led to significant improvements in the prediction of tidal patterns as well as to a better understanding of the ways in which sound waves can be used to study the properties of ocean water.

Over the past decade, Munk has been especially interested in problems of global climate change. Many scientists suspect that carbon dioxide released by human activities on Earth is causing an increase in the average annual temperature of the planet. But verifying the relatively small temperature increases that may be occurring has been a difficult task. Munk and his colleagues have suggested one way for dealing with this problem in a project known as Acoustic Thermometry of Ocean Climate (ATEC).

ATEC makes use of low-frequency acoustic speakers placed at various depths underwater in locations throughout the Pacific Ocean. The audio signals sent out by these speakers can be picked up thousands of kilometers away with detectors placed off the coast of California. Minute changes in water temperature affect the pattern in which sound waves travel from speaker to detector, thus permitting scientists to measure very small temperature changes that are difficult to determine by other means.

Munk has received many honors and awards throughout his career. He was elected to the National Academy of Sciences in 1956 and to the Royal Society in 1976. In 1993, he received the Vetlesen Prize, one of the highest honors given in the field of earth sciences.

Music and Water

Water and music are interrelated in two major ways. First, some musical instruments use water as a means for producing sound. Examples of such instruments are the water gong, water drum, and water organ. Second, water has been taken as a theme by many different composers throughout history. In addition, water-related events are the major theme of sea shanties—traditional songs sung by sailors as they work.

Musical Instruments

According to composer Tan Dun, winner of the 2001 Oscar for best original score, there are more than thirty different ways of using water in some form or another to play music. Dun has made use of many of these ways in his 1999 composition Concerto for Water Percussion and Orchestra in Memory of Toru Takemitsu ("Composer Tan Dun to Perform in Taipei"). Some water instruments have been used since the earliest days of human civilization and are still popular in many cultures.

Water Gong

The water gong is the name given to a traditional gong or tam-tam that is struck and then submerged into water or, alternatively, struck while in water and then lifted out. In either case, changing from an air to a water environment or vice versa changes the pitch of the sound produced.

Water gongs are not a traditional part of classical scores, but they have become popular among modern composers. In addition to Tan's concerto, the instrument has been employed by composers such as Michael Pratt ("Dancing on the Wall," 1990), John Cage ("First Construction," 1939, and "Double Music," 1941), Annea Lockwood ("Water Gong," 1967), Garvin Bryars ("Four Elements," 1990), and George Crumb ("Mundus Canis," 1998).

Water Drum

Water drums have been popular musical instruments among Indian tribes in North and South America, Africa, and New Guinea for hundreds of years. The most common type of water drum is made from a small hollow log, barrel, kettle, crock, or other container covered with animal skin. Water is added to the container, and sound is produced by drumming on its surface. The pitch produced depends on the amount of water in the container.

A second type of water drum, especially popular in New Guinea, consists of a hollow object, such as a gourd, floating in a large container of water. Sound is produced by drumming on the object. Different pitches can be produced by using containers of different sizes and shapes.

Water Organ

A water organ is an instrument in which water is used to compress air, which is then forced through pipes to make sounds. Although the terminology used by various writers differs, two types of water organs can be distinguished: the *hydraulic organ* and the *hydraulis* (or *hydraulicus*). The hydraulic organ is an automatic instrument, while the hydraulis requires a human performer to produce sounds. Although no water organs exist today, a great deal of information about them is still available.

The hydraulic organ was described by the Greek writers Ktesibios (about 250 B.C.), Philo (third century B.C.) and Hero (A.D. 62).

It was one of many similar inventions capable of replicating the sounds of trumpets, harps, birds, and other objects, and was widely popular among the Greeks. The principle of the water organ later became known to the Arabs. A treatise by one Banu Musa in the ninth century, for example, described an elaborate instrument constructed in the shape of a man with an organ pipe in his hand and a reed pipe in his mouth, both powered by water-driven compressed air.

Hydraulic organs first appeared in Europe during the thirteenth century, where they took a variety of shapes. In some cases, inventors simply reproduced or adapted instruments familiar to the Greeks and Arabs. In other cases, they designed new and more imaginative devices based on the hydraulic organ principle. For example, many members of the upper class had hydraulic organs built into gardens, grottos, or other natural formations where they were powered by waterfalls, streams, or other sources of water. By the end of the seventeenth century, however, interest in hydraulic organs had begun to wane; therefore, no new instruments were being built, and those in existence were allowed to fall into disrepair.

A water organ described by Vitruvius in about A.D. 20 illustrates the principles on which this instrument works. A vessel (such as a wooden box) is filled with air and suspended inside a larger container filled with water. The air vessel has holes in the bottom through which water from the surrounding container can flow in and out. A pipe leading into the air vessel is supplied with air from an exterior pump.

When the exterior pump is in its "up" position, air is pushed into the interior vessel, forcing water out of the vessel and into the surrounding container. When the pump is in its "down" position, air escapes from the interior vessel, allowing water from the surrounding container to flow into it. As water flows into the air vessel, it compresses the air and forces it out the top of the vessel into a series of pipes. A performer may then release air through one or more of the pipes by pushing on a key that releases air through a valve into the pipe.

Musical Themes

Water has been a theme in all types of musical compositions over many hundreds of years. The following examples are only suggestive of the great variety of classical, folk, popular, and other musical categories in which the mention of water occurs. In some cases, the motivation for using water as a musical theme is very clear. In other cases, however, there is no clear connection between the title of a musical composition and the subject of water itself.

Classical

Water Music (1717), by George Frideric Handel (1685–1759), was composed to be played during a cruise up the River Thames from London to Chelsea on a royal barge. The piece was composed in honor of and played for the enjoyment of King George I, who had acceded to the throne in 1714. A historian of the time wrote that the king's approval was "so great that he caused it to be played three times in all, twice before and once after supper, even though each performance lasted an hour" (Musical Heritage Society).

La Fille mal gardée (1786), by Louis-Joseph-Ferdinand Hérold (1791–1833), contains one of the earliest sections in which orchestral instruments are used to represent the sound and fury of the lightning and thunder that accompany a storm. Similar examples found in later compositions include those in Christoph Willibald Gluck's (1714–1787) *Don Juan* (1761), Gioacchino Rossini's (1792–1868) *William Tell* (1829), Gaetano Donizetti's (1797–1848) *Lucia di Lammermoor* (1835), Ludwig van Beethoven's (1770–1827) *Creatures of Prometheus*

(1800) and the Sixth Symphony (1808), and Richard Strauss's (1864–1949) *Alpine Symphony* (1915).

Fingal's Cave (*Hebrides* Overture; 1830–1832), by Felix Mendelssohn (1809–1847), pays tribute to a spot located on the Island of Staffa, off the west coast of Scotland. It is one of the most famous caves and most famous seafront locations in the British Isles, a place where crashing waves present a spectacular sight that is captured musically by Mendelssohn in this work. A year earlier, Mendelssohn had written another sea piece, *Calm Sea and Prosperous Voyage*, based on two poems by the German poet Goethe.

The Gondoliers (or *The King of Barataria*; 1889), *H.M.S. Pinafore* (1878), and *The Pirates of Penzance* (or *The Slave of Duty*; 1879) are three operettas written by Sir W. S. Gilbert (1836–1911) and Sir Arthur Sullivan (1842–1900) based on nautical themes. In *Gondoliers*, two recently married gondoliers are told that one of them—though it is not known which—is actually the King of Barataria. Both leave their occupation as gondolier to return to that country to subdue a rebellion and bring peace to the nation. *Pinafore* satirizes nautical life in the Royal Navy, from the lowliest seaman (a handsome sailor named Ralph) to the "leader of the navy" (Sir Joseph Porter). In typical Gilbert and Sullivan style, the two men were actually exchanged at birth, making Ralph a member of the nobility and Sir Joseph a commoner. *Penzance* tells the story of pirate life off the coast of Cornwall, again complicated by the fact that a member of the pirate crew had been assigned to the ship accidentally. His nursemaid had arranged to have him apprenticed to a "pirate" ship rather than a "pilot" ship.

Scheherazade (1888), by Nicolai Rimsky-Korsakov (1844–1908), tells the story of the Sultan Shahriar, who, convinced that all women are unfaithful, resolves to murder each of his wives after their first night together. The Sultana Sheherazade foils this plan by telling her husband a series of tales, the ending of which is delayed until the following night. The Rimsky-Korsakov work is in four movements: "The Sea and Sinbad's Ship," "The Story of the Kalender Prince," "The Young Prince and the Young Princess," and "Festival at Baghdad—The Sea—The Ship Goes to Pieces against a Rock Surmounted by a Bronze Warrior." In spite of the titles, the composer apparently had little or no programmatic story line in mind when he wrote the piece, and the music describes the titles only in the broadest sense.

The Swan of Tuonela (1893), by Jean Sibelius (1865–1957), is one of four tone poems in the composer's *Four Legends*, or *Lemminkainen* Suite, which tells of the adventures of the Finnish national hero by that name. *The Swan* was reworked from material originally intended as the introduction to a grand opera (never written) called *Building of the Boat*, based on Finland's national epic the *Kalevala*. *The Swan* describes a magical swan as it swims in the black waters of the Kingdom of Death.

Sea Pictures (1899), by Sir Edward Elgar (1857–1934) is a collection of five songs based on poems by five different writers, one of whom was his wife Alice. The songs are entitled "Sea Slumber Song," "In Haven—Capri," "Sabbath Morning at Sea," "Where Corals Lie," and "The Swimmer."

Sea Drift (1903), by Frederick Delius (1862–1934), is written for baritone, chorus, and orchestra, and based on Walt Whitman's romantic poem "Out of the Cradle Endlessly Rocking."

La Mer (1905), by Claude Debussy (1862–1918), contains three "symphonic sketches" based on sea-related phenomena: "From dawn to midday on the sea," "Play of the waves," and "Dialogue of the wind and the sea."

Sea Symphony (1903–1909), by Ralph Vaughn Williams (1872–1958), is a powerful expression of life and death on the oceans, based on Walt Whitman's "Leaves of Grass." The work's themes are based on Whitman's major sections, "A Song for All Seas, All Ships," "On the Beach, Alone," "The Waves," and "The Explorers." The symphony is scored for soprano, baritone, chorus, and orchestra.

Fountains of Rome (1917), by Ottorino Respighi (1879–1936), wrote that his composition was intended to "give expression to the sentiments and visions suggested to him by four of Rome's fountains, contemplated at the hour in which their character is most in harmony with the surrounding landscape." The four movements describe "The Fountain of the Villa Giulia at Dawn," "The Triton Fountain in the Morning," "The Fountain of Trevi at Midday," and "The Villa Medici Fountain at Sunset."

Mississippi Suite (1924), by Ferde Grofé (1892–1972), was written for band leader Paul Whiteman and introduced at New York City's Carnegie Hall in 1925. The work makes use of Native American Indian, jazz, and African American themes. Its four movements are entitled "Father of the Waters," "Huckleberry Finn," "Old Creole Days," and "New Orleans at Carnival Time."

Suite from *The River* (1937), by Virgil Thomson (1896–1989), was written for a documentary film produced under the auspices of the WPA (Works Progress Administration) and directed by Pare Lorentz. The four movements are entitled "The Old South," "Industrial Expansion in the Mississippi Valley," "Soil Erosion and Floods," and "Finale."

Four Sea Interludes (1945) are musical selections that introduce, separate, and connect the seven scenes in Benjamin Britten's (1913–1976) opera *Peter Grimes*. The opera tells the story of Peter Grimes, a fisherman whose goal in life is to wed the widowed schoolmistress, Ellen Orford. The interludes are entitled "Dawn," which links the prologue and first scene; "Storm," which appears in the middle of act 1; "Sunday Morning," the prelude to act 2; and "Moonlight," the prelude to act 3.

Victory at Sea (1952), by Richard Rodgers (1902–1979), was written to accompany a twenty-six-part television documentary about the part played by the U.S. Navy during World War II. The score consists of a number of suite-like compositions with titles such as "Guadalcanal March," "Theme of the Fast Carriers," "Fire on the Waters," "D-Day," "Song of the High Seas," "Hard Work and Horseplay," and "The Pacific Boils Over." The work was highly successful, resulting in the production of three soundtrack albums and one CD with twelve of the most popular themes.

The Death of Klinghoffer (1991), by John Adams (1947–), is based on a historic event—the hijacking of the cruise ship *Achille Lauro* in 1985. During the opera, Islamic terrorists who have taken control of the ship explain the reasons behind their actions, while their captive passengers comment on the situation in which they find themselves. The only casualty of the hijacking, Klinghoffer, is killed offstage while his wife is singing about disability, illness, and death.

Folk and Popular Music

Songs that refer to lakes, rivers, streams, oceans, rain, and other water-related phenomenon go back to some of the earliest music of which we know. The following are only a few of the countless examples that could be cited.

"Logan Water" is a Scottish folk song that first appeared in print in 1709. The song has had different lyrics throughout its history, one set composed by Robert Burns (1759–1796) in 1793. The current lyrics tell of a woman whose lover has left her to "face his faes, far, far, frae me and Logan braes."

"The Water Is Wide" is a Scottish folk song originally called "Waly, Waly." It was first published in 1724 and remains one of the most popular and frequently performed folk songs of the day. It tells a sad story of an unfaithful lover who has deserted the songwriter, although he vows not to complain because "Some day, I hope to love again."

"On the Banks of the Allan Water" is a Scottish folk song of indeterminate age that tells of a "miller's lovely daughter" whose soldier lover was false so that "left alone was she." The song ends telling of the daughter's grave on the banks of Allan Water, a river in Scotland.

"The Lakes of Pontchartrain" is a traditional song about a traveler from New Orleans who falls in love with a Creole girl "by the lakes of Pontchartrain." When he asks her to marry him, she declines because she already has a lover "and he was far at sea."

"Old Folks at Home" is probably better known by the first line of its lyrics: "Way down upon the Suwannee River." The song was composed by Stephen Foster (1826–1864) in 1851 and expresses the sentiments of a black man thinking back on his happy days on the plantation: "Sadly I roam, still longing for de old plantation, and for de old folks at home."

"A Hard Rain's A-Gonna Fall," by Bob Dylan, was first released in 1963. The song uses the analogy of a "hard rain" falling to represent a variety of human woes, including "the song of a poet who died in a gutter," "one man who was wounded in love," and a place "where the pellets of poison are flooding their waters."

"Rain" (1966) is one of the Beatles most famous songs, although it was originally recorded on the B side to "Paperback Writer." Professor Gary Kendall of Northwestern University, an expert on the group, writes that "John's [Lennon] lyrics are influenced by his reading of Buddhism. 'Rain' and 'shine' represent a fundamental dualism. John is saying that you don't have to respond to these in an unthinking way like other people. There is greater awareness in stepping beyond . . . these dualisms" (http://www.northwestern.edu/musicschool/classes/beatles/projects/rain).

"Fire and Rain," written in 1970 by James Taylor, has become a theme for this songwriter-singer's life and the title of his biography, published in 2000. The theme of the song is expressed in the chorus: "I've seen fire and I've seen rain. I've seen sunny days that I thought would never end. I've seen lonely times when I could not find a friend."

"Purple Rain" was the hit song and defining story of a 1984 Warner Brothers film by the same name, performed by the artist Prince. The film and song are reputedly based on autobiographical themes about the artist's childhood. In the song, the singer tells his lover that he "only wanted to one time see you laughing. I only wanted to see you laughing in the purple rain."

"November Rain," a Guns N' Roses 1991 song, tells of two lovers trying to work through their relationship, a "love restrained." The singer acknowledges that "nothing lasts forever and we both know hearts can change and it's hard to hold a candle in the cold November rain."

Sea Shanties

Sea shanties (also *sea chanties*) are work songs sung by sailors on sailing ships while carrying out their routine tasks, such as raising or setting an anchor or raising or lowering sails. Shanties were sung not just as a form of amusement or to pass the time of day, but to coordinate routine tasks that required cooperation among a group of men.

The word *shanty* (or *chantey*) is thought to have been derived from the French word *chanter*, meaning "to sing."

The shanty was led by one person, the shantyman, who chose the song and set the tempo for the job. Specific emphases were placed on certain words corresponding with specific parts of the task, such as pushing, pulling, or lifting. Different rhythms were used for different types of tasks. For example, halyard shanties were sung while raising and lowering sails, while short drag shanties were used with more difficult tasks, such as raising a masthead. In spite of their basic association with work, shanties were also sung during free time and during special ceremonies, such as crossing the equator.

Examples of some common sea shanties include "Blow the Wind Westerly," "Common Sailors," "Drunken Sailor," "Heave Away," "Homeward Bound," "One More Day," "Strike the Bell," and "Roll the Woodpile Down."

Further Reading

Colcord, Joanna C. *Songs of American Sailormen*. 1938. Reprint, New York: W. W. Norton, 1967.

"Composer Tan Dun to Perform in Taipei," *China Post*, 15 May 2001. http://th.gio.gov.tw/show.cfm?news_id=8953.

DeSolla Price, Derek. "Automata in History: Automata and the Origins of Mechanism and Mechanistic Philosophy." *Technology and Culture*, winter 1964, 9–23.

Farner, Henry George. *The Organ of the Ancients*. London: William Reeves, 1931.

Hugill, Stan. *Shanties and Sailors' Songs*. New York: F. A. Praeger, 1969.

Musical Heritage Society, http://www.musicalheritage.com/CLASSICAL/digsinfo.sp?RECORD_NUMBER=5356&.

N

Naiads

See Nymphs

Nargile

See Hookah

Nargilla

See Hookah

National Ice Center

The National Ice Center (NIC) is a collaborative agency representing the U.S. Navy, the National Oceanic and Atmospheric Administration (NOAA), and the Coast Guard. The purpose of the center is to provide data and forecasts on the size, location, and movement of large ice masses in the Arctic and Antarctic Seas, the Great Lakes, and the Chesapeake Bay. The information is made available to the armed forces of the United States and allied nations, the Departments of Commerce and Transportation, the general public, and U.S. and international government agencies. Data are primarily collected through aerial surveys conducted by per-

sonnel affiliated with the Naval Ice Center and from a variety of satellites.

The beginnings of the National Ice Center can be traced to 1956 when NOAA and U.S. Fleet Weather Central, a division of the U.S. Navy, began to collaborate on ice research and reporting. With the growth of data produced by satellite observations, this collaboration grew and expanded to include other government agencies. The collaboration was formally acknowledged as the Joint Ice Center in 1976. In 1995, the center's name was officially changed to the National Ice Center.

Over the years, the center has been involved in some important projects, including the attempted recovery of the submarine USS *Thresher* in 1964 and the Apollo 12 and 13 lunar missions. Today, the center provides important data used in the study of global climate change in ocean properties.

Further Information

National Ice Center
Federal Building #4
4251 Suitland Road
Washington, DC 20395
(301) 457-5305
URL: wysiwyg://right/63/http://www.natice.noaa.
    gov/mainone.htm

National Oceanic and Atmospheric Administration

The National Oceanic and Atmospheric Administration (NOAA) was created within the U.S. Department of Commerce on 3 October 1970 by the Reorganization Plan Number 4 of 1970. NOAA was assigned responsibilities previously held by a number of other federal agencies, primarily the National Ocean Service (NOS), the National Weather Service (NWS), and the National Marine Fisheries Service (NWFS). The administration was later assigned additional responsibilities in the Coastal Zone Management Act of 1972; the Marine Mammal Protection Act of 1972; the Marine Protection, Research, and Sanctuaries Act of 1972; the Weather Modification Reporting Act of 1972; the Endangered Species Act of 1973; the Offshore Shrimp Fisheries Act of 1973; the Fishery Conservation and Managements Act of 1976; and the Land Remote Sensing Act of 1984.

NOAA's mission covers a number of areas, such as:

- exploring, mapping, and charting the ocean and its living resources, and the management, use, and conservation of those resources;
- describing, monitoring, and predicting conditions in the atmosphere, ocean, Sun, and space environments;
- issuing warnings against impending destructive natural events;
- assessing the consequences of inadvertent environmental modification over several scales of time;
- managing and disseminating long-term environmental information.

The administration has defined seven strategic goals through which it carries out the above mission. These seven goals are

1. providing advance short-term weather warnings and forecast services;
2. implementing programs of seasonal to interannual climatic forecasts;
3. assessing and predicting decadal to centennial climatic changes;
4. promoting safe navigation;
5. building sustainable fisheries;
6. helping the recovery of protected species;
7. sustaining healthy coastal ecosystems.

For additional information on specific NOAA programs and activities, refer to the agency's home page, which is divided into a number of "theme" pages. These theme pages describe in more detail NOAA's work in areas such as the Arctic, atmospheric modeling and mapping, aviation weather, drought, fire weather, El Niño and La Niña, hurricanes, lightning, marine forecasts, ozone levels, river forecasts, volcanic ash, and weather emergencies.

Further Information

National Oceanic and Atmospheric Administration
14th Street & Constitution Avenue, N.W., Room 6013
Washington, DC 20230
Telephone: (202) 482-6090
Fax: (202) 482-3154
URL: http://www.noaa.gov
e-mail: answers@noaa.gov

National Water Research Institute

The National Water Research Institute (NWRI) is Canada's largest freshwater research institute. Its mission is to develop scientific information on which government policies and programs, public decision making, and early identification of environmental problems can be based. The institute works with both international and other Canadian agencies to carry out this mission.

Two major areas of research currently being conducted are the Canadian Aquatic Biomonitoring Network and the Endocrine Disrupting Substances Groundwater study. NWRI works closely with a number of other organizations interested in water-quality issues, including the Canadian Glacier Information Center, the National Laboratory for Environmental Testing, and the *Water Quality Research Journal of Canada*. NWRI's work is carried out at two locations, listed below.

Further Information

Canada Centre for Inland Waters
P.O. Box 5050
867 Lakeshore Road
Burlington, ON L7R 4A6
Telephone: (905) 319-6900

National Hydrology Research Centre
11 Innovation Boulevard
Saskatoon, SK S7N 3H5
Telephone: (306) 975-5717
URL: http://www.cciw.ca/nwri

Nekton

See Plankton

Neptune

See Poseidon

Nereids

See Nymphs

1998 International Year of the Ocean

In 1993, the Intergovernmental Oceanographic Commission (IOC) of the United Nations Educational, Scientific, and Cultural Organization (UNESCO) passed a reso-

lution calling for the creation of an International Year of the Ocean (IYO). The resolution was adopted by the UN General Assembly, and the year 1998 was set aside for this event.

The IYO was established with three general objectives in mind:

- to raise awareness of the oceans and coastal areas as finite-sized economical assets;
- to obtain commitments from governments to take action, provide adequate resources, and give the oceans the priority which they deserve;
- to emphasize that it is only through global scientific cooperation that we can begin to improve our understanding of how the oceans work (http://www.unesco.org/opi/eng/98iyo/iyo-98.htm).

Some of the themes around which IYO activities are organized include maritime transportation; national security; ocean resources; marine environmental quality; recreation and tourism; and weather, climate, and natural hazards.

Planning and implementation of IYO activities was under the general direction of the IOC. The commission was established by UNESCO in 1960 to "develop, recommend, and coordinate international programmes for scientific investigation of the oceans and to provide related ocean services to Member States" (http://www.unesco.org/opi/eng/98iyo/ioc.htm). A number of other intergovernmental, national, scientific, and other organizations worked with IOC in carrying out IYO programs. These included the United Nations Food and Agricultural Organization, the International Atomic Energy Agency, the International Marine Organization, the United Nations Environment Programme, and the World Meteorological Organization. In the United States, the

National Oceanic and Atmospheric Administration took the lead role in promoting the IYO in this country.

A large number and variety of activities were carried out as part of the IYO. The most prominent may have been EXPO '98, held in Lisbon, Portugal, from 22 May to 30 September 1998. The central theme of the exposition was "The Oceans: A Heritage for the Future." The United Nations and many of the other 145 participants in the exposition chose ocean-related topics as themes for their pavilions. Overall, more than 15 million tourists visited the exposition.

The IYO also served as an impetus to the writing and publication of "The Ocean Charter," a statement of principles about the oceans developed at a conference held in St. John's, Newfoundland, in September 1997. The charter can be accessed on the IYO page at http://www.unesco.org/opi/eng/98iyo/charter.htm. Other activities included in the ocean year were educational programs for children and the general public; an assessment of the best scientific information currently available about the oceans; the printing and distribution of special ocean-related stamps; a variety of special awards to individuals and organizations who have made contributions to our understanding of the ocean; at least thirty scientific conferences on topics such as marine pollution, coastal erosion, ocean circulation, and the problems of urban growth; more than a dozen research and training cruises, many of which included stops at ports to raise public awareness on ocean issues; and a host of publications on the oceans, including a multivolume encyclopedia entitled *Humanity and the Sea* and an electronic document, *United Nations Interactive Atlas of the Oceans*. A complete list of activities and events carried out in conjunction with the IYO can be found at http://www.unesco.org/iyo/toc.htm.

Further Reading

Home page for the 1998 International Year of the Ocean is http://www.unesco.org/opi/eng/98iyo.

A list of some major publications resulting from the IYO can be found at http://ioc.unesco.org/iyo/activities/publications.htm.

A good overview of IYO can be found in *International Year of the Ocean: Fact Sheets, Guide to Additional Resources*. Washington, D.C.: U.S. Department of Commerce, National Oceanic and Atmospheric Administration, 1998.

Nymphs

The nymphs are a group of minor goddesses prominent in Greek mythology and, in similar forms, the mythology of a few other cultures. The nymphs were said to populate the mountains, trees, plants, woods, and waters, and to have relationships with both deities and humans. They are generally described as beautiful naked women, who remain young in their appearance no matter how long they live.

The water nymphs can be divided into three large categories—Naiads, Nereids, and Oceanids—according to the type of water in which they were found. The Naiads lived in freshwater and were further subdivided according to the type of water in which they could be found. For example, those who lived in fountains were called the Crinaeae; those found in springs were the Pegaeae; those who populated the lakes were the Limnatides; those who lived in rivers were the Potameides; and those found in the marshes were the Eleionomae. In most stories, Oceanus and Tethys are said to be the parents of the Naiads.

The Nereids were saltwater nymphs, the daughters of the deities Doris and Nereus. The Oceanids were daughters of Tethys and Oceanus and were found in the open seas.

The attributes possessed by the nymphs varied somewhat from story to story. They were generally not thought to be immortal, but lived only as long as the location in which they were found survived. Plutarch, for example, claims that the average lifespan of a Naiad was 9,620 years, and that the goddesses died when their water source dried up or was displaced.

Nymphs were often thought to have supernatural powers that allowed them to make prophecies and to provide protection for plants, animals, and humans in their special environment. The Nereids, for example, were thought to provide special protection for sailors, with whom they often became lovers.

Scholars do not agree as to the exact number of nymphs of each category, but the names of no more than about ten Oceanids and as many as fifty Nereids have been tabulated.

Many stories in Greek mythology tell of the love relationships between nymphs and humans and deities. Apollo, in particular, seems to have been associated with a number of nymphs. The nymphs are generally depicted as kind, loving, and compassionate figures although, when aroused, they are also described as having the capacity to bring down the power of natural punishments, such as thunder, lightning, and storms, on their adversaries.

The Rusalki are nymphlike figures found in both Slavonic and Russian mythology. They are water nymphs generally thought to be the spirits of drowned girls. In some cases, they are depicted as beautiful young girls who try to lure unsuspecting travelers into the water by singing a magical song to them. In other cases, they are described as ugly hags who attempt to seize unsuspecting travelers and drag them into the river, where they are drowned. *See also* Gods and Goddesses of Water

Further Reading

Larson, Jennifer. *Greek Nymphs: Myth, Cult, Lore*. Oxford: Oxford University Press, 2001.

O

Oceanids

See Nymphs

Oceanography

Oceanography is the study of the oceans, focusing on the structure and properties of the ocean bottoms, the physical and chemical properties of seawater, the marine life found in the oceans, and the uses to which ocean resources can be put.

History

Given their importance as water highways and as a source of fish, shellfish, and other types of food, it is hardly surprising that the oceans have been the subject of intense study from the earliest days of humankind. Such early research, however, was of an entirely practical nature, with relatively little concern for theoretical issues. The early Greeks were probably the first humans to consider abstract questions, such as the origin of the ocean and the cause of the tides. Such questions were of purely philosophical concern, however, and were associated with little or no actual experimentation.

Most of the first experimental research in oceanography was carried out as part of the great explorations that began in the six-teenth century. During his trip around the world, for example, the Spanish explorer Ferdinand Magellan (ca. 1480–1521) attempted to determine the depth of the Pacific Ocean using a rope more than a kilometer (a half mile) in length. The rope, however, was not long enough to reach the ocean bottom.

Another famous explorer, Captain James Cook (1728–1779), charted many parts of the Pacific and Atlantic Oceans; mapped the general outlines of both oceans; carried out studied of wind, currents, and water temperature; studied a number of coral reefs; and conducted extensive astronomical studies on his voyages.

One of the most challenging problems facing early oceanographers was determining the depth of the ocean and the general shape of the ocean bottom. The first successful research on this problem came in 1818 when British explorer Sir John Ross (1777–1856) measured the depth of one portion of Baffin Bay off Greenland as 1,050 fathoms (6,300 ft). Sir John was also able to collect a sample of mud from the ocean bottom at this point. Ross's nephew, Sir James Clark Ross (1800–1862), improved on his uncle's research and obtained soundings of 2,425 fathoms (14,500 ft) in the South Atlantic and 2,677 fathoms (16,060 ft) off the Cape of Good Hope.

The person credited with being the Father of Modern Oceanography is American naval officer Matthew Fontaine Maury (1806–1873). Maury compiled analyses of wind and current patterns, published maps of those patterns, and authored the first textbook on oceanography, *The Physical Geography of the Sea*.

An important factor in the early development of oceanography as a science was a series of extensive scientific expeditions. The first of these, the U.S. Exploring Expedition, had as its objective a survey and exploration of the Pacific Ocean for the purpose of increasing commercial prospects for U.S. companies. The expedition took three years and covered nearly 90,000 miles. Unfortunately, most of the scientific information collected on the expedition was either lost or misplaced.

Probably the most famous and most productive expedition of the nineteenth century was the Challenger Expedition, sponsored by England's Royal Society and the British Royal Navy. Under the direction of Sir Charles Wyville Thomson (1830–1882), the expedition lasted four years, traveled 68,890 miles, and circumnavigated the globe. The expedition traveled on the steamship *Challenger*, which carried a crew of 243 and a scientific team of 6 members. The team completed 492 deep soundings and 133 dredgings, and collected data from 362 oceanographic stations. More than 4,700 new species of marine life were discovered, and the deepest part of the ocean, the Marianas Trench at 4,475 fathoms (26,850 ft), was discovered.

Later Advances

By the 1920s, oceanographers had been provided with a new technique for the study of the ocean bottom: echo sounding. German physicist Alexander Behm (1880–1952) invented a method for bouncing sound waves off the ocean bottom. The time of transit of the sound waves provided a method for determining the depth of the ocean bottom. Behm first used this procedure in 1920 to determine the depth of various parts of the North Sea.

Behm's method was later put to use by oceanographers in all parts of the world. In 1922, for example, the U.S. destroyer escort *Stewart* traveled across the Atlantic Ocean, making more than nine hundred echo soundings. Three years later, the most extensive study of the ocean floor to date was conducted by the German expedition *Meteor*. The *Meteor* traveled back and forth across the South Atlantic thirteen times between 1925 and 1927, making 67,400 soundings and conducting studies of ocean currents, salinity, temperature, and oxygen content. In addition, the expedition carried out studies of marine life, atmospheric properties, and geologic phenomena.

The development of modern oceanography has also been dependent on the development of major research institutions, such as the Scripps Institute of Oceanography in La Jolla, California, founded in 1903, and the Woods Hole Oceanographic Institute, established in 1930.

Improved technology over the past half century has led to a number of important discoveries about the oceans and the ocean bottoms. In 1930, for example, American naturalist Charles William Beebe (1877–1962) invented a device for deep-sea diving known as the *bathysphere*. In 1934, Beebe and his companion Otis Barton descended to a depth of 505 fathoms (3,028 ft) in a bathysphere, breaking all previous records for deep-sea diving by humans. The bathysphere was a large steel sphere suspended by a steel cable from a ship at the surface.

Nearly two decades later, a modification of Beebe's bathysphere was developed by Swiss physicist Auguste Piccard (1884–1962). Piccard had already made his fame in the use of balloons to study the upper atmosphere when he decided to extend his studies in the opposite direction, toward the ocean's bottom. In the late 1940s, he developed a bathysphere-type vessel that was able to travel under its own power underwater, a vessel he named the *bathyscaphe*. In 1954, two French naval officers used Piccard's bathyscaphe to descend to a depth of 2,214.5 fathoms (13,287 ft), or about 2½ miles, off the Mediterranean coast of Africa. By 1960, Piccard and his companion Don Walsh used an improved version of their bathyscaphe, the *Trieste*, to dive to the bottom of the Challenger Deep, more than seven miles below the water's surface.

Subject Matter

Oceanographers are interested in every aspect of Earth's oceans and ocean bottoms. They study current speed and direction; wave height, length, and direction; temperatures, light intensity, oxygen content, and other properties at different depths and different parts of the ocean; composition of the ocean bottom; and plant and animal life found in the oceans.

Some idea of the scope of oceanography can be gained by a review of the types of courses professional oceanographers are expected to complete, or may select to take, during their formal education. In addition to basic courses in physics, chemistry, biology, and earth sciences, these courses include topics such as general oceanography, marine pollution, physical oceanography, chemical oceanography, geological oceanography, biological oceanography, marine microbiology, aquatic ecology, geophysical fluid dynamics, waves, tides, physical chemistry of seawater, organic geochemistry, subduction zones, global paleoclimatology, plankton paleoecology, marine stratigraphy, deep-sea sediments, marine geophysics, paleomagnetism and geomagnetism, phytoplankton taxonomy, phytoplankton physiol-

ogy, marine bio-optics and remote sensing, zooplankton, marine fish ecology and production, fish population dynamics, costal marine ecosystems, animal communication, and marine zooplankton ecology. *See also* Cook, James; Magellan, Ferdinand; Maury, Matthew Fontaine; Oceans and Seas; Submarines and Submersibles

Further Reading

Duxbury, Alison, Alyn C. Duxbury, and Keith A. Sverdrup. *Fundamentals of Oceanography*. New York: McGraw-Hill, 2001.

Garrison, Tom. *Oceanography: Invitation to Marine Science*. Belmont, CA: Brooks/Cole, 2001.

Nybakken, Jim, ed. *Encyclopedia of Oceanography and Marine Science*. London: Fitzroy Dearborn, 2001.

Schlee, Susan. *The Edge of an Unfamiliar World: A History of Oceanography*. London: Hale, 1975.

Oceans and Seas

The oceans and seas are large bodies of salt water covering about 71 percent of Earth's surface. At one time, mariners referred to "the seven seas," by which they meant the Atlantic, Pacific, Indian, and Arctic Oceans; the Mediterranean and Caribbean Seas; and the Gulf of Mexico. Today, oceanographers tend to regard all oceans and seas as part of three major systems: the Atlantic, Pacific, and Indian oceans. Each ocean system includes a number of smaller tributaries. Tributaries associated with the Indian Ocean, for example, include the Red Sea, Persian Gulf, Arabian Sea, Bay of Bengal, Andaman Sea, and Great Australian Bight.

Characteristics

The chart below summarizes some major characteristics of the three ocean systems. The chart includes the Arctic and Antarctic

Characteristics of Three Ocean Systems

Ocean	Area (10^6 km^2)	Volume (10^6 km^3)	Mean Depth (m)
Arctic	14,090	17.0	1,205
North Atlantic	46,772	153.6	3,285
South Atlantic	37,364	152.8	4,091
North Pacific	83,462	322.0	3,858
South Pacific	65,521	254.9	3,891
Antarctic	32,249	120.3	3,730

Sea	Ocean System	Area (10^6 km^2)	Volume (10^6 km^3)	Mean Depth (m)
Mediterranean Sea	North Atlantic	2,516	3,758	1,494
Caribbean Sea	North Atlantic	2,754	6,860	2,491
Gulf of Mexico	North Atlantic	1,543	2,332	1,512
Hudson Bay	North Atlantic	1,232	158	128
Arabian Sea	Indian Ocean	3,863	10,561	2,734
Bay of Bengal	Indian Ocean	2,172	5,616	2,586
Bering Sea	North Pacific	2,304	3,683	1,598
Okhtosk Sea	North Pacific	1,590	1,365	859
Coral Sea	South Pacific	4,791	11,470	2,394
Arafura Sea	South Pacific	1,037	204	197
South China Sea	North & South Pacific	3,685	3,907	1,060
Banda Sea	North & South	695	2,129	3,064

Oceans and divides both the Atlantic and Pacific Oceans into northern and southern portions. It also lists some of the major tributaries to each of the oceans.

Ocean Environments

The marine environment can be divided into two major categories: benthic and pelagic. The *benthic* environment refers to the physical conditions and plant and animal life on the ocean bottoms. The *pelagic* environment refers to all of the ocean waters overlying the seafloor. The pelagic environment can, in turn, be divided into two major subdivisions: the neritic zone and the oceanic zone. The *neritic zone* includes the portion of the ocean from the shore to about 200 meters (650 ft) out. The neritic zone essentially overlies the continental shelf. The *oceanic zone* includes all the ocean beyond the neritic zone, and can be divided into three horizontal layers. The *bathyal zone* includes the layer between about 200 and 2,000 meters (650–6,500 ft); the *abyssal zone* lies below the bathyal zone and extends to a depth of about 6,000 meters (20,000 ft); the *hadal zone* lies below the abyssal zone at the ocean's bottom.

The benthic region (seafloor) is divided into zones in much the same way as the pelagic region. The uppermost zone is called the *littoral zone* and extends from the high-tide line to a depth of about 200 meters (650 ft). Because so much plant and animal life is found in this zone, it is further divided into three sections, depending on the conditions available for the support of life. The *supralittoral zone*, for example, consists of the shallowest portions of the ocean—regions that may be dry at least part of the time. Organisms that live in this region must be able to survive in both dry and wet conditions and must be able to withstand the action of tides, waves, and storms.

The *eulittoral* (or *intertidal*) zone extends from dry land to a depth of about 50 meters (150 ft). Plants and animals living in this area must also be able to survive the action of waves and, in some cases, have adapted to those conditions by constructing burrows in which they spend part of their lives. Finally, the *sublittoral zone* extends from the eulittoral zone to a distance of about 400 meters (1,300 ft). At this distance from shore, relatively few plants and animals are able to survive.

Beyond the littoral zone, the benthic region is divided into three major sections corresponding to those of the pelagic region: bathyal, abyssal, and hadal. Low temperatures, high pressures, and lack of light make it difficult for many forms of life to survive in these three zones. For example, the average amount of biomass found in waters within a few dozen meters of shore is anywhere from 100 to 5,000 grams per square meter. At distances of 50 to 200 meters, that value drops to about 200 grams per square meter. On the ocean floor at average depth, there is no more than about 5 grams of biomass per square meter. And in the deepest parts of the ocean, so little life survives that the amount of biomass is less than 0.001 gram per square meter.

The marine environment is often divided according to the amount of life received. In the upper zone, or *euphotic zone*, for example, enough sunlight penetrates the water to permit photosynthesis to occur. The euphotic zone extends from the ocean's surface to a depth of about 50 meters (160 ft). In the next layer down, the *dysphotic zone*, sunlight can be detected with measuring instruments, but the levels are too low to permit photosynthesis. The dysphotic

zone extends from the bottom of the euphotic zone to a depth of about 1,000 meters (3,300 ft). Finally, the deepest zone, the *aphotic zone*, is one in which sunlight cannot be detected at all, and plants and animals living in this region must adapt to life based on some process other than photosynthesis.

Seafloor Structure

All oceans have a similar horizontal structure. In moving seaward from the shore, one encounters a nearly flat or gently sloping stretch of ocean bottom known as the *continental shelf*. To an observer, the continental shelf usually appears flat, with an average worldwide slope of about 0°07'. Continental shelves are a significant geological formation, covering about a sixth of Earth's surface. By common agreement among oceanographers, a continental shelf is an area that extends outward into the ocean to a maximum depth of about 300 fathoms (1,800 ft). In many cases, this depth is reached at a distance of less than 200 kilometers (120 mi) from shore.

Continental shelves end rather abruptly at their outer edge, plunging down toward the ocean bottom at a sharp angle. The point at which this change occurs is called the *shelf break* (or *shelf edge*) and leads into the next zone, the *continental slope*. The continental slope is a much steeper structure with angles of descent of about 5°. In the most extreme cases, however, the slope may drop off at as much as a 20° angle.

After a distance of a few dozen kilometers, the angle of descent of the ocean floor levels off once again, forming the *continental rise*. The continental rise, like the continental shelf, slopes very gently toward the ocean bottom, with an average angle of descent of about 0.5°. The continental rise extends from the base of the continental slope to a distance of about 800 kilometers (500 mi) from the shore. The continental rise ends when it has reached the deepest part of the ocean floor, a region known as the *ocean basin*.

The ocean basin is one of the two major topographic features of Earth, the other being continents. Ocean basins cover about 70 percent of the total sea area and about half of the planet's total surface area. For thousands of years, scientists knew far less about the ocean basins than they did about the rest of Earth or, for that matter, than they did about the surface of the Moon. Today, we know that the ocean basins are complex regions, marked by ridges, trenches, fracture zones, abyssal plains and hills, and other structures not unlike those found on the continents themselves.

Ridges are underwater mountain ranges comparable to those found on the continents. One such range, the Mid-Atlantic Ridge, begins at the northern tip of Greenland, runs down the center of the Atlantic Ocean, and ends at the southern tip of Africa. The Central Indian Ridge runs beneath the Red Sea, dividing Africa from Asia; through the center of the Indian Ocean; and then westward toward the Mid-Atlantic Ridge and eastward toward the Pacific-Antarctic Ridge. The eastern end of the Pacific-Antarctic Ridge splits off the coast of Chile and runs northward as the East Pacific Rise and southeasterly as the Chile Ridge. After a break, the East Pacific Rise reappears off the coast of Washington State as the Juan de Fuca Ridge. Because of the interconnections among these ranges, some scientists suggest that a single oceanic ridge exists encircling Earth and covering a distance of more than 65,000 kilometers (40,000 mi).

In most locations, the tops of oceanic ridges lie at least 2,000 meters (6,500 ft) below sea level. These underwater mountains vary widely in their height but are, on average, about 2,500 meters (8,250 ft) tall. Running through the middle of oceanic ridges is a deep crevice known as a *rift*. The rift that cuts through a ridge may be as deep as 2,000 meters (6,500 ft). Scientists believe that rifts are formed when two adjacent tectonic plates pull away from each other, allowing magma to flow from the mantle onto the seafloor.

Trenches are narrow, canyonlike structures usually found adjacent to a continent. Scientists believe that trenches are formed when a portion of the earth's crust beneath the oceans collides with a portion of the crust underlying a continent. Since the oceanic crust is slightly less dense that the continental crust, it is forced downward, forming a depression in the ocean floor—the oceanic trench.

For example, the Nazca plate that underlies the eastern Pacific Ocean is in contact with the South American plate that underlies South America. As the South American plate moves westward, it forces the leading edge of the Nazca plate downward, forming the Peru-Chile Trench, which drops to a maximum depth of about 8 kilometers (5 mi).

The major oceanic trenches are found around the perimeter of the Pacific Ocean: the Peru-Chile, Cascadia, Aleutian, Kuril, Japan, Marianas, Tonga, and Java Trenches. The Marianas Trench is 11,034 meters (36,201 ft) deep at its maximum point, the deepest trench known. It is more than 2,000 meters (7,000 ft) deeper than Mount Everest is tall. The deepest point in the Indian Ocean is the Java Trench, with a maximum depth of 7,449.9 meters (24,440 ft).

Fracture zones are cracks in the ocean floor caused by earthquakes or volcanic activities. Fracture zones are commonly found on oceanic ridges, oriented at right angles to the length of the ridge. Some of the largest fracture zones are found along the eastern edge of the Pacific Ocean. The Clipperton and Clarion Fracture Zones, for example, originate along the western coast of Mexico and extend more than 5,000 kilometers (3,000 mi) to the west. The maximum width of these zones is about 50 kilometers (30 mi), and their maximum depth, about 3,000 meters (10,000 ft).

Abyssal plains and hills are topographic features that correspond to comparable structures found on the continents. The plains are relatively flat regions that extend for many kilometers in all directions, with a vertical rise of less than 0.1 percent. Scientists believe that the flatness of abyssal plains is a consequence of the deposition of eroded materials carried from rivers and streams into the oceans. Any irregularities in the ocean floor that may once have existed have long since been covered over by this deposition of sediment.

Abyssal plains are more extensive and more widespread in the Atlantic and Indian Oceans than in the Pacific Ocean. The explanation for this observation is that the majority of Earth's rivers flow into the Atlantic and Indian Oceans, providing them with a much larger supply of sediment.

Abyssal hills are irregular structures on the ocean floor with an average height of about 250 meters (825 ft). Similar structures of larger size are known as *seamounts*. A seamount is a steep-sided protrusion that may reach 1,000 meters (300 ft) in height. One of the largest seamounts is Greater Meteor Seamount, found in the northeastern Atlantic Ocean. It has a diameter of 110 kilometers (70 mi) at its base and an elevation of more than 4,000 meters (13,000 ft).

Another variation of an abyssal hill is a *guyot*. A guyot is simply a seamount with a flattened top. Scientists believe that the tops of guyots were once above sea level and were worn off by wave action. It seems likely that abyssal hills, seamounts, and guyots have all been formed by earthquakes and volcanic action.

Ocean Currents

An ocean current is a horizontal flow of seawater on the ocean's surface. Ocean currents are produced when wind systems exert a force on water at the ocean's surface, driving it in the same direction as prevailing wind patterns in the atmosphere. The flow of energy from the tropics to the mid-latitudes and on to the polar regions also drives and sustains ocean currents. The general pattern of most ocean currents is at least partly determined by the continental masses between which they flow.

Five major ocean current patterns can be found in the world's three major oceans. These patterns have roughly elliptical shapes with water flowing in a clockwise direction in the Northern Hemisphere and a counterclockwise direction in the Southern Hemisphere. The five systems are as follows:

North Atlantic: Gulf Stream, North Atlantic Drift, Canary Current, and North Equatorial Current

South Atlantic: South Equatorial Current, Brazil Current, South Atlantic Current, and Benguela Current

North Pacific: North Pacific Current, California Current, North Equatorial Current, and Kuroshio Current

South Pacific: South Equatorial Current, East Australian Current, South Pacific Current, and Peru Current

Indian: South Equatorial Current, West Australian Current, South Indian Current, and Agulhas Current

Specific ocean currents differ from each other to some extent in physical characteristics, such as width, depth, and velocity of water movement. For example, the Gulf Stream has a maximum width of about 200 kilometers (120 mi) and a maximum velocity of about 2.25 meters per second. It transports an average of about 100 million cubic meters per second, more water than flows through all the rivers in the world. By contrast, the North Equatorial Current in the Pacific Ocean flows no faster than about

0.20 meters per second, moving about half as much water as in the Gulf Stream.

Subsurface currents also exist in the oceans. These currents flow much more slowly than do surface currents, and are caused by differences in temperature and salinity of seawater. Most subsurface currents are associated with a single large system that begins off the coast of Newfoundland and flows down the eastern side of North and South America, and eastward across the Antarctic Ocean. The current then splits, with one segment flowing north along the east coast of South Africa, where it returns to the surface. The second segment continues eastward through the Antarctic Ocean, flowing northward east of Australia. This current surfaces off the east coast of Japan.

Origin of the Oceans

Scientists now believe that Earth was formed about 4.5 billion years ago—with the solar system's other planets—when a massive dust cloud condensed to form the solid bodies that now make up that system. By about 4 billion years ago, Earth's surface had cooled sufficiently to allow water to condense from the vapor phase, in which it had previously existed, to the liquid phase. That water probably came from two main sources: outgassing from rocks that made up the planet, and collisions with comets.

As liquid water formed, it filled whatever depressions were present at the time. Scientists believe that, at its earliest stages, Earth may have consisted largely of one massive land body and one massive body of water. The former has been given the name *Pangaea*, and the later, *Panthalassa*.

This state was not stable, however, largely because of the churning action of magma underlying Earth's crust. Over many millions of years, this action, and perhaps other forces, resulted in the breakup of Pangaea into the continents we know today. As the continents formed, the oceans filled in the space between them, forming the Atlantic, Pacific, and Indian Oceans and other oceans and seas as they now exist. *See also* Aquatic Plants and Animals; Estuaries; Gulf Stream; Oceanography; Submarines and Submersibles

Further Reading

EAO Scientific Systems. *Introduction to Earth's Oceans.* Halifax, NS: EOA Scientific Systems, Inc., 2001.

Monahan, Dave, ed. *World Atlas of the Oceans.* Westport, CT: Firefly Books, 2001.

Prager, Ellen J., with Sylvia A. Earle. *The Oceans.* New York: McGraw-Hill, 2000.

The Random House Atlas of the Oceans. New York: Random House, 1991.

Schneider, David A., and Glenn Zorpette. *Scientific American Presents: The Oceans.* New York: Scientific American, 1998.

Ocean Thermal Energy Conversion (OTEC)

See Energy from Water

Ocean Waves and Currents (as power sources)

See Energy from Water

Osmosis

Osmosis is the movement of solvent molecules through a semipermeable membrane from a region of low solute concentration to one of higher solute concentration. In solution terminology, a *solute* is a substance that is dissolved in a second substance, known as the *solvent*. A *semipermeable membrane* is a membrane that allows the passage of small molecules but not the passage of larger molecules. Osmosis occurs when any two solutions of different concentration are separated by a semipermeable membrane, but the cases involving aqueous solution are most common and of most interest.

Mechanism

Suppose that a sac made of a semipermeable membrane is filled with a solution of sugar dissolved in water and then immersed in a beaker filled with pure water. In such a case, water molecules move back and forth between the sugar solution and pure water through the semipermeable membrane. Sugar molecules in the sugar solution, however, are too large to pass through the membrane. The movement of water molecules in one direction (pure water to solution) is more rapid that in the other direction (solution to pure water).

Over time, two changes can be observed in this type of system. First, there is a net transfer of water molecules from the pure water into the sugar solution. That is, the sugar solution becomes more dilute. Second, the rate at which water moves from outside the membrane into the membrane decreases. As the sugar solution becomes more dilute, the rate at which water molecules move *out of* the solution begins to approach the rate at which they move *into* the solution. At some point, the rate at which water molecules move into and out of the solution is equal, and the system is said to be in *dynamic equilibrium*. This term means that water molecules continue to move through the membrane in both directions but at equal rates, so that no net change in the volume of the sugar solution takes place.

Osmotic Pressure

The process of osmosis can be stopped by applying pressure on the solution within the semipermeable membrane. The amount of pressure that must be applied to prevent osmosis from occurring with a solution suspended in a pure solvent is called that solution's *osmotic pressure*. Conversely, osmotic pressure can also be defined as the pressure exerted by a solution when it has reached dynamic equilibrium.

The osmotic pressure exerted by a solution is a function of two variables: molarity and temperature. Molarity (M) is a measure of solution concentration measured in moles per liter. In the equation representing osmotic pressure shown below, temperature (T) is measured in degrees kelvin (on the absolute temperature scale), and the letter R represents the ideal gas constant, whose value is 0.082 L-atm/K-mol. The symbol π is commonly used to represent osmotic pressure.

$$\pi = MRT$$

As an example, the osmotic pressure exerted by a 4.0M solution of sucrose dissolved in water at 25°C (298K) is:

$$\pi = 4.0 \ mol/L \times 0.082 \ L\text{-}atm/K\text{-}mol \\ \times 298K = 98 \ atm$$

Osmotic pressure can be determined empirically with an instrument known as an *osmometer*. An osmometer is simply a device used to apply pressure on the top of a solvent so as to prevent osmosis from taking place. The pressure needed to prevent osmosis is then read directly off the instrument.

Terminology

When two solutions (or a solution and a pure solvent) are separated by a semipermeable membrane, the more concentrated solution is said to be *hypertonic* in comparison to the less concentrated solution. Conversely, the less concentrated solution is said to be *hypotonic* compared to the more concentrated solution. These two conditions can also be defined in terms of the relative osmotic pressure of the two solutions. The solution with the higher osmotic pressure is hypertonic, while that with the lower osmotic pressure is hypotonic. Two solutions with equal concentration or osmotic pressure are said to be *isotonic* to each other.

When osmosis occurs with living cells, water may pass into or out of the cell, depending on its surrounding environment. If the cell is hypertonic to its surrounding environment, water will move into the cell, causing it to swell and, in some cases, burst. When this process occurs with red blood cells it is known as *hemolysis*. By contrast, a red blood cell that is hypotonic in relation to the surrounding environment will tend to lose water and shrivel up, a process known as *crenation*.

Biological Significance

Osmosis plays an important function in many biological systems. To a considerable extent, osmotic effects can be explained because a living cell can usually be thought of as a hypertonic solution enclosed in a semipermeable membrane surrounded by a hypotonic solution. Water tends to flow from the surrounding environment into the cell unless some other force acts to prevent that flow.

Osmosis in Plant Cells

Consider the fact that some kinds of green plants (grasses, for example) are able to stand upright even though they lack the woody tissue that supports trees, shrubs, and other kinds of plants. This ability to remain upright is one effect of osmotic pressure in the plant. That pressure is first exerted in the cells that make up the root hairs of the plant. The interior of those cells is hypertonic in comparison to water in the soil that surrounds them. There is a constant flow of water, therefore, from the soil into the cells.

This inward flow of water increases the pressure inside the cell, tending to cause them to burst. It is only the rigid cell wall

that surrounds these cells that prevents their destruction. As water continues to flow into root hair cells (and cells higher up in the plant), plant tissues retain a firm, upright structure. The pressure arising from the inward flow of water into cells is called *turgor*.

If adequate water is not supplied to the plant, osmotic pressure is reduced and the plant loses its turgidity. On a cellular level, the lose of water from plant cells is called *plasmolysis*, and we describe the overall effect on the plant as *wilting*. Plasmolysis and wilting also occur when the environment surrounding root hair cells is increased. For example, adding salt to the soil around a plant increases the concentration of groundwater, and that water may become hypertonic with respect to the interior of the plant cell. In that case, water flows out of the plant cells into the ground, producing plasmolysis and wilting. The same effect can be produced by adding excessive amounts of fertilizer to the soil, causing a wilting effect in plants sometimes described as *burning*.

Osmosis in Animal Cells

One example of the way osmosis functions in animals can be found in the exchange of water (and other substances) that occurs between the capillaries and the interstitial fluid that surrounds them. Capillaries are small blood vessels that connect arterioles (small arteries) to venules (small veins). The interstitial fluid is hypotonic with respect to the fluid inside the capillaries, so water tends to flow *into* the capillaries. The osmotic pressure in such cases amounts to about 28 mmHg (millimeters of mercury).

But other forces are present within the capillaries, most important of which is the hydrostatic pressure caused by the beating of the heart. In the arteries, arterioles, and capillary ends attached to the arterioles (where blood is being forced outward by the heart), this hydrostatic pressure tends to be relatively high, amounting to about 36 mmHg. The net effect of the hydrostatic pressure pushing water *out of* the capillaries (36 mmHg) compared to the osmotic pressure pushing water *in* (28 mmHg) results in the loss of water from the capillaries to the interstitial fluid.

At the opposite end of a capillary, however, the situation is reversed. Hydrostatic pressure caused by heart contractions in this part of the circulatory system is lower, about 21 mmHg. In this region, then, the osmotic pressure (28 mmHg) is greater than the hydrostatic pressure (21 mmHg), resulting in a flow of water back into the capillaries.

The significance of this phenomenon is that water does not travel into and out of the capillaries as a pure solution. Instead, it tends to transport other substances that have been dissolved in it. At the arteriole end of a capillary, for example, the flow of water from the capillary to the interstitial fluid carries with it oxygen, glucose, and other substances needed for cells to function normally. At the venule end of the capillary, water flowing back into the capillary carries with it ammonia, carbon dioxide, and other waste products formed as a result of cell metabolism.

Reverse Osmosis

Normally, the flow of water in an osmotic system is from the solution with a lower concentration to the one with a higher concentration. However, the direction of flow can be changed by applying pressure to the hypertonic solution. For example, the osmotic pressure of normal seawater is about 25 atmospheres. If a pressure of 25 atmospheres is applied to a container of seawater separated from pure water with a semipermeable membrane, no movement of water will occur from the latter to the former. Furthermore, if a pressure *greater than*

25 atmospheres is applied to the seawater, water will flow against the normal pressure gradient, out of the seawater into the pure water. This process is known as *reverse osmosis*.

Reverse osmosis has some important commercial applications. It is sometimes used, for example, for the purification of waste water. One of the largest such plants in the world is located in Yuma, Arizona, and operated by the U.S. Bureau of Reclamation. The plant was built to reduce the concentration of salts present in waste water leaving irrigation systems in the area. The plant produces about 275 million liters (72 million gal) of desalted water per day and reduces the salinity of water that passes through the plant from more than 3,000 ppm (parts per million) to about 28 ppm. The plant was originally authorized as part of an agreement between the United States and Mexico to guarantee that water leaving the United States would be suitable for use in Mexican irrigation systems.

Reverse osmosis has also been used to produce potable water by desalinating seawater. The principle on which such plants operate is the same as that described in the preceding paragraph. The world's largest reverse osmosis desalination plant is located at Ghar Lapsi, Malta. The plant began operation in 1983 and produces 20,000,000 liters of pure water per day from a supply of brackish water. Water introduced into the plant has a salinity of 36,500 ppm, while the salinity of product water is only 380 ppm. *See also* Desalination

Further Reading

Sourirajan, S. *Reverse Osmosis*. New York: Academic Press, 1970.

Thain, J. F. *Principles of Osmotic Phenomena*. London: Royal Institute of Chemistry, 1967.

P

Pack Ice

See Icebergs

Pagaeae

See Nymphs

Panama Canal

See Canals

Patrick, Ruth (1907–)

Ruth Patrick is one of the world's premier limnologists. She has made important contributions in the use of diatoms to detect and measure the amount and type of pollution present in freshwater sources.

Personal History

Patrick was born in Topeka, Kansas, on 26 November 1907. She was granted her bachelor's degree from Coker College in 1929, and her master's and doctoral degrees in botany, both from the University of Virginia in 1931 and 1934, respectively. At a time when paid positions for women research scientists were scarce, she began her academic career as a volunteer at the Academy of Natural Sciences in Philadelphia. She also taught at the Pennsylvania

School of Horticulture and worked as a laboratory technician at Temple University.

Professional Accomplishments

In 1937, Patrick was appointed to her first paid position at the Academy of Natural Sciences, as curator of the academy's Leidy Microscopical Society, a post she held for the next decade. During that period, she concurrently held the position of associate curator of the academy's Department of Microscopy. In 1947, Patrick was appointed curator and chair of the Department of Limnology at the academy. In 1973, she was promoted to the Francis Boyer Research Chair.

Patrick's interest in the relationship between diatoms and water properties dates to her doctoral thesis in which she studied the way in which these organisms survive and thrive in various kinds of water. In the late 1940s, she gave a paper on this topic at a professional meeting at which William B. Hart, an oil company executive, was present. Hart was sufficiently impressed with the potential use of diatoms for the monitoring of water pollution that he arranged for Patrick to receive a grant to pay for further research on the topic. She used the money to carry out an extensive survey of pollution in Pennsylvania's Conestoga Creek, a classic study that established important baseline data about diatom populations in polluted waters. The study also brought Patrick's work to the attention of other scientists around the world.

Patrick has carried out studies of fresh-water resources in many parts of the world. By her own estimate, she has worked in and around more than eight hundred different rivers during her research. In 1987, Patrick co-authored an important book on ground-water resources, *Groundwater Contamination in the United States*. The book summarized much of what was then known about groundwater resources, the threats posed to those resources, and possible ways of dealing with these threats.

Among Patrick's many honors and awards have been the Gimbel Philadelphia Award in 1969, the Pennsylvania Award for Excellence in Science and Technology in 1970, the Eminent Ecologist Award of the Ecological Society of America in 1972, and the Tyler Ecology Prize in 1973. Patrick has also assumed an active role in lobbying and advising governmental agencies and was involved in the drafting of the Clean Water Act.

Further Reading

"Women Environmental Leaders: Ruth Patrick (1907–)," http://www.mtholyoke.edu/proj/cel/patrick.html.

pH

The expression known as *pH* is a measure of the acidity of a solution. The pH concept was invented by the Danish chemist Søren Peter Lauritz Sørenson (1868–1939) in 1909. Sørenson developed the concept to help understand the effect of hydrogen ion concentration (acidity) in enzymatic processes. The symbol *pH* represents the phrase "potential of hydrogen."

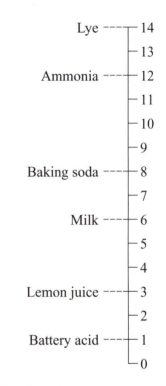

The pH values of some common substances.

Mathematically, the pH of a solution is the negative logarithm of the hydrogen ion concentration of a solution, or

$$pH = -log\,[H^+].$$

More accurately, pH is expressed in terms of the *hydronium ion* concentration, the ion formed when a hydrogen ion becomes attached to a water molecule, or

$$pH = -log\,[H_3O^+].$$

The pH scale is usually thought to extend from 0 to 14, with 0 being a very acidic solution, and 14 being a very basic solution. In fact, there are no theoretical limits to the pH scale, and while the vast majority of solutions have a pH between 0 and 14, some solutions have negative pHs and some have pHs of more than 14.

The diagram on the previous page illustrates the pH of some familiar solutions.

The concept of pH is important and widely used in biological systems, industrial operations, environmental studies, and other situations. For example, most biological reactions take place only within a narrow pH range. Living systems have developed methods for protecting such reactions against large changes in pH; such systems are known as *buffers*.

Similarly, some chemical reactions proceed readily at some pH levels but more slowly or not at all at other pH levels. Industrial chemists must understand the pH conditions needed for the operations for which they are responsible, and design systems to ensure that those pH levels are maintained. *See also* Aqueous Solution

Phytoplankton

See Plankton

Plankton

Plankton is the general name given to marine and freshwater plants, animals, and protists that float passively in the water or are only weakly motile (able to swim). The term *plankton* comes from the Greek word for "drifting." Plankton can be distinguished from other aquatic organisms, such as the *benthos*, which live in or on the ocean bottom, and the *nekton*, which are strong swimming organisms, such as fish.

Types of Plankton

Plankton can be classified into two large categories: the plants, or *phytoplankton*, and the animals, or *zooplankton*. Aquatic protists such as bacteria and fungi are also abundant.

Plankton can also be categorized according to their size. Those with a length of more than 1 mm (millimeter) are called *macroplankton* and are primarily zooplankton. Organisms with a length of less than 1 mm but greater than 0.05 mm are called *microplankton* and typically include both phytoplankton and zooplankton. The smallest form of plankton are the *nannoplankton*, with a length of less than 0.05 mm. Nannoplankton consist primarily of phytoplankton. Macroplankton can generally be collected with coarse-mesh nets, while nets of fine cloth are needed to capture microplankton. Nannoplankton are able to pass through both types of nets.

Phytoplankton

Two of the most abundant forms of phytoplankton are the *diatoms* and the *dinoflagellates*. Diatoms are algae members belonging to the class Bacillariophyceae, composed of beautiful symmetrical shells made of silica. They are typically no longer than about a micron (micrometer) in length. When diatoms die, they settle to the bottoms of lakes and oceans, where they may form deposits known as *diatomaceous ooze*. Dinoflagellates are so named because they have two appendages, or *flagella*, extending from their skeleton. They are members of the class Phytomastigophora and generally range from about 10 to 200 microns in length. A third type of phytoplankton found most commonly in tropical waters is *coccolithophorid*, a one-celled type of golden-brown algae also containing two flagella. The outer surface of coccolithophorid is covered with interlocking plates made of calcite. They range in size from about 5 to 50 microns. When these organisms die, they settle to the bottom of the ocean and form *coccolith ooze*.

Phytoplankton are thought to be responsible for production of about two-thirds of all food on Earth, as much as 160 billion tons per year. They constitute the lowest step in a food pyramid on which nearly all ocean life—and much land-based life—is based.

Phytoplankton are the main food source for zooplankton, which, in turn, are the primary food source for small fish, including herring, anchovies, sardines, and mackerel, as well as larger fish, such as sharks, tuna, and swordfish, and even mammals, such as dolphins and whales. Indeed, the world's largest mammals—the great whales—feed directly on phytoplankton and zooplankton, often consuming three tons of plankton every day.

Phytoplankton also perform another essential function for the planet. By some estimates, they produce more than half of all the oxygen consumed by animals on Earth.

Zooplankton

Two kinds of zooplankton are sometimes distinguished. *Holoplankton*, or *temporary zooplankton*, are eggs and larvae of the benthos and nekton that float on or near the water's surface during their early stages of life. As they mature, they develop the ability to swim on their own and are no longer classified as zooplankton.

By contrast, *meroplankton*, or *permanent zooplankton*, are organisms that spend all their lives floating and drifting on or near the surface of an ocean or lake. Nearly every group of marine organism is represented among the zooplankton. They range from a few microns to many meters in size and include organisms such as the *foraminifera, radiolaria*, and *crustacea*.

Foraminifera and Radiolaria occur widely throughout the world's oceans. Their bodies are covered with shells made of calcite, which, when the organism dies, settle to the bottom of the ocean at intermediate depths (in the case of foraminifera) or greater depths (in the case of radiolaria). Indeed,

very large areas of the world's ocean floors are covered with the remains of these organisms.

The most common form of zooplankton are the crustacea, the Copepoda and Euphausiacea in particular. The Copepoda are the primary food of a number of commercial fish and are thought to be the most numerous animal on Earth. The Euphausiacea, commonly known as *krill*, are also found everywhere in Earth's oceans and are the major food of whalebone whales, such as the great blue and finback whales.

In contrast to phytoplankton, zooplankton can grow to substantial size. The jellyfish known as *lion's main*, for example, may develop tentacles more than 30 meters (100 ft) in length.

Life Cycles

Like land-based plants, phytoplankton are photosynthetic organisms that use solar energy to convert carbon dioxide and water into organic material. In addition to sunlight, they require only nutrients to survive. Some of these nutrients are carried into lakes and oceans through runoff from the land. But the vast majority of nutrients in the oceans come from the decay of other plants and animals that have lived in the oceans, died, settled to the bottom, and decayed.

Phytoplankton can never migrate very far from the narrow band of the ocean near the surface, called the *euphotic zone*, where adequate supplies of sunlight are available to make photosynthesis productive. They are forced to rely on the churning action of the oceans to bring them the nutrients they need to survive and grow. The extent of this churning action depends on latitude and season.

During the summer, solar energy effectively warms the upper layer of the ocean to a depth of about ten meters. But sunlight does not penetrate much farther, and deeper parts of the ocean remain cold. During this time of the year, the nutrients needed by phytoplankton remain near the ocean's bottom.

In the winter, the amount of solar energy is greatly reduced, and the ocean's upper layers are no longer much warmer than its lower layers. Nutrients can much more readily flow upward to the upper reaches of the ocean, where they are available to phytoplankton. In addition, winter storms tend to produce waves and vertical currents in the ocean that assists the flow of nutrients upward.

By spring, the necessary combination of nutrients and solar energy is now available for the growth of phytoplankton. These organisms begin to grow and spread very rapidly in a phenomenon known as *spring bloom*. The explosion of phytoplankton growth is accompanied by a parallel growth of zooplankton, who feed on them.

As spring changes into summer, the ocean's upper layer again becomes much warmer than its lower layers, the flow of nutrients is reduced, and the spring bloom comes to an end. The cycle of phytoplankton growth begins again.

Latitude is also a factor in the life cycle of phytoplankton. Near the equator, the difference between upper and lower layers of the ocean is less pronounced. The winter upwelling that brings so many nutrients to phytoplankton living in higher latitudes is much less apparent in equatorial regions. As a result, the equatorial regions of Earth's oceans are a less abundant source of phytoplankton (and, thus, zooplankton) than are the higher latitudes.

The productivity of a region is also dependent upon ocean currents. In some parts of the world, currents produce regular upwelling of water for most of the year. These upwellings bring an even greater supply of nutrients than in high-latitude regions. Regions such as these can be found along the edge of the Gulf Stream, for example, accounting for the extraordinary fishing productivity of the area known as the Grand Banks. Ocean currents may also carry nutrients washed off the land, as by the Peru Current off the western coast of South America and the Benguela Current off the west coast of southern Africa.

Applications

While it is clear that plankton are a rich source of food for aquatic animals, some authorities believe that they might also be a possible food source for humans. Their protein content is greater than 50 percent, with a suitable balance of essential amino acids, which makes them an especially attractive food source for nations unable to produce other protein-rich foods, such as meat. Humans, however, are unable to digest plankton protein effectively, so some sort of mechanism would need to be devised to convert them to usable foods. *See also* Aquatic Plants and Animals; Limnology; Oceanography

Further Reading

Hallegraeff, Gustaaf M. *Plankton: A Microscopic World*. Leiden: E. J. Brill, 1988.

Newell, G. E., and R. C. Newell. *Marine Plankton: A Practical Guide*. London: Hutchinson, 1977.

Raymont, John E. G. *Plankton and Productivity in the Oceans*. 2d ed. New York: Pergamon Press, 1983.

Further Information

Journal of Plankton Research
Oxford University Press
Great Clarendon Street
Oxford OX2 6DP
United Kingdom
Telephone: +44 (0) 1865 267907
Fax: +44 (0) 1865 267485
URL: http://planktloupjournals.org
e-mail: jnl.info@oup.co.uk

Plumbing

The term *plumbing* applies to all of the pipes, fixtures, and other apparatus used for the distribution and use of water and wastes in a building. The word *plumbing* comes from the Latin term *plumbus*, for "lead," of which pipes were made in the Roman civilization.

A plumbing system consists of three primary components: (1) a method for delivering potable (safe for drinking) water to a building, home, or other structure; (2) sinks, toilets, showers, bathtubs, and other appliances in which water is used; and (3) a method for the removal of wastewater.

Plumbing systems vary in size and complexity depending on the size of the structure they are designed to serve and the variety of functions needed within the structure.

Water Supply

The water needed for a plumbing system is generally derived from one of three sources: one or more wells; a cistern or other reservoir designed for the storage of water; or a pipe or system of pipes (the water *main*) that carries water from a river, stream, or other natural source to a community.

All of these delivery methods were already well developed in ancient times. In the Minoan civilization that flourished in Crete between about 3000 and 1500 B.C., for example, plumbers designed elaborate systems to deliver water from elevated regions to towns and cities. Archaeologists have found systems of pipe made of baked clay, terra-cotta, and other ceramic materials buried underground up to depths of 11 feet in some cases.

The first aqueduct system for the delivery of water to towns and cities is thought to have been built by order of the Assyrian king Sennacherib (704–681 B.C.). The system consisted of a 10-mile-long canal that carried water to his capital of Nineveh from mountain sources. The canal was made of hardened earth lined with flagstones and waterproofed with bitumen.

Ancient Egyptians relied on canals to carry water from the Nile River for many of their domestic needs. King Menes, for example, is said to have ordered the construction of an extensive systems of dikes and canals to bring water from the Nile to his capital city of Memphis. That system remained in use until the Roman occupation of Egypt that lasted from about 30 B.C. to A.D. 641. Early Egyptians also dug wells to obtain a supply of pure water. One of the most famous of those wells, the Well of Joseph, was dug through 300 feet of solid rock near the Pyramids of Giza in about 3000 B.C.

Today, most homes and other buildings obtain their water from centralized distribution plants that force water through water mains with pumps. In some cases, the water is first pumped into an elevated storage tank and then allowed to flow by gravity downward through the community's water pipes. In communities that obtain their water from mountainous regions, pumping may not be necessary—the force of gravity alone may be sufficient to drive water through the community's water-supply system.

Plumbers have always used the strongest, most durable, most noncorrosive materials available for the construction of water pipes. In ancient times, these materials included bitumen, baked clay, and other types of ceramics. In rare cases, pure metals or alloys were used to make pipes. The Romans constructed their city's extensive plumbing system from lead, a metal that gave its name to the practice and probably had significant and tragic effects on the health of Roman citizens. Lead is a toxic metal, and small amounts must have constantly been absorbed by water running through the city's water pipes. Thus, every Roman citizen was exposed to a constant, if low, amount of lead in their drinking water.

Pipes used in modern plumbing systems are generally made of copper, brass, steel, or plastic. Water usually enters a home or other building through a water meter, which measures the amount of water used in the structure. It then flows through a cold-water system to toilets, sinks, tubs, washing machines, dishwashers, and other appliances in the building. It also flows to water heaters, where it is heated and then transported through a hot-water system to appliances that use hot water.

Fixtures

Some early civilizations developed primitive water-using appliances, although such devices were generally available only to royalty and the nobles. In Egypt prior to 2500 B.C., for example, artisans had learned to make bathtubs and toilets with plastered interiors and metal connections. Water emptied from such devices into containers below them; wastes flowed through copper pipes that were fitted with lead stoppers.

Most homes and work structures in developed countries today contain at least one or more toilets (also known as *water closets*), kitchen sinks, bathroom sinks (*lavatories*), and bathtubs or showers. Many homes and buildings also have other water-using appliances, such as dishwashers, garbage disposals, ice makers, drinking fountains, and urinals. The surfaces of these appliances with which water comes into contact must be nonabsorbent, durable, and easy to clean. Some of the most common materials used for this purpose are stainless steel; some type of ceramic material, such as vitreous china or porcelain; or cast iron covered with an enamel finish. After water has been used in one of these appliances, it drains out into the waste system.

Waste System

Generally speaking, ancient civilizations were not as concerned about nor as effective in dealing with the problems of wastewater disposal than they were with developing efficient water-supply systems. In many cases, wastes were simply thrown into the nearest street, discarded into a nearby stream or river, or carried away in buckets to an empty lot. One historian reports, for example, that ancient Babylonians used the city's streets as refuse dumps. From time to time, wastes that had accumulated on a street were simply covered over with a layer of clay. Over time, streets became so high that ladders had to be constructed to allow residents to climb down from the street into their homes (The History of Plumbing—Babylonia).

Wastewater disposal was one of the great public-health problems in the history of human civilization. It was not until the mid–nineteenth century that scientists realized that human wastes contained microorganisms that could cause cholera, diphtheria, typhoid, and many other diseases. At that point, health authorities realized that the disposal of human wastes posed quite a different issue than the disposal of dirty wash water.

In today's plumbing systems, wastes flow from toilets, bathtubs, showers, sinks, and other appliances through waste pipes (usually made of cast iron, steel, or copper) into a municipal sewer pipe or a septic tank. Layout of the wastewater pipes must be carefully designed to ensure that wastes always flow downward into the pipe and that there is no chance that they can back up and mix with incoming potable water.

Wastewater systems also include a second element, the venting system. As wastewater flows downward and out of a building, it tends to leave areas of reduced pressure behind it. Without a venting system, the wastewaters may back up and flow back into an appliance. The most common venting system consists of pipes that lead from various parts of the drainage system to one or more stack vents on the roof of the building. This system ensures that pressure within the drainage system is equalized and that back-ups will not occur.

Another protective element included in the wastewater disposal system is the *U-trap* found in many appliances. The U-trap is a pipe in the shape of the letter U located beneath a sink, toilet, or other appliance. The trap remains partially filled with liquid at all times, preventing sewer gas and pests from entering the building by way of the appliance.

Plumbing Codes

The installation of all kinds of plumbing systems is now closely regulated by rules and regulations called *plumbing codes*. These codes differ from locality to locality and type of structure, but cover a wide variety of features of the plumbing system. Some aspects typically covered by a code include the way pipes are laid in the ground, fed into a building, attached to appliances and other fixtures, and placed in relation to gas and electric service; the size and placement of drain pipes; the design of venting systems; and the installation of toilets, water heaters, and other appliances. In general, most plumbing codes require that systems be installed by licensed plumbers. *See also* Aqueducts; Fountains; Water Purification

Further Reading

The Complete Guide to Home Plumbing. Minnetonka, MN: Creative Publishing International, 2001.

Complete Home Plumbing. Menlo Park, CA: Sunset Books, 2001.

Kardon, Redwood, Michael Casey, and Douglas Hansen. *Code Check Plumbing: A Field Guide to the Plumbing Codes*. Newtown, CT: Taunton Press, 2000.

An extensive history of plumbing can be found at http://www.theplumber.com

Polubarinova-Kochina, Pelageya Yakovlevna (1899–1999)

Polubarinova-Kochina was one of modern Russia's greatest mathematicians whose interest in fluid dynamics has led to some important discoveries in and contributions to the field of hydrology. Perhaps her best known work in that field was a 1952 book, *Theory of Ground Water Movement*, which is still a standard reference in the field. According to the English translator of the book, the work contains "over thirty of her original and significant contributions on the hydromechanics of porous media (groundwater and oil flow)." Polubarinova-Kochina's theories have been applied to problems as diverse as the construction of dams, irrigation systems, soil drainage, and the theory of tides.

Personal History

Pelageya Yakovlevna Polubarinova-Kochina was born in Astrakhan, Russia, on 13 May 1899. The family moved to St. Petersburg at least partially to find better schools for their four children. After graduating from the Pokrovskii Women's Gymnasium in 1916, Polubarinova-Kochina enrolled in the Bestudzevskii women's program because, at the time, women were not allowed to pursue the same college programs as men. One benefit of the 1917 Revolution, however, was that this policy was revoked, and the Bestudzevskii program was absorbed into the University of Petrograd, from which Polubarinova-Kochina received a bachelor's degree in pure mathematics in 1921.

Professional Accomplishments

After graduation, Polubarinova-Kochina held a series of academic posts, including staff member at the Institute of Transportation, assistant lecturer at Leningrad University, staff researcher at the Leningrad Institute of Civil Aviation Engineering, and professor at Leningrad University. From 1935 to 1959, Polubarinova-Kochina was associated with the Division of Mechanics at the Steklov Mathematics Institute, which later became the Division of Hydromechanics, Institute of Mechanics at the University of Moscow.

In 1959, Polubarinova-Kochina volunteered to move to Siberia to establish a branch of the Soviet Academy of Sciences at Novosibirsk. She served as director of the Department of Applied Hydrodynamics and head of the Department of Theoretical Mechanics there until 1970, when she returned to Moscow. Upon her return to Moscow, she was appointed director of the Section for Mathematical Methods of Mechanics at the Institute for Problems in Mechanics in Moscow. She remained active until very late in her life, publishing her final paper on fractional linear transformation at the age of 100. She died on 3 July 1999 in Moscow, nearly two months after her 100th birthday.

In addition to her work in mathematics and hydrology, Polubarinova-Kochina has been interested in the history of mathematics and, in 1948, published the first extensive biography of the great Russian mathematician Sofya Kovalevskaya.

Further Reading

"Pelageya Yakovlevna Polubarinova-Kochina," http://www.physics.ucla.edu/~cwp/Phase2/Polubarinova-Kochina,_P._Ya@921234567.html (with references to other sources of information about Polubarinova-Kochina).

Poseidon

Poseidon is one of the twelve major gods in Greek mythology. He had dominion over the oceans and was responsible for the occurrence of earthquakes. In the latter capacity,

A fifth-century B.C. sculpture of the Greek god of the sea, Poseidon. *Vanni Archive/Corbis.*

he was sometimes called *Enosigaios* or *Enosichton*, meaning "Earth-shaker." The Romans later adopted Poseidon as their god of the seas and renamed him Neptune.

Poseidon was the son of Cronus and Rhea. According to one legend, he and his siblings were swallowed by their father, but were eventually rescued by Poseidon's brother, Zeus. Once released, the brothers made war on their father and eventually dethroned him. They then rolled dice to decide how control of the natural world would be divided. In this contest, Zeus was given dominion over the sky, a second brother, Hades, was given control over the underworld, and dominion over the sea was awarded to Poseidon.

Poseidon was married to the Nereid Amphitrite, with whom he had three children, Antaios, Orion, and Polyphemos. However, he was well known as a womanizer and had affairs with many other women. One of the most famous affairs was with the Gorgon Medusa, with whom he sired the winged horse Pegasus. He is said to have fathered many other horses also.

One of the most famous tales involving Poseidon was his contest with Athena for the allegiance of Athens. Poseidon initiated the contest by sinking his trident into the Acropolis in Athens, causing water to burst forth from the temple. But Athena contested his claim to the city by planting the first olive trees around the city. Poseidon pursued his conquest of the city by offering the Athenians the horse, the first time humans had seen the animal. In the end, the gods decided to award the city of Athens to Athena.

Greek mariners and travelers made sacrifices to Poseidon to ensure safe ocean voyages. He was also honored at horse races in Corinth, and in Argos, where horses were sacrificed to him by drowning in a whirlpool.

Poseidon is usually depicted as an older man with a long beard, carrying a three-pronged trident. In paintings and sculptures, he is often portrayed as riding a horse, a bull, or a dolphin. *See also* Gods and Goddesses of Water

Further Reading

Classical Myths: Poseidon: Texts, http://web .uvic.ca/grs/bowman/myth/gods/poseidon_ t.html.

Mythology Biofile, http://www.york.cuny.edu/ ~clip/m_biofiles/balanta.html.

Poseidon, http://www.messagenet.com/myths/ bios/poseidon.html.

Precipitation

The term *precipitation* refers to the forms in which liquid or solid water fall to the earth from the atmosphere. The most common forms of precipitation are rain, sleet, glaze, snow, mist, drizzle, hail, rime, and graupel. Dew and white frost are sometimes considered forms of precipitation also.

How Precipitation Occurs

Meteorologists have long puzzled over the question as to how droplets of water and tiny ice crystals in clouds coalesce to form particles large enough to fall to the earth as rain, snow, or some other form of precipitation. The fundamental problem is that cloud droplets are very small, about 10–20 μm (micrometers) in diameter, while an average rain drop is about 2,000 μm in diameter. By what mechanism do the thousands of tiny cloud particles combine with each other to produce a particle large enough to survive the fall to earth without evaporating?

The Bergeron Process

One answer to this puzzle was posed by the Swedish meteorologist Tor Bergeron. In 1933, Bergeron pointed out that clouds at a temperature of about −10°C contain a mixture of water droplets and tiny ice crystals. Water molecules in an ice crystal are held tightly in a distinct geometric structure, while water molecules in water are attracted to each other only weakly. Thus, water molecules can escape from water droplets more readily than they can from ice crystals.

In a mixture of water droplets and ice crystals, then, the tendency will be for water molecules to escape from the droplets, condense on the ice crystals, and, thereby, increase the size and mass of the ice crystals. Over time, ice crystals in a cloud will grow larger and larger by collecting water from water droplets around them.

Eventually, the ice crystals will become heavy enough to begin to fall in the form of snow crystals, or snowflakes. The simple act of falling toward the surface of the earth may then augment the coalescence process described above. Falling snowflakes may collide with and collect water droplets as they fall through a cloud. And the snowflakes may break apart, with each new flake acting as a condensation nucleus for the collection of even more water drops.

The process described here generally occurs when temperatures are still below the freezing point of water, 0°C (32°F). Thus, precipitation produced by the Bergeron process starts out as snow. As that snow falls through lower elevations, however, it may melt and change into rain or some other form of precipitation.

The Collision-Coalescence Process

One problem with Bergeron's explanation of the precipitation process is that it fails in cases where precipitation begins in warm clouds with temperatures above −10°C. To explain precipitation in such cases, the collision-coalescence process has been developed.

According to this explanation, precipitation can begin in a warm cloud only when relatively large water droplets begin to form. Such droplets must have diameters at least twice as large (50μm) as typical cloud droplets. Water droplets of this size are large enough to begin to fall toward the earth's surface. By comparison, water droplets of 10–20 μm are so light that they are buoyed up by updrafts in clouds. Larger water droplets can form when unusually large condensation nuclei (such as larger crystals of salt) are present in a cloud.

As large water droplets fall through a cloud, they collide with smaller water droplets suspended in the cloud. As they collide, the smaller water droplet may become attached to the larger water droplet, increasing its mass and the velocity at which it falls. Eventually, larger water droplets reach the earth's surface as rain, snow, or some other form of precipitation.

The collision-coalescence process is actually more complex than described above. For example, larger water droplets may be torn apart by friction as they fall through a cloud. The smaller droplets thus produced may then themselves become nuclei for the formation of even more droplets large enough to fall as precipitation.

Forms of Precipitation

Various forms of precipitation can be distinguished from each other primarily on the basis of their physical state and the size of particles of which they are made. As an example, rain is made of liquid droplets, while snow consists of solid crystals. Mist consists of particles no more than about 50μm in diameter, while hail particles are generally at least 5 mm in diameter.

The form in which precipitation reaches the earth's surface depends primarily on the temperature of the layers of the atmosphere through which it passes before reaching the ground. For example, suppose that snow crystals formed by the Bergeron process in a cloud fall through the air whose temperature is never above 0°C. In that case, the snow crystals will reach the ground in the form of snow.

But suppose those same snow crystals leave a cloud and pass through lower layers of the atmosphere with temperatures greater than 0°C. In that case, the precipitation may reach the ground in the form of rain. It may also occur as mist or drizzle, depending on the size of the water droplets that reach the earth. Generally speaking, mist consists of particles between 5 and 50 μm in diameter, drizzle consists of particles that range from 50 to 500 μm in diameter, and rain particles range in size from 500 μm to 5 mm in diameter.

Sleet is produced when precipitation falls through three atmospheric layers. The highest layer in which the precipitation is formed must be cold enough for snowflakes to form by the Bergeron process. The temperature of the next lower layer, then, is above freezing, at which point the snow changes to rain. In the lowest layer, near the ground, the temperature must once again be less than 0°C, such that the water droplets formed in the middle layer once again freeze, forming sleet. Sleet, then, consists of frozen raindrops.

Glaze (also called freezing rain) is produced by a process similar to that for sleet. The difference is that the raindrops formed in the middle layer don't freeze while they pass through the lowest, cold layer. They strike the ground as liquid droplets, but freeze immediately to produce a coating of ice on solid objects they strike.

Hail particles form only in tall cumulonimbus clouds where updrafts may reach more than 150 km/hr in velocity. Falling ice crystals formed by the Bergeron process may be caught in one of these updrafts and carried upward into the cloud again. As they travel upward, these ice crystals collide with and collect more water droplets. Eventually, they become heavy enough to start falling again. On their way down, they may encounter another updraft and be carried upward one more time. This process can be repeated a number of times, with the hailstones formed growing in size each time. Large hailstones can be cut in half to find concentric shells that show the number of up-and-down trips they have made through a cloud.

Most hailstones grow no larger than about a centimeter. However, hailstones the size of golf balls are not unknown, and the largest hailstone on record was recovered in Coffeyville, Kansas, on 3 September 1970. The stone had a diameter of 44.5 cm (17.5 in.) and a mass of 766 g (1.67 lb) (Recorded Weather Extremes, http://www.infoplease .com/ipa/A0001431.html).

Dew and White Frost

Dew and white frost (or just *frost*) are forms of condensation rather than precipitation. That is, they involve the change of water from a vapor state into the liquid state (in the case of dew) or the solid state (in the case of white frost) at the earth's surface rather than within a cloud. Dew or frost forms when objects close to the ground have radiated sufficient heat to cool the surrounding air below the dew point. Water vapor then condenses as liquid droplets or as solid ice crystals on those objects. Dew or frost forms depending on the surrounding temperature. At temperatures above 0°C, water vapor condenses as a liquid; at temperatures below 0°C, it condenses in the form of ice.

Precipitation Extremes

The amount of precipitation varies with both season and geographical location. The regions with the greatest average yearly rainfall are said to be Cherrapunji and Mawsynram, India, and Mt. Waialeale, Kauai, Hawaii, with 11.43, 11.87, and 11.68 meters of rain annually. By contrast, the driest region in the world is probably the area around Arica, Chile, with a fifty-nine-year average of 0.08 cm (0.03 in.). The region had no rainfall at all during one fourteen-year period.

In the United States, the driest area is Death Valley, California, with an average yearly rainfall of 4.14 cm (1.63 in.). The region with the longest period of no measurable rain is Bagdad, California, which went 767 days without rain from 3 October 1912 to 8 November 1914 (Recorded Weather Extremes, http://www.infoplease .com/ipa/A0001431.html). The largest single snowstorm recorded in North America occurred at the Mt. Shasta Ski Bowl, California, from 13–19 February 1969, during which 480 cm (189 in.) of snow fell. The largest single snow season occurred in 1998–99 at Mt. Baker, Washington, when 2,895.6 cm (1,140 in.) of snow was recorded. *See also* Clouds; Humidity; Hydrologic Cycle; Ice

Further Reading

Lutgens, Frederick K., and Edward J. Tarbuck. *The Atmosphere: An Introduction to Meteorology.* 8th ed. Upper Saddle River, NJ: Prentice Hall, 2001.
"Precipitation: Online Meteorology Guide," http://ww2010.atmos.uiuc.edu/(Gh)guides/mtr/cld/prcp/home.rxml.
Reynolds, Ross. *Cambridge Guide to the Weather.* Cambridge, UK: Cambridge University Press, 2000.
Upgren, Arthur R. *Weather: How It Works and Why It Matters.* Cambridge, MA: Perseus Publishing, 2000.

Properties of Water

Water is a chemical compound composed of 88.8 percent oxygen and 11.2 percent hydrogen by weight. Its chemical formula is H_2O, which can also be written HOH. The chemical formula expresses the fact that water is composed of molecules containing two hydrogen (H) atoms and one oxygen (O) atom each.

Shape of the Water Molecule

The atoms that make up a water molecule are not arranged linearly, as in the structure shown on the left below. Instead, they take on a fork-shaped arrangement, as shown on the right. In this structure, the two hydrogen-oxygen bonds form an angle of 104.5 degrees with each other.

The shape of the water molecule is significant because of the consequences it has for many physical and chemical properties of water. In the arrangement shown above, the oxygen atom has a larger nucleus and, hence, a greater attraction for the two electrons that make up the oxygen-hydrogen bonds. By contrast, the smaller nucleus of the hydrogen atoms has less attraction for those electrons. The two oxygen-hydrogen bonds are, therefore, polar, with the oxygen end being more negative than the hydrogen end in each case.

Because of this fact, the water molecule as a whole is also polar. The "upper" end of the molecule (as shown in the diagram) is more negative than the "lower" end.

Because water molecules are polar, they tend to attract each other. The positive end of one water molecule tends to be attracted to the negative end of a second molecule. This force of attraction between two adjacent molecules is known as *hydrogen bonding* and accounts for many of the properties described below.

Physical Properties

Water is a clear, colorless, odorless, tasteless liquid with a boiling point of 100°C (212°F) and a freezing point of 0°C (32°F). Its density is 1.00 grams per cubic centimeter (62.3 pounds per cubic foot), and its refractive index is 1.333. Its viscosity is 0.01002 poise (at 20°C). The specific heat of liquid water is 1 calorie per gram, and of ice, 80 calories per gram. The latent heat of vaporization of water at 100°C is 540 calories per gram, one of the largest such values of any common material.

These physical constants are all affected by the presence of hydrogen bonding in water. For example, boiling is the process that occurs when heat is applied to a liquid. The heat energy added to the liquid is used to increase the speed with which the molecules of the liquid are moving. Eventually, those molecules move fast enough to escape from the liquid and form a vapor.

In the case of water, additional heat is needed for this process to occur. Some of the heat added to liquid water is first used to reduce the hydrogen bonding between adjacent molecules. Once water molecules are free to move independently, additional heat can then be used to convert them to the gaseous state.

Water is also one of the best solvents known and, in fact, is sometimes, if inaccurately, called the *universal solvent*. The vast majority of inorganic compounds as well as many organic compounds dissolve in water—many to an appreciable degree. The polar character of the water molecule is largely responsible for the liquid's effectiveness as a solvent. When ionic or polar compounds, such as table salt (sodium chloride; NaCl) or sugar (sucrose; $C_{12}H_{22}O_{11}$) are added to water, the electrical nature of the water molecule attracts ions or charged molecules in the solute (the material being dissolved). These ions or charged molecules are then torn away from the bulk of the material and solvated by water molecules.

In solvation, the ion or charged molecule is surrounded by water molecules. During the process of solution, then, the solute is eventually reduced entirely to atomic-sized particles that are surrounded by water molecules and made invisible to the eye.

Chemical Properties

Water molecules ionize very slightly. In this process, the negatively charged oxygen atom of one molecule is able to pull strongly enough on the positively charged hydrogen atom of a second molecule to remove that hydrogen atom from its molecule entirely. The two species that remain are a water molecule with an additional hydrogen ion (a *hydronium ion*; H_3O^+) and a water molecule with one less hydrogen ion (a *hydroxide ion*; OH^-). That reaction can be represented by the following chemical equation:

$$H_2O + H_2O \leftrightharpoons H_3O^+ + OH^-$$

The double-arrow ($\leftrightharpoons$) in this equation indicates that the reaction can proceed in either direction. In fact, the right-to-left reaction is strongly preferred, and in a collection of water molecules, the vast majority will tend to be in the form of water (H_2O) molecules and only a very small number in the form of hydronium (H_3O^+) and hydroxide (OH^-) ions. The proportion of water molecules that do ionize is about one out of ten million.

Generally speaking, water is a relatively inert substance. The chemical bond that holds the oxygen and hydrogen atoms together in a water molecule is very strong. Relatively few chemical species are active enough to break this bond and react with water chemically.

Perhaps the best known example of water's chemical activity is its behavior with the active metals in group one of the periodic table. For example, when sodium metal is added to water, sodium atoms replace hydrogen atoms in the water molecule, forming sodium hydroxide and hydrogen gas.

$$2Na + 2HOH \rightarrow 2NaOH + H_2$$

This reaction occurs so vigorously, in fact, that enough heat may be evolved to ignite the hydrogen gas formed in the reaction.

Water reacts very slowly or not at all with most other elements and compounds. For example, the vast majority of metallic elements, such as iron, copper, tin, lead, and nickel, oxidize very slowly when placed into water. The oxidation occurs because of oxygen gas dissolved in water as well as oxygen atoms present in the water molecule. *See also* Capillary Action; Electrolysis of Water; Hydrate; Hydrogen Bonding; pH; Surface Tension

Further Reading

Franks, Felix. *Water*. London: Royal Society of Chemistry. 2d ed., 2000.

Leopold, Luna B. *Water: A Primer*. San Francisco: W. H. Freeman, 1974.

Psychrometer

See Hygrometer

Purification by Water

See Biblical References to Water

Purification Rites

See Biblical References to Water

R

Rain

See Clouds; Precipitation

Ray, Dixy Lee (1914–1994)

Dixy Lee Ray was a marine biologist, educator, and government official who was perhaps best known to the general public for her strong views on important energy-related issues. Some of Ray's most significant research was carried out while she was chief scientist for the International Indian Ocean Exploration in 1964. The purpose of this expedition was to study the physical and biological characteristics of the Indian Ocean, a region that had been largely ignored by scientists up to that time.

A 1977 photo of Washington State Governor Dixy Lee Ray. *Bettmann/Corbis.*

Personal History

Dixy Lee Ray was born in Tacoma, Washington, on 3 September 1914. As a child, Ray spent her summers on Puget Sound and became interested in marine biology. She attended Mills College in Oakland, California, from which she received her bachelor's degree in 1937 and her master's degree in 1938. She then took a job as a science teacher in the Oakland Public Schools.

In 1942, Ray left this job to begin graduate studies in zoology at Stanford University, where she was awarded her doctoral degree in 1945. In that same year, she was appointed instructor in zoology at the University of Washington. Over the next twenty-seven years, she rose to become assistant and then associate professor at the university. From 1960 to 1962, Ray served as a consultant in the field of biological oceanography to the National Science Foundation.

Professional Accomplishments

In 1965, Ray was appointed director of the newly created Pacific Science Center in

Seattle. The center was established in a group of buildings left over from the Seattle World's Fair of 1962. Under Ray's direction, the center became one of the most effective programs of science education for children and adults.

In 1972, Ray was asked to serve on the Atomic Energy Commission (AEC) by President Richard Nixon. A year later, she was appointed chair of the commission. In that post, she became actively involved in promoting the development of nuclear energy as an alternative to the world's limited supply of fossil fuels.

After leaving the AEC, Ray continued her interest in politics and ran successfully for the governorship of Washington. She was elected in 1977 and reelected to a second term in 1979. After leaving the governor's office, she became an engineering consultant.

Ray was well-known for her strong positions on political and scientific issues and her willingness to express those positions in public speeches and in print. Toward the end of her life, she became particularly upset by what she saw as the extremism of some environmentalists and environmental groups. She wrote two books, *Trashing the Planet* (1990) and *Environmental Overkill* (1993), in which she took environmentalists to task for arousing government leaders and the general public about unimportant and nonexistent environmental issues. She argued in the latter book, for example, that the threat of global climate change from human activities had been greatly exaggerated, and that such exaggeration could only lead to unnecessary restrictions on the economic activity of the world's nations.

Ray was awarded the Clapp Award in Marine Biology in 1958, the Seattle Maritime Award in 1967, the Frances K. Hutchinson Medal for Service in Conservation in 1973, the United Nations Peace Medal in 1973, and the Francis Boyer Science Award in 1974. Ray died of respiratory failure on 3 January 1994.

Recreation

See Aquatic Sports

Reservoirs

A reservoir is a place where some material, such as water, is collected and kept in storage for future usage. Probably the earliest known water reservoirs were those built by the ancient Egyptians in the fifth millennium B.C. These reservoirs held water diverted from the Nile River and kept in reserve for the irrigation of crops. Archaeological evidence suggests that the ancient Chinese and Babylonians also developed reservoirs to maintain their irrigation systems. One of the most advanced of these systems was that created under Babylonian king Hammurabi (1792–1750 B.C.).

Uses

Reservoirs can be built in a variety of ways, but the most common involve the diversion of water from a river or stream or the damming of a river or stream. The water stored in reservoirs is used for many different purposes, including:

- irrigation
- hydroelectric power
- recreation activities, including boating, swimming, and fishing
- dilution of wastewater and reduction of salinity
- emergency purposes, such as firefighting
- protection for fish and other wildlife during periods of drought
- recharge of aquifers that supply residential, industrial, and agricultural wells

A somewhat different function of water reservoirs is in flood control. Reservoirs provide a place for excess rainwater to collect and reside, protecting communities, farms, and other facilities in an endangered area from flood damage.

Environmental Impact

Concerns have been expressed that the construction and use of water reservoirs may have certain harmful environmental effects on an area. Some examples include:

- contamination of water supplies taken from the reservoir and from downstream areas by reservoir water use, such as boating and other forms of recreation
- destruction of wildlife habitat from the dammed river or stream
- loss of land submerged by the reservoir
- increase in susceptibility of loss to life and property in case of earthquakes

- alteration in riparian systems, usually resulting in the loss of long rivers and streams in favor of many smaller rivers and streams
- damage to aquatic life adapted to previously existing riparian systems, including loss of genetic diversity of species living in the area

Large Reservoirs

The capacity of reservoirs is usually given in one of two measures: cubic meters or acre feet. An acre foot is the amount of water needed to flood a piece of land one acre in size to a depth of one foot.

The largest reservoir in the world is located at Owens Falls, Uganda. It was built in 1954 and has a capacity of 204,800 billion cubic meters (166 billion acre feet; 7,231,500 billion cu ft). The next five largest reservoirs in the world and the five largest reservoirs in the United States are listed below. *See also* Dams; Energy from Water; Irrigation

The Largest Reservoirs in the World

Rank	Name	Location	Capacity (million m³)	Built
2	Kariba	Zambia/Zimbabwe	181,592	1959
3	Bratsk	Russian Federation	169,270	1964
4	High Aswan	Egypt	168,900	1970
5	Akosombo	Ghana	147,960	1965
6	Daniel Johnson	Canada	141,851	1968

The Largest Reservoirs in the United States

Rank	Name	Location	Capacity (million m³)	Built
1	Lake Mead	Arizona/California Nevada	23,135	1936
2	Lake Powell	Utah	21,900	1964
3	Lake Sakakawea	North Dakota	19,400	1956
4	Lake Oahe	South Dakota	18,920	1962
5	Fort Peck	Montana	15,330	1940

Further Reading

Henderson-Sellers, Brian. *Reservoirs*. London: Macmillan, 1979.

LaBounty, James F. *Lakes, Ponds, and Reservoirs: What They Are and How to Study Them*. Denver, CO: Bureau of Reclamation, U.S. Department of the Interior, 1985.

Lagler, Karl F., ed. *Man-Made Lakes: Planning and Development*. Rome: United Nations Development Programme, Food and Agriculture Organization, 1969.

United Nations Environment Programme. *The Pollution of Lakes and Reservoirs*. Nairobi: United Nations Environment Programme, 1994.

Revelle, Roger (1909–1991)

Roger Revelle is known for his many contributions to the field of oceanography and environmental science, as well as his work to educate the general public about important issues in science education and environmental science. Of all his scientific research, the project for which Revelle was probably best known to the general public was his work on the Deep Sea Drilling Project. The purpose of this project was to study the structure and composition of the ocean floor, including the boundary between the earth's crust and upper mantle in these regions. One result of this research was the discovery of unexpectedly high heat radiation in certain parts of the ocean, later to be found to be regions where tectonic plates were in motion relative to each other. This research provided fundamental support for the theory of plate tectonics.

Personal History

Revelle was born in Seattle on 7 March 1909, but grew up in Southern California. He earned his bachelor's degree in geology from Pomona College in 1929 and then continued his graduate studies, first at Pomona and later at the University of California at Berkeley. It was during his time at Berkeley that Revelle was offered a research position at the Scripps Institute of Oceanography (SIO) in La Jolla, California. He continued his research at SIO while working in a doctoral program in oceanography, a degree he received in 1936.

Professional Accomplishments

Upon graduation, Revelle was offered a teaching post at SIO, where he remained for the rest of his academic career. He began as an instructor and, after World War II, was appointed director of the institute, a post he held until 1964. During the war, Revelle was assigned to the Hydrographic Office as director of the Bureau of Ships. He remained with the government as chief of the geophysics sector of the U.S. Office of Naval Research. During this period, he was involved in research tests of the nation's nuclear weapons capabilities.

In the later 1950s, Revelle became interested in the issue of anthropogenic effects on the earth's atmosphere. He argued that the release of carbon dioxide by human activities was altering the composition of the atmosphere in such ways that climate changes were likely to occur. This view was one of the earliest expressions of concern about possible global warming due to human activities. In an effort to learn more about this question, Revelle was involved in the organization of the International Geophysical Year of 1957–58. As part of this program, Revelle was instrumental in the creation of a carbon dioxide measuring station near the Mauna Loa volcano on the island of Hawaii, a station that has since produced some of the most important data about global climate change.

Always interested in government service, Revelle accepted an appointment as science advisor to Secretary of the Interior

Stewart Udall from 1961 to 1963. During this period, he visited the Indus River basin of Pakistan to study issues of water management. An unexpected consequence of this assignment was Revelle's exposure to the economic and social problems existing in Third World nations. He returned to the United States with a new interest in problems of population growth. In 1964, he asked for a leave of absence from his position at SIO to help found the Center for Population Studies at Harvard University. He remained at the center for a decade, while he taught, wrote, and spoke about population issues. It was during this period that he edited and authored four important books on population issues: *The Survival Equation: Man, Resources, and His Environment* (1971, co-edited with Ashok Khosla and Maris Vinovskis), *Consequences of Rapid Population Growth* (1972), *Population and Social Change* (1972, co-edited with David Glass), and *Population and Environment* (1974). He returned to California in 1975, where he remained until his death in San Diego on 15 July 1991.

Further Reading

Dorfman, Robert, and Peter P. Rogers, eds. *Science with a Human Face: In Honor of Roger Revelle*. Boston: Harvard School of Public Health, 1997.

Morgan, Judith. *Roger: A Biography of Roger Revelle*. La Jolla, CA: Scripps Institute of Oceanography, 1996.

Rivers and Streams

A river is a natural flow of water with a significant volume of water. Rivers are a major factor shaping the topology of the earth's land surfaces. They represent an important natural resource for humans, other animals, and plants, and are significantly affected by human use.

River Development

Rivers can originate from a variety of different sources. One of the most common sources is a spring—a supply of water that issues from underground. Springs are formed when the water table reaches the earth's surface and water issues forth from holes or cracks in the surface.

Other important sources of rivers are lakes and marshes. Lakes and marshes themselves are formed by different factors; they may be filled by rivers and streams that flow into them, or they may obtain their water primarily from rainfall and other forms of precipitation. Finally, rivers may be formed from runoff in high mountain areas.

Whatever their source, rivers tend to develop through a relatively uniform and predictable series of steps. Those steps can be described by considering what happens when precipitation falls onto a flat surface on the side of a mountain. At first, that precipitation either sinks into the ground and becomes groundwater or evaporates back into the atmosphere. Of course, both processes can and do occur simultaneously.

At some point, the ground becomes saturated with water and additional precipitation begins to flow downward across the surface of the ground. This water is called *runoff*. In an ideal situation, runoff flows downward across the ground in a thin sheet. Ideal situations seldom if ever occur, however. Instead, the earth's surface is usually covered with rocks, stones, plants, and other objects that change the flow of water. In addition, the type of ground over which water flows may be very different, consisting of very hard bedrock, more fragile sandstone or limestone, sand, gravel, or soil. All of these factors affect the way water flows over the earth's surface. They divert water flow and change it from a uniform thin sheet into a series of *rills*—very small streams of water. Rills continue to carry water downward along paths that are determined by the land's contour.

At some point, the paths of rills are likely to intersect with each other. A group of two, three, or more rills may join together to form a larger body of flowing water, called a *brook*. At this point, the system of flowing water has begun to take on the characteristics of almost all river systems: a *dendritic* pattern. The term *dendritic* means "treelike," and it describes the appearance of the river system as seen from the air. In this pattern, rills look like the tiniest branches of a large tree that are joined to somewhat larger branches, the brooks.

As water continues to flow downhill, this pattern continues. Brooks begin to intersect and flow into each other, forming a *stream*, and streams begin to intersect and flow into each other, forming a *river*. The terms used here are somewhat difficult to distinguish from each other in terms of the size and volume of waterways involved. However, they do serve to outline the development of the mature river as it grows from very small water flows into larger and larger waterways.

The dendritic pattern of the river system is now complete. In a mature system, the major river is represented by the trunk of a tree, while the smaller waterways (streams, brooks, and rills) are represented by the largest, smaller, and smallest branches that grow off the main trunk. A river system of this type may cover hundreds, thousands, or, in the case of major rivers, millions of square kilometers. The rills, brooks, and streams that flow into the major river are known as its *tributaries*, and the overall pattern is referred to as the river's *drainage basin*.

Rivers sometimes empty into deserts, from which their water eventually evaporates back into the atmosphere. But the ultimate repository of nearly all running water is the oceans. In many cases, rivers run into major lakes, such as the Great Lakes, but those lakes are usually only temporary repositories. They, in turn, empty into other rivers that ultimately flow into the oceans.

Types of Rivers

Rivers can be classified into three general categories depending on their regularity of flow. Most rivers contain water all the time, although the level of water may vary depending on season and rainfall. Such rivers are called *permanent* rivers. Other rivers, *periodic* rivers, may dry up at one or more times of the year. Such rivers can be found in parts of the world where rainfall is relatively low and where the rate of evaporation from the river may exceed the rate at which precipitation falls. Finally, *episodic* rivers carry water at only certain times of the year or, in some cases, only once every few years. Episodic rivers are found in regions where rainfall is quite low, amounting to no more than a few inches per year. Permanent and periodic rivers tend to flow into the oceans or into large lakes, and while episodic rivers may also do so, they may also drain into deserts.

Gravitational Effects and the Aging of Rivers

The force that drives the flow of water in rills, brooks, streams, and rivers is gravity. In the absence of gravity, droplets of water that fall on higher elevations would simply remain on the earth's surface or evaporate back into the air. But because gravity exerts a force on all objects toward the earth's center, those droplets of water tend to flow downward across the ground.

Gravitational effects are strongly affected by the topography of the land. On a surface that is perfectly flat, gravity can have no effect at all since there is no direction in which water can move. On land with a small slope, the force of gravity is largely "diluted" by the flatness of the ground's angle, and water can flow only very slowly. But in areas where the slope is significant, as on the side of a mountain, water is free to flow downward quite easily and does so with considerable velocity.

The effects of gravity on the flow of water explain in large part the formation of rivers at different elevations. In high mountains, water flows through channels with a relatively high velocity, scraping along channel bottoms with significant force. The force of running water is able to erode away rocks, stones, pebbles, sand, gravel, and other material on the channel bottom. In this part of the river, the force of running water operates primarily on the channel bottom but also, to a lesser extent, on the sides of the channel.

Because running water at high elevations tends to exert a force primarily downward, the valleys formed in these regions tend to have a V shape. A V-shaped valley is representative of a young river, one found at high elevations with rapidly running water.

As rivers reach lower elevations, this pattern of erosion changes. Gravitational effects become less important at these elevations. Eventually, a point is reached at which running water cuts downward to only a moderate degree and begins to cut sideways into the walls of its channel. As this process continues, the river valley experiences an overall change in shape. It changes from a V-shaped valley to a U-shaped valley.

If the land were perfectly flat without any kind of obstacle, a mature river valley would look like a long, straight trough, like the gutter on the roof of one's house. But land is never perfectly flat or lacking in obstacles. As a consequence, the flow of

water in a river is pushed one way or another by slight changes in elevation, the presence of a large boulder, a section of soft rock in the ground, or some other factor. In other words, the river course changes into an undulating pattern that is determined by the type and elevation of ground over which it flows.

The flatter the ground over which a river flows, the more extreme becomes its undulation or *meanders*. In extreme cases, meanders can become very significant, taking on the shape of huge bows or loops. At some points, the river can cut into the bank of a meander and form a new channel to another part of the meander, cutting off a section of the river in the process. These cut-off portions of the river then take on the appearance of half-moon-shaped lakes, called *oxbow lakes*.

Erosion and Deposition

Running water is a major factor in changing the earth's surface. Rivers and streams erode rocks and soil from one part of the land and deposit them in other parts. In general, the effect of running water is to carry materials from higher elevations to lower-elevations, wearing mountains down and building up lower-lying areas. Many of the most prominent landforms, such as valleys, canyons, and deltas, are produced by the action of running water.

Running water causes erosion in three directions: downward, into the streambed; sideways, along the walls of the waterway; and forward, in the direction of river movement. The relative extent to which each type of erosion occurs depends on many factors.

One of the most important of these factors is slope. The greater the slope of a waterway, the more rapidly water runs; the greater the kinetic energy it has; and the more efficiently it is able to erode soil, rocks, and other material. As an example, a river with a slope of about 1 percent is able to transport about 6,000 kg of sand per day. By comparison, a river with a slope of about 2 percent can carry about 21,000 kg of sand per day.

Elevation can also be an important factor in determining the rate of erosion. At high elevations, gravitational forces have more effect than they do at lower elevations. Many mountain rivers are efficient erosional forces both because they tend to have steep slopes and because they are at high elevations. In such cases, most erosion occurs downward and in a forward direction. At lower elevations, by contrast, the greatest amount of erosion occurs in a direction normal to that of water flow—that is, against the walls of the river or stream.

Climate is another factor in determining the rate of erosion. In drier regions, soil tends to be less consolidated and more fragile than in moist regions. Running water tends to cause a greater amount of erosion in such areas than in more temperate climates.

The amount of soil carried by running water depends on the velocity of the water. For example, a stream flowing at the rate of 1,000,000 liters per day can carry about 14,000 kg of sand per day. But a stream flowing twice as fast can carry 2.5 times as much sand.

When running water slows down, it loses its ability to carry suspended solids, and those solids are precipitated out of the water. This effect occurs most commonly when rivers and streams empty into large lakes or the oceans or when they flow onto a flat arid region, such as a desert. Deposits that settle out of rivers and stream in lakes or the oceans are known as *deltas*, while those that form in deserts are called *alluvial fans*.

Large Rivers

Rivers occur in all sizes and shapes, determined primarily by the topography and climate of a region. The world's two longest rivers are the Nile, at about 6,500 km (4,040 mi), and the Amazon, at about 6,520 km (4,050 mi). The Nile River rises in Burundi and flows northward through Egypt and into the Mediterranean Sea. The Amazon has its source in the Andes Mountains in Peru,

River Systems of the World

River	Length (in km/mi)	Area of Drainage Basin (in thousand km²/mi²)	Rate of Discharge (in m³/sec)
Amazon	6,520/4,050	7,180/2,270	180,000
Nile	6,500/4,040	2,881/1,170	1,584
Mississippi/Missouri	6,019/3,740	3,221/1,247	17,545
Yangtze	5,800/3,600	1,970/698	35,000
Ob'	5,570/3,460	2,975/1,154	12,600
Yenisei-Angara	5,550/3,480	2,605/996	19,600
Yellow	4,845/3,010	745/290	1,365
Congo	4,700/2,900	3,822/1,440	42,000

Source: "River Systems of the World," http://www.rev.net/~aloe/river.

Note: Various sources provide different data for the numbers shown in this table.

about 100 miles from the Pacific Ocean. The chart on page 258 lists data for these and other major rivers of the world.

Environmental Issues

Rivers are so common in many parts of the world that it is easy to take them for granted and to assume that they will always provide an abundant supply of clean water for all human needs. Such is, however, not the case. A number of problems created by human activities have begun to threaten the health of rivers around the world. A "healthy" river is one that is able to supply water of a quality that allows its use for various human purposes. That definition differs depending on the kind of use to which the water must be put. Water for agriculture or hydroelectric power use, for example, may carry more contaminents than water intended for human consumption or recreational purposes. The two most important of these problems are disruption of natural flow and pollution.

Disruption of Natural Flow

The term *natural flow* refers to the regular cyclic movement of water into and through a river system, from source to final outlet. Natural flow is important in a river because it provides a diverse and complex habitat for a variety of plants and animals. In addition, it provides a dependable and safe supply of safe water for a variety of human needs.

In today's world, a number of factors threaten the natural flow of many rivers. Some of these factors are described:

Urban development: Growing urban areas usually increase the amount of water taken from rivers faster than it can be replenished by natural sources. As a result, the natural flow in the river may be changed. In addition, the infrastructure that makes up an urban area—including parking lots, malls, housing developments, industrial areas, and paved roads—change the watershed surrounding a river. The ground in such areas becomes less impervious, and rainfall tends to run off the land more rapidly rather than soaking into the soil. Drainage patterns are changed and rivers may become more contaminated with soil washed away with the runoff water.

Logging may remove trees and other plants that normally help soil hold water. As a consequence, runoff and soil erosion may increase in logged areas.

Dams dramatically alter the natural flow of rivers. They change the rate, volume, temperature, and other characteristics of water flowing through a river. Aquatic plants and animals that have adapted to conditions in a natural river are often not able to survive after dams have been built and these conditions changed.

Diversion systems that capture water for agricultural purposes and channelization projects that are carried out to improve navigation or control flooding may have similar results on aquatic ecosystems.

Pollution

Almost any human activity carried out along a river or stream or within a river's drainage basin has the potential for polluting that waterway to some extent or another. For example, factories, power-generating plants, and other industrial facilities often release toxic chemicals, heavy metals, radioactive wastes, and other hazardous materials into a river system. Cooling water returned to rivers from such facilities can significantly increase the temperature of the river water and therefore act as a pollutant.

Active and abandoned mines can also be the source of significant pollution of waterways. Mercury, lead, cadmium, and other toxic and carcinogenic metals can be carried away into a river system, through which they can reach and harm humans, other animals, and plants. Sand, gravel, and other sediments disturbed during the mining process can also be washed into a river, causing siltation.

Lumbering activities can also increase the risk of river pollution. The extensive root systems of trees is an important factor in absorbing water and preventing erosion. When trees are cut down, that protection is lost and soil is washed easily into the nearest waterway.

Farms and dairies also produce a variety of materials that can contaminate rivers. These materials include livestock wastes, fertilizers and pesticides, and bacteria and other disease-causing organisms. Communities ranging in size from large urban areas to small rural towns also produce a host of different kinds of wastes that can damage water quality in a river. These wastes include detergents, garbage, humans wastes, oil and grease, plastics, and lawn and garden pesticides.

In its 2000 report the U.S. Environmental Protection Agency surveyed 23 percent of the nation's 3.6 million miles of rivers and streams. The agency found that the water quality in 55 percent of those waterways could be classified as "good," 10 percent as "good, but threatened," and 35 percent as "impaired." The study found that the major single type of pollution threatening rivers was siltation, which impacted 40 percent of all waterways studied. Siltation affects water quality in a number of ways, suffocating fish eggs and other bottom-dwelling organisms, modifying aquatic habitats, interfering with drinking-water treatment processes, and limiting recreational uses of a river or stream.

The next most important pollutants and the percent of rivers affected were pathogens (primarily bacteria; 38%), nutrients such as nitrogen and phosphorus (30%), oxygen-depleting substances (24%), metals (20%), pesticides (16%), habitat alterations (15%), and thermal modifications (12%).

The most important single cause of river pollution, according to the study, was agriculture, accounting for detrimental conditions on 60 percent of the river mileage studied. Agricultural activities were responsible for most of the major pollutants, including siltation, pesticides, nutrients, and oxygen-depleting substances. The next most important causes of water pollution in rivers and streams were hydromodification, such as channelization, dredging, and dam construction (20% of all river mileage studied); urban runoff and storm sewers (10%); municipal point sources (9%); resource extraction (8%); forestry (5%); land disposal (5%); and habitat modification (5%).

Efforts are now under way to prevent further pollution of rivers and to reduce pollution in those waterways that have already been damaged. The leading federal agency in this effort is the Environmental Protection Agency, operating under the authority of the Clean Water Act of 1972 and its later amendments. The original Clean Water Act provided comprehensive protection for the nation's waterways against industry and municipal effluents, while the 1977 amendments focused on the problem of toxic wastes in waterways.

Many private groups are also working to preserve, protect, and rehabilitate the nation's rivers. For example, the organizations American Rivers and River Network both work to educate the general public about problems relating to the nation's rivers and to organize and support programs for the protection and rehabilitation of rivers. A comprehensive list of public and private organizations involved in river quality issues can be found in the *National Directory of River and Watershed Conservation Groups,* published by River Network and the Rivers, Trails & Conservation Assistance Program of the National Park Service (additional information at jhamilla@rivernetwork.org).

See also Hydrologic Cycle; Limnology; Water Pollution

Further Reading

Boon, P. J., Bryan Davies, and Geoffrey E. Petts, eds. *Global Perspectives in River Conservation*. New York: John Wiley & Sons, 2000.

National Geographic Society Staff. *Exploring the Rivers of North America*. Washington, D.C.: National Geographic Society, 2000.

Patrick, Ruth. *Rivers of the United States*. New York: John Wiley & Sons, 2001.

Penn, James R. *Rivers of the World: A Social, Geographical, and Environmental Sourcebook*. Santa Barbara, CA: ABC-CLIO, 2001.

"The Quality of Our Nation's Waters: A Summary of the National Water Quality Inventory: 1998 Report to Congress," Washington, D.C.: Office of Water, United States Environmental Protection Agency, June 2000.

Waters, Thomas F. *Wildstream: A Natural History of the Free Flowing River*. St. Paul, MN: Riparian Press, 2000.

Further Information

American Rivers
1025 Vermont Ave., N.W., Suite 720
Washington, DC 20005
Telephone: (202) 347-7550
Fax: (202) 347-9240
URL: http://www.americanrivers.org/contactus/default.htm
e-mail: amrivers@amrivers.org

River Network
520 SW 6th Avenue, #1130
Portland, OR 97204
Telephone: (503) 241-3506 or (800) 423-6747
Fax: (503) 241-9256
URL: http://www.rivernetwork.org
e-mail: info@rivernetwork.org

Rowing

See Aquatic Sports, Rowing

S

Sacred Waters

The term *sacred waters* refers to rivers, streams, lakes, wells, springs, and other sources of water thought to have magical powers. The association of water with supernatural events can probably be traced to the earliest days of human culture. Water itself has often been regarded as one of the fundamental elements of life, the source from which all life on earth began, and a substance with particularly strong powers to cure the body and the soul.

Water resources earn their reputation for holiness for a variety of reasons. In some cases, it is thought that a god, goddess, or spirit created the water source for the particular benefit of humankind. In other cases, the water source may be blessed because of some historic and significant event that took place at that location.

A list of all the sacred waters revered by humans over the centuries would be very long indeed. The following is a summary of some of the best known of these sources.

The Holy Rivers of India

All rivers are revered in the Hindu religion. For example, a grand celebration was held on 6 September 1996 in honor of the birth of A.C. Bhaktivedanta Swami Prabhu-pada, founder of the International Society for Krishna Consciousness (the Hare Krishna movement). At this ceremony, water taken from 1,008 holy rivers and waters was used to wash a statue of the prophet.

Still, some rivers are thought to be more holy than others. Some authorities list seven especially sacred rivers, and others count only five. Among these rivers are the Cauvery, Ganga (Ganges), Godavari, Indus, Jumna, Narmada, Saraswathi, Sindhu, and Yamuna.

Perhaps the best known of these rivers to many Westerners is the Ganga River. Hindus believe that bathing in the river washes away all their sins. Any person who dies near the river, according to tradition, will be transported directly to heaven. The river is said to possess its holiness because it flows from the toe of Vishnu. The Ganga is intimately involved in many mythical tales from Hindu history. On many occasions, it has assumed the form of a god or goddess or a human and performed wonderful acts of healing and protection for humans.

To some people, other rivers are at least as important as the Ganga. People in the region of Amarkantak, for example, argue that the river Narmada is even more powerful than the Ganga. In proof, they offer the claim that a person must bathe in the waters of the Ganga, say prayers for seven days on the banks of the Yamuna, or offer devotions to the Saraswati for three days to receive blessings but need only have a view of the Narmada to receive purification.

The River Jordan

The River Jordan is regarded as holy by many Christians because it is the site at which John the Baptist baptized Jesus. The specific site where this event was to have occurred is the region where the river leaves the Sea of Galilee. Many Christian pilgrims visit the river to obtain its blessings. They may also take some of the river water home with them to be used as holy water in their own residence or local church.

Holy Wells and Springs

Certain wells, springs, and other natural sources of water throughout the world are regarded as holy by specific religions or individuals. In many cases, the reason that such spots are thought to be sacred is no longer known—they simply have centuries of tradition behind them. In other cases, some hint as to the possible source of the supernatural force of the site may be available. The Chalice Well in Glastonbury is an example. The well produces a continuous supply of reddish brown water with a resemblance to blood. For centuries, pilgrims have come to the well to seek both physical and spiritual healing.

One of the most famous of all sacred springs is found in the grotto at Lourdes. In 1858, a fourteen-year-old girl by the name of Bernadette Soubirous reported seeing the Virgin Mary in a grotto outside the small town of Lourdes in southwestern France. She said that the Virgin had told her to drink of the water in the grotto, although there was nothing other than muddy rainwater to be had. After Bernadette scooped out and drank some of the water, a stream began to appear. From that point on, the grotto continued to

issue forth with its new source of water. Both the grotto and the stream became holy places, and Bernadette went on to become Saint Bernadette. In 1876, a basilica was built at the grotto. It has become a popular destination for pilgrims, with about 3 million of the faithful visiting every year. *See also* Holy Water

Further Reading

1008 Holy Waters Bathing of Prabhupada, http://www.afn.org/~centennl/centkals.htm.
Sacred Places: Water and the Sacred, http://www.arthistory.sbc.edu/sacredplaces/water.html.
Tales from the Indian Epics, http://www.investindia.com/newsite/kids/stories/indepic.htm.

Salinity

The term *salinity* refers to the total quantity of salts dissolved in seawater, expressed in parts per thousand (ppt). Thus, a sample of seawater that contains one gram of dissolved salts in 1,000 grams of water has a salinity of 1 ppt. The term is also used in a more general sense to describe a property of soil in which salts have been deposited from running water or some other source. Thus, one might say that the salinity of the soil in an area has increased because of extensive use of irrigation water on the soil. As the irrigation water evaporates, it leaves some dissolved salts behind, increasing the salinity of the soil. This entry discusses the first definition of salinity only.

Measuring Salinity

Two methods are commonly used to determine the salinity of water. In the first, a weighed sample of seawater (m_Σ) is allowed to evaporate and the mass of the dissolved salts left behind (m_s) is weighed. The salinity of the sample can then be calculated by dissolving the mass of the dissolved salts by the mass of the sample and multiplying by 1000, or:

$$S = \frac{m_s}{m_\Sigma} \times 1000$$

The more common method for determining salinity is by measuring the electrical conductivity of a water sample. The conductivity of seawater depends, in a dilute solution, on the concentration of salts present. Pure water itself does not conduct an electric current, but adding salts to the water makes it conductive, with the conductivity increasing as the amount of salt added is increased. Electrical conductivity is a more popular method for determining salinity than measurement of dissolved solids because it is easier and quicker.

Salinity is expressed in mg/L (milligrams per liter) when measured by the first method described and in μS/cm (microsiemens per centimeter) when measured by electrical conductivity. There is no direct relationship between these two measures, but the conversion factor commonly found on conductivity meters and in the literature is:

2 μS/cm = 1 ppm (parts per million), or
2 μS/cm = 0.001 ppt

Definitions

Seawater is usually defined as water containing 18–35 ppt of dissolved solids. *Brackish water* is like seawater in that it contains dissolved solids, though a lower concentration of them—typically about 1–4 ppt. An intermediary type of seawater, called *salted water*, is sometimes identified. Salted water is generally defined as a form of seawater with 4–18 ppt of dissolved solids.

Brackish water and salted water form in regions where seawater comes into contact and mixes with freshwater. Coastal estuaries and salt marshes typically contain brackish water, as do lakes and ponds into which

ocean water can flow, as during a severe storm. Brackish water also occurs in inland areas where a balance between freshwater input by way of precipitation or running water and evaporation results in a body of water with a dissolved salt concentration of 1–4 ppt.

Effects of Salinity on Living Organisms

A number of organisms have evolved the traits needed to survive in saline water. Some plants do well in brackish water but not seawater, while others have adapted to an environment (such as the oceans) in which salt concentrations can reach 35 ppt. Such plants have evolved mechanisms by which the salt in their aquatic environments is filtered out by specialized cells.

By contrast, freshwater plants are not very tolerant of saline conditions. Most plants that live in or near rivers, lakes, ponds, and other bodies of freshwater tolerate no more than 5 ppt of dissolved salts, and many begin to die off at much lower levels of salinity. Even plants that have adapted to survival in saline conditions can experience deleterious responses when the concentration of salts in water increases.

A large variety of animals have also adapted to salt water with varying concentrations of dissolved salts. Many ocean fish, for example, can tolerate saline concentrations as high as 40 ppt, while some highly tolerant species, such as the shelduck, can survive at concentrations up to 100 ppt. Most mammals who live in salt water tolerate levels of up to about 35 ppt.

Again, by contrast, freshwater animals are much more sensitive to high concentrations of dissolved salts. Most freshwater fish, for example, begin to die off at 10 ppt or less, a level that also begins to affect freshwater birds. The tolerance level for frogs and reptiles is less well known, but is thought to be about 9 ppt for the former and 5 ppt for the latter.

Humans and domestic animals are not very tolerant of elevated levels of dissolved salts in water. In general, water with a salt content of less than 0.4 ppt is regarded as safe for drinking, irrigation, and livestock consumption. Water with a salt content of 0.4 to 1.5 ppt can be used by humans and for livestock, but is suitable for irrigation only if special precautions are taken to provide adequate drainage on suitable soils for certain types of plants. Water with a salt content of up to 3.0 ppt can be used as drinking water for pigs and chickens, while most other domestic animals can tolerate levels of up to 5.0 ppt. Water with a salt content of 5.0 ppt can be used for irrigation with certain salt-tolerant crops, but water with a higher salinity can be used only in emergency situations.

Finally, very saline water (over 5.0 ppt) should not be used for human consumption or irrigation, and is suitable only for beef cattle (whose limit is about 8.5 ppt) and adult sheep (whose limit is 11.5 ppt). The only safe use for such saline water is for flush toilets.

Salt Lakes

All naturally occurring bodies of water contain at least some dissolved solids. As rivers and streams flow over the ground, they dissolve minerals. These dissolved minerals then contribute to the salinity of the lakes and oceans into which those rivers and streams flow. The salinity of most lakes remains relatively constant because water (and the dissolved solids it contains) that enters at one location drains out at one or more other locations. The salinity of the oceans gradually increases over time because, of course, they have no outlet. But the volume of water in the oceans is so great that this increase in salinity is imperceptible over even relatively long periods of time.

Some bodies of water, however, do not fit the pattern described for lakes above. Such bodies may have a source of water—such as one or more rivers and streams and, of course, rain—but have no outlet. Water that enters these bodies of water leaves only through evaporation. As a consequence, the concentration of dissolved solids in these lakes gradually increases over time. Such bodies of water, known as *terminal lakes*, can be found throughout the world. Their evolution to more and more saline conditions is similar to that which occurs in the oceans. The difference is that terminal lakes have small enough volumes that the increase in salinity can be observed and measured relatively easily.

Two of the best known terminal lakes in the world are the Great Salt Lake in Utah and the Dead Sea, which lies on the border between Israel and Jordan. The Great Salt Lake is a remnant of an even larger body of water, Lake Bonneville, that was created by melting glaciers more than thirty thousand years ago. At its largest point, Lake Bonneville was about 520 kilometers (325 mi) long and 215 kilometers (135 mi) wide, covering an area of more than 52,000 square kilometers (20,000 sq mi) in western Utah. At its deepest point, the lake was more than 300 meters (1,000 ft) deep. At the time, water drained from the lake by way of a number of rivers and streams.

About twelve thousand years ago, significant topographic and climatic changes began to occur. The ground beneath Lake Bonneville subsided, and nearly all outflowing streams and rivers were cut off. At the same time, the climate in Utah became warmer and drier. Water continued to flow into the lake, as it does today, but at a much reduced volume, and was lost from the lake through evaporation only. Over the last ten thousand years, the volume of the lake—the largest portion of which is now known as the Great Salt Lake—has continued to diminish, while its salinity has continued to increase. On average, the lake is currently losing about 3 million acre-feet of water a year.

The volume and salinity of the lake change from year to year, depending on the amount of precipitation and average annual temperature. The salinity of the lake water also varies in different parts of the lake. The lowest salinity ever measured in the lake was about 5 percent, only slightly greater than that of seawater. The highest salinity ever measured was nearly 27 percent, just less than the highest concentration of dissolved salts water can hold. The lake currently contains about 5 billion tonnes (5 billion tons) of dissolved salts, distributed as shown below.

The Dead Sea was formed about a million years ago when huge earth movements created an enormous north-south crevice in the earth's surface now known as the Syrian-African Rift. The body of water into which the Jordan River and other rivers and streams then emptied—the modern Dead Sea—sank to the bottom of that crevice. The lake's original outlets to the Mediterranean Sea all disappeared. Today, the Dead Sea is the lowest point on the earth's surface, 400 meters (1,300 ft) below sea level.

Water has continued to flow into the Dead Sea from the Jordan River and other sources, as well as from precipitation. Its only outlet, however, is by means of evaporation. As with other terminal lakes, therefore, the only natural factors affecting lake volume and salinity are the amount of annual precipitation (usually quite low) and average annual temperature (usually quite high). In recent years, however, the volume and salinity of the Dead Sea have been significantly affected by another factor—the diversion of water from the Jordan and other rivers for irrigation and hydroelectric power by both Jordan and Israel.

As a result of both natural and human factors, the salinity of the Dead Sea has reached the maximum concentration possible based on its salt content of about 33 percent. Plans have long been discussed for connecting the Dead Sea with the Mediterranean Sea (the "Med-Dead" connection), or with the Red Sea (the "Red-Dead" connection), with the difference in water heights being used to generate hydroelectric power. For a number of economic and political reasons, however, that project has never gone beyond the planning stage.

Many other terminal lakes exist in the Middle East, China, and other parts of the world. For example, Sevier Lake in northern Utah is another remnant of ancient Lake Bonneville. It currently covers an area of about 42,000 square kilometers (16,200 sq mi) and has an average salinity of about 18 percent. In western Nevada, Walker, Pyramid, and Winnemucca lakes are all terminal lakes formed when ancient Lahontan Lake (like Lake Bonneville) began to dry up about ten thousand years ago. *See also* Desalination

Further Reading

French, Richard H. *Salinity in Watercourses and Reservoirs: Proceedings of the 1983 International Symposium on State-of-the-Art Control of Salinity*. Boston: Butterworth, 1984.

Chemical Composition of Three Salt Lakes

Source	Sodium	Potassium	Magnesium	Calcium	Chloride	Sulfate
Great Salt Lake	32.8	2.0	3.3	0.2	54.5	7.2
Seawater	30.8	1.1	3.7	1.2	55.5	7.7
Dead Sea	12.3	2.3	12.8	5.3	67.2	0.1

Source: http://www.ugs.state.ut.us/online/PI-39/PI39PG9.HTM.

Poljakoff-Mayber, Alexandra, and J. Gale, eds. *Plants in Saline Environments*. Berlin: Springer-Verlag, 1975.

Talsma, T., and J. R. Philip, eds. *Salinity and Water Use*. London: Macmillan, 1971.

Scripps Institution of Oceanography

The Scripps Institution of Oceanography (SIO) is one of the oldest, largest, and most prestigious institutions devoted to the study of oceanography in the world. It was founded in 1903 by University of California professor William E. Ritter and a group of citizens of San Diego, California. The institution was created "to carry on a biological and hydrographic survey of the waters of the Pacific Ocean adjacent to the coast of Southern California; to build and maintain a public aquarium and museum; and to prosecute such other kindred undertakings as the Board of Trustees may from tine to time deem it wise to enter upon." At the time, the institution was known as the San Diego Marine Biological Station and was sponsored by the Marine Biological Association of San Diego (MBASD).

In 1905, the institution was moved from Coronado, its original home, to its present home in La Jolla, California. Seven years later, MBASD became a part of the University of California system and was renamed the Scripps Institution for Biological Research. It was given its current name in 1925.

Today, SIO's program of research covers a wide range of topics, including physical, chemical, geological, and geophysical studies of the oceans; research on the atmosphere and the land beneath the seas; and studies on marine biology. Researchers use a variety of instruments, tools, and methods that involve ground- and ship-based studies as well as remote sensing by satellite. The institution currently occupies sixty-seven buildings on 230 acres, with satellite campuses at Point Loma and Mount Soledad, both in California. It also owns and operates four research vessels and one research platform. There are about 1,200 staff and 190 graduate students.

Examples of the research groups working at SIO include Acoustic Thermometry of Ocean Climate, Antarctic Circumpolar Wave, California Coastal Ocean Observation System, Electrical Conductivity, Experimental Climate Prediction Center, Hydraulics Laboratory, Indian Ocean Experiment, Ocean Bottom Seismograph Facility, Satellite Marine Gravity, Southwest Fisheries Science Center, and Whole Earth Geophysics.

The institution publishes technical reports on research carried out under its auspices as well as news briefs and other information intended for the general public. The publication the *SIO Research Overview* is available online at the institution's web site or in print form.

Further Information

Scripps Institution of Oceanography
Public Information MC 0233
UC San Diego
9500 Gilman Drive
La Jolla, CA 92093
Telephone: (858) 534-3624
Fax: (858) 534-5306
URL: http://www.sio.ucsd.edu
e-mail: siocomm@sio.ucsd.edu

Scuba

The term *scuba* (also *SCUBA*) is an acronym for *self-contained underwater breathing apparatus*. A popular form of scuba equipment is known by the trade name Aqua-Lung. Scuba systems are of two general types: open-circuit and closed-circuit. In an open system, the diver breathes in compressed air or oxygen stored in a cylinder

Scuba diver with elkhorn and staghorn corals. *OAR/National Undersea Research Program (NURP), NOAA.*

that he or she carries. Waste gases are then breathed out into the surrounding water. In a closed system, exhaled air is recycled through a mechanism that absorbs carbon dioxide and allows oxygen to be reused by the diver.

History

Inventors have been searching for ways to provide divers with self-contained breathing systems for nearly two hundred years. One of the earliest scuba systems was designed in 1825 by the English inventor William James. James's invention consisted of a cylindrical belt containing compressed air worn around the diver's waste. It is unclear as to whether this device was ever tested or used.

One of the first closed-circuit devices was invented by the English merchant seaman Henry A. Fleuss. In Fleuss's system, air exhaled by the diver passed over a rope soaked in caustic potash (potassium hydroxide). Carbon dioxide in the exhaled air reacted with and was absorbed by the potash. The apparatus was successfully used at depths of up to 25 feet for periods of up to three hours.

By the early twentieth century, the French, British, and American navies were all beginning to show interest in research on scuba systems. In the 1920s, for example, the U.S. Navy was studying the use of helium-oxygen mixtures for use in both scuba and surface-to-diver systems. One of the most successful inventors of the time was the French naval captain Yves Le Prieur. In 1933, Le Prieur tested an open-circuit system in which the diver carried a tank containing pressurized air on his back. The amount of air supplied to the diver was controlled by a *demand regulator* (or *demand valve*) that functioned according to the depth to which the tank was carried. The device was adopted for use by the French navy in 1935 and a year later, Le Prieur organized the first scuba diving club, the Club of Divers and Underwater Life.

In spite of the apparent usefulness of the demand regulator, it fell out of use with almost all scuba equipment for a decade. Instead, air or oxygen was supplied by means of a continuous-flow system or through a manual on/off switch controlled by the diver. In the early 1940s, however, French naval officer Jacques-Yves Cousteau and French engineer Emile Gagnan rediscovered the demand regulator and designed a scuba system based on its use. They successfully tested the device and, in 1943, received a patent for it under the name Aqua-Lung. Aqua-Lung was soon marketed commercially in France (1946), Great Britain (1950), Canada (1951), and the United States (1952).

Operation

Scuba systems operate using either compressed air or compressed oxygen mixed with some inert gas, such as helium. The number and size of tanks carried by a diver depend on the depth of the dive and the time it is expected to take. For depths of maximum depth and time, three cylinders, each of about seven liters (1.6 gal) in capacity are used.

In both open- and closed-circuit systems, two pressure regulators are provided. The regulators reduce the pressure of the gas in the tank to the pressure surrounding the diver in two steps. Supply tanks also contain a warning device that notifies the diver when the supply of air or oxygen is low. At that point, the diver can use a changeover valve to turn on a reserve supply of air. *See also* Cousteau, Jacques-Yves

Further Reading

Graver, Dennis. *Scuba Diving*. 2d ed. Champaign, IL: Human Kinetics Publishers, 1999.

Griffiths, Tom. *Sport Scuba Diving in Depth: An Introduction to Basic Scuba Instruction and Beyond*. Princeton, NJ: Princeton Book Company, 1991.

Sea Giants

See Jotuns

Sea Monsters

Sea monsters are huge animals believed to have lived—or be living—in lakes, rivers, oceans, and other bodies of water. Stories about sea monsters extend very far back into history. One of the most famous tales from Greek mythology, for example, was that of the fight between Hercules and the Hydra. The Hydra was an enormous snakelike sea creature with many heads. Each time Hercules cut off one head, two more would grow in its place.

The Loch Ness Monster

Arguably the most famous sea monster of them all is the Loch Ness monster, a giant dragonlike water creature thought to inhabit the Loch Ness in northern Scotland. Sightings of the monster go back to at least A.D. 565, when Saint Columba is said to have discovered the remains of an animal devoured by the monster on the shores of the lake. It was said that Saint Columba actually saw the monster himself.

Although tales of the monster continued to be told for many centuries in the region of the lake, it began to receive international attention only in 1933, when tourists in the area reported seeing splashing in the middle of the lake, apparently caused by some enormous creature. The world's press began to take an interest in the story, and the monster soon became a worldwide celebrity.

At least three major scientific expeditions have been conducted in the last quarter century in an effort to locate the monster, but none of these efforts produced verifiable proof of the monster's existence. While many laypersons continue to believe in its existence, most scientists are highly dubious, at the very least, that such a creature exists.

Other Examples of Sea Monsters

Many large bodies of water around the world have been identified as the homes of sea monsters. For example, a creature referred to as Champ was first identified in Lake Champlain between New York and Vermont in 1609 by the French explorer Samuel de Champlain. The creature was reputed to look like a twenty-foot-long snake with the head of a horse. At one time, P. T. Barnum offered a reward of $50,000 for capture of the monster, but no one was able to claim the reward.

Another famous lake monster in North America is Ogopogo, whose home is reputed to be Lake Okanagan in British

Depictions of sea dragons from the sixteenth-century *Historiae Animalium. Academy of Natural Sciences of Philadelphia/Corbis.*

Columbia. Native American folklore of the area includes tales of the creature, known by the name of Natiaka, or Lake Monster, at the time. During the 1920s, a number of people living in the area claimed to have seen the monster, including a group of more than fifty individuals witnessing a baptism in the lake in 1926.

Similar stories have been told about a monster in Lake Manitoba by the name of

Manipogo. Again, the first mention of such a creature can be found in folk tales told by Native Americans living in the area. In 1962, two men reported seeing the monster and taking its picture, but the picture was too blurry to provide convincing evidence of the monster's existence.

The open seas have been the location of many sea-monster sightings. One of the most famous occurred in December 1848 when the crew of the sailing ship *Pekin* found what it thought to be a sea monster off the Cape of Good Hope. Excitement was high at first because another ship, the HMS *Daedalus*, had reported a similar sighting only two months earlier. When a search boat from the *Pekin* reached the "monster," however, it was found to be nothing other than a very large piece of seaweed.

Similar excitement surrounded the discovery of an enormous carcass in April 1977 by the Japanese trawler *Zuyio Maru*. The crew was unable to bring the carcass back to port for examination, but it did save some tissue for later examination. Scientists were finally able to determine that the animal had been a basking shark, *Cetorhinus maximus,*—a very large member of the shark family. The "sea monster" turned out to be nothing other than a well-known marine animal.

Explanations for Sea Monsters

Stories about supposed sea monsters often remain in circulation long after most scientists have concluded that such animals do not exist. For example, radio commentator Dennis L. Finnan recently based one of his regular programs on sea monsters, claiming that the numerous sightings reported were evidence that the theory of evolution was incorrect, and that the biblical explanation of creation was correct.

Genesis describes the existence of such "monsters," he claimed, so humans should not be surprised today to find that they still exist. Finnan even used the *Zuyio Maru* discovery to support his case, telling his listeners that the creature "was totally unknown and could not be classified." He said that scientists thought that the monster was probably a close relative of the dinosaurs, which, according to scientists, had all vanished from the earth.

For most believers, sea monsters have less cosmic significance, representing nothing more than interesting mysteries that have not been resolved to the satisfaction of the believer. As such, they have provided subject matter for endless newspaper articles, television documentaries, radio reports, and other media features.

Most scientists suggest that there are relatively simple explanations for sea-monster sightings. In many cases, the creatures that people have seen are probably nothing other than very large marine animals. The blue whale, for example, is the largest animal in the world, with a weight of more than 200 tons and a length of up to 100 feet. It would not be surprising for mariners to mistake a breaching blue whale for a sea monster or for shore dwellers to misinterpret the partially decayed carcass of one on the beach.

In other cases, sharks, octopus, squid, seaweed, or other plants or animals may be confused for sea monsters. Such confusion can occur easily on water, where the reflection and refraction of light can distort the image of an otherwise familiar object.

Further Reading

Ellis, Richard. *Monsters of the Sea*. New York: Knopf, 1994.

Hillier, Chris, and Owen Hill. *The Devil and the Deep: A Guide to Nautical Myths and Superstitions*. Dobbs Ferry, NY: Sheridan House, 1997.

Spaeth, Frank. *Mysteries of the Deep*. St. Paul, MN: Llewellyn Publications, 1998.

Suckling, Bob. *The Book of Sea Monsters*. New York: Overlook Press, 1998.

SeaWeb

SeaWeb is a multimedia education project designed to raise awareness about the world's oceans and the life within them. The project is concerned with eight major ocean problems: discharge of oil, chemical contaminants, and other debris into the ocean; effects resulting from commercial fishing operations; discharge of nutrients into the ocean; loss and disruption of coastal habitats; introduction of fish and other organisms into environments where they do not occur naturally; disruption and degradation of estuarine environments; reduction in stratospheric ozone; and global climate change.

SeaWeb has chosen a number of specific issues through which to deal with these general problems. Those issues are declines in swordfish and tuna populations, trawling and long-line fishing practices, shrimp and salmon farming, algal blooms, marine sanctuaries and marine zoning, shark finning, Florida Bay, and land-based toxic pollutants.

The organization uses a variety of tools to carry out its objectives. For example, it sponsors the Spokesteam project, which arranges for teams of scientists to speak with journalists around the United States about ocean problems. SeaWeb also sponsors Ocean Citations, a list of many articles dealing with ocean issues; and Oceans Briefing Book, an ongoing collection of scientific articles dealing with marine environmental issues.

SeaWeb also publishes *Ocean Update Newsletter*, a monthly publication featuring recent news, views, and events concerning marine and coastal environment and wildlife. It also provides an Ocean Events Calendar, listing ocean-related activities taking place around the world. The SeaWeb Aquaculture Clearinghouse also collects and makes available information on fish farming in North America.

Further Information

SeaWeb
1731 Connecticut Ave., N.W., 4th Floor
Washington, DC 20009
Telephone: (202) 483-9570
URL: http://www.seaweb.org
e-mail: seaweb@seaweb.org

Seaweed

See Aquatic Plants and Animals

Ships

See Boats and Ships

Shipwrecks

A shipwreck is an accident in which a boat or ship is damaged or destroyed with consequent loss of human life and/or cargo. The loss of ships at sea or on large lakes has been a possibility ever since humans first started building them. The oldest known shipwreck was discovered off the coast of Turkey by Dr. George Bass, of Texas A&M University, in 1984. The remains of the ship were dated to about 1300 B.C. The oldest shipwreck in deep water was found in 1999 by Dr. Robert Ballard of the Woods Hole Oceanographic Institute. This site contains two boats, believed to have been Phoenician cargo ships traveling from Tyre to Carthage in about 750 B.C. Their remains rest at a depth of about 500 meters (1,600 ft).

Shipwrecks are caused by a number of factors, some of which are natural events and some of which result from human error. In the first category are storms, shallow water or reefs undetected by the ship's crew, and fires caused by natural events, such as lightning. The second category includes, faulty design and/or improper construction or maintenance of the vessel, and careless navigation. Some possibly unavoidable causes include an old ship or aging equipment, which can result in malfunction of engines, steering equipment, sails (if used), or other apparatus on the ship.

Shipwrecks are of great interest to archaeologists. Each ship can be thought of as a small community at least partially characteristic of life at the time the ship was destroyed. The discovery and study of such ships gives modern scholars a glimpse of the kind of materials used at the time of the disaster, written records of life at the time, relationships among the crew, and other sociological phenomena. In addition, many wrecked ships hold cargo, ranging from food and building supplies to art objects and jewelry, that provide further information about a specific historical period.

Some famous shipwrecks in history are listed below. Those dated prior to 1833 are known primarily from underwater archaeological research conducted on ship remains. Those dated later are based on contemporary records of the events.

Fifteenth Century: A cargo ship from the fifteenth century was discovered and studied in 1999 by a research team led by Dr. Mensun Bound of the Oxford University Department of Marine Archaeology. The ship's remains were found in 70 meters of water off the Hoi An coast of Vietnam. The ship's hold contained more than 244,000 artifacts.

Sixteenth Century: In 1992, researchers from the Florida Bureau of Archaeological Research discovered the oldest shipwreck yet found along the coast of Florida—a Spanish galleon from the fleet commanded by Tristán de Luna Lunay Arellano (1510–1573). The ship was destroyed by a hurricane in 1559 and now lies at the bottom of Pensacola Bay. Thus far, about three thousand artifacts have been recovered.

Seventeenth Century: In June 1629, the VOC (Dutch East India Company) vessel *Batavia* was wrecked off the coast of Western Australia. The wreck, the second oldest in Australian waters, was first discovered in modern times by crayfisherman Dave Johnson in 1963. An account of the disaster and discoveries made in its remains can be found in *Islands of Angry Ghosts*, by Hugh Edwards.

Eighteenth Century: On 21 November 1996, Mike Daniels of Intersal, Inc., located a shipwreck in 7 meters (23 ft) of water off the coast of Beaufort, North Carolina, believed to be the *Queen Anne's Revenge*, once captained by the notorious pirate Blackbeard. Investigators found a treasure of artifacts, providing them with one of the best insights on the everyday life of pirating during the period. They concluded that the pirate had intentionally grounded the ship in an attempt to reduce the number of men then working for him.

1833: The English passenger ship *Lady of the Lake* struck an iceberg and sank 250 miles east of Cape St. Francis, St. Johns, Newfoundland. The ship was en route from Belfast to Quebec with about 275 passengers, of whom at least 215 were lost in the disaster.

1865: On 27 April, a boiler in the steamboat *Sultana* exploded, killing an estimated 1,547 passengers. The disaster was the worst shipwreck in American history. Most of those who died in the explosion were Union prisoners of war who had just been released from Confederate prison camps, such as Andersonville, and were on their way home. The ship carried no lifeboats or life jackets, so chances of survival were slim at best.

1904: An aging paddle-wheel excursion steamer, the *General Slocum*, caught fire in the East River off Manhattan, resulting in the death of 1,021 passengers. The boat had been declared safe for use only five weeks earlier by the U.S. Steamboat Inspection Service. More than five hundred of the passengers who had embarked at the Third Street Dock were under the age of twenty.

1912: The British luxury liner *Titanic* struck an iceberg on the evening of 14 April 1912 on its maiden voyage from Southampton to New York. The ship carried the most up-to-date engineering and

The April 1967 wreck of the tanker *Torrey Canyon*. Hulton-Deutsch Collection/Corbis.

safety equipment and was thought to be "unsinkable." An estimated 1,513 people died in what has arguably become the most famous shipwreck in history.

1915: The cruise ship *Eastland* capsized at the dock in Chicago as passengers boarded for an all-day excursion trip to Michigan City, Indiana. More then seven thousand employees of the Western Electric Company had been invited to the picnic, and early arrivals dashed to board the newest and finest of the six cruise ships chartered for the trip. The captain's efforts to prevent the ship from listing first to starboard, then to port, as passengers swarmed on board were eventually unsuccessful, and the ship rolled over, killing 812 people.

1942: On a voyage carrying military personnel from New York to England, the *Queen Mary* collided with and cut in half a British anti-aircraft cruiser. The cruiser, along with a number of other military craft, had been maneuvering back and forth across the bow of the *Queen Mary*, and it was unable to move quickly enough to avoid being struck. Three hundred thirty-eight British sailors were killed in the disaster.

1948: An unidentified Chinese troopship sank near the city of Yingkow, east of Beijing, while attempting to evacuate Nationalist troops from their position in Manchuria. An estimated six thousand persons died in the accident.

1956: On the night of 25 July 1956, in heavy fog, the Swedish-American passenger liner *Stockholm* collided with the Italian luxury liner *Andrea Doria*. The *Stockholm*'s bow, reinforced for the purpose of breaking through ocean ice, sliced into the starboard side of the *Andrea Doria*, causing the ship to capsize and sink almost exactly twelve hours later. Only fifty-two passengers and crew from the Italian ship were unaccounted for, and more than sixteen hundred were saved, thanks largely to the presence of the *Ile de France* in the vicinity of the accident.

1983: The Nile steamer *10th of Ramadan* caught fire and sank in Lake Nasser near the city of Aswan. Two hundred seventy-two passengers and crew died in the accident, and another seventy-five were missing.

1987: The passenger ferry *Dona Paz* collided with the oil tanker *Victor* off Mindoro Island, about 100 miles south of Manila. The ferry was licensed to carry eighteen hundred passengers, but more than three thousand people were on board at the time of the collision. Of that number, only twenty-four are known to have survived.

1993: On 17 February 1993, the Haitian passenger ferry Neptune capsized off the southwestern town of Jeremie. More than one thousand people were believed drowned in the disaster, with another three hundred surviving.

See also Boats and Ships

Further Reading

Delgado, James P. *Lost Warships: Great Ship-wrecks of Naval History*. New York: Facts on File, 2001.

Edwards, Hugh. *Islands of Angry Ghosts*. New York: Morrow, 1966.

Gould, Richard A. *Archaeology and the Social History of Ships*. Cambridge: Cambridge University Press, 2000.

Marx, Robert F., with Jenifer Marx. *The Search for Sunken Treasure: Exploring the World's Great Shipwrecks*. Toronto: Key-Porter Books, 1993.

Throckmorton, Peter, ed. *The Sea Remembers: Shipwrecks and Archaeology from Homer's Greece to the Rediscovery of the Titanic*. New York: Weidenfeld & Nicolson, 1987.

Further Information

Shipwrecks (magazine)
Society Secretary
BMRS
Flat 5 Trevone House
6 Kent Road
Southsea
Hants PO5 3EN
England
URL: http://www.geocities.com/shipwrecks_magazine
e-mail: shipwrecksworld@aol.com

An excellent general reference on shipwrecks is the web page of the Florida State University Program in Underwater Archaeology at http://www.anthro.fsu.edu/uw/links/directory_files/archprojects.html.

Sirens

See Mermaids and Mermen

Skeleton

See Aquatic Sports, Skeleton

Sling Psychrometer

See Hygrometer

Snow

See Precipitation

Softening of Hard Water

See Hardness of Water

Steam

Steam is the name given to water in the vapor state. The term is somewhat ambiguous, however, as steam may occur in a variety of impure and pure forms.

At room temperature, water exists in the liquid state. As heat is added to liquid water, its temperature rises. The addition of 1 calorie of heat to 1 gram of water causes an increase of 1°C. When the temperature of liquid water reaches 100°C (212°F), the addition of more heat does not cause a corresponding rise in temperature, but provides sufficient energy for some fraction of the water molecules present to escape from the liquid state and become gaseous (water vapor). This process is known as *boiling*.

As more heat is added to the liquid water that remains, more water molecules escape into the vapor phase. That is, more of the liquid water begins to boil. Eventually, all of the liquid water is converted to the vapor phase. The amount of heat needed to convert a sample of liquid water completely to the vapor phase at any temperature is known as the *latent heat of vaporization*. The latent heat of vaporization for water at 100°C is 540 calories per gram. That is, it takes 540 calories to completely convert 1 gram of liquid water at 100°C to the vapor phase.

The latent heat of vaporization of water is very high compared to that of other substances. The latent heats of a few other substances at their boiling points are listed below for comparison.

Latent Heat of Vaporization

Substance	Latent Heat of Vaporization (cal/g)
ammonia	
(liquid)	327
benzene	94
chloroform	65
ethanol	
(ethyl alcohol)	204
lead (liquid)	206
mercury	70
methanol	
(methyl alcohol)	263

States

At or near 100°C, the vapor phase above a sample of liquid water consists of two parts. One part consists of water vapor, or steam, formed by the evaporation (boiling) of the liquid water. The other part consists of liquid droplets that have escaped from the liquid into the vapor phase. The mixture of gaseous and liquid water is known as *wet steam*. The composition of the vapor phase is often referred to as its *quality*, the mass percentage of liquid to gaseous water. For example, if 100 grams of the vapor phase contains 98 grams of gaseous water and 2 grams of liquid water, the system would have a quality of 98 percent.

By contrast, *dry steam* is steam that contains no liquid water. It has a quality of 100 percent. Dry steam is formed when a sample of liquid water has been heated to the point that every water molecule has been converted to the vapor phase. Dry steam is also known as *saturated steam*. Saturated steam is in an unstable state in that the loss or gain of any heat energy converts it to another form. For example, if heat is removed from a sample of saturated steam,

some of the steam will condense to form liquid water. If heat is added to saturated steam, however, it remains in a gaseous state although its temperature begins to increase. Dry steam with a temperature greater than 100°C is known as *superheated steam*. Small increases in the heat added to dry steam causes relatively significant increases in its temperature. For example, the addition of 1 calorie of heat to steam causes an increase of about 2°C, about twice that attained with liquid water.

Applications

Steam is one of the most widely used industrial chemicals, and it owes its popularity to a number of physical and chemical properties. First, it is produced from perhaps the most abundant chemical on earth, water. When boiled, water is converted to a colorless, odorless, tasteless, nontoxic gas—steam. After use, steam can be condensed and recovered as pure water, which can then be reused, usually with little or no loss.

Also, steam can be transported easily from the place where it is produced to sites where it is to be used. Since water and steam are relatively inactive chemically, the latter can be transported through pipes and other conduits that require little if any special coating or preparation.

Finally, a given mass of steam contains a relatively large amount of heat energy. As shown above, few commonly available materials release as much heat in changing from the gaseous to the liquid phase (latent heat of condensation/vaporization) as does steam. Although the cost of converting liquid water to steam is relatively high because of the heat needed to bring about the change of state, that heat can be recovered easily from relatively small quantities of steam.

Steam has a great variety of industrial, commercial, residential, and other applications. For example, it is used to heat homes, office buildings, and other structures. It is also used to transfer heat energy or provide heat for many different industrial processes. Steam has been widely used for more than two centuries in steam engines. In the simplest such engine, the force of steam is used to push a piston upward in a cylinder. When the steam condenses, a partial vacuum is left behind in the cylinder and the piston returns to its original position.

Steam is also used in a number of chemical processes. It reacts with coke and air, for example, to produce water gas—a mixture of carbon monoxide, hydrogen, carbon dioxide, and nitrogen. Steam also reacts with hydrocarbons to form hydrogen gas, along with other by-products. Steam is also used as a cleaning agent, in the vulcanization of rubber, and in the secondary recovery of petroleum. *See also* Steam Engine; Steam Pressure Sterilizer

Further Reading

Wagner, Wolfgang, and Alfred Kruse. *Properties of Water and Steam*. Berlin: Springer-Verlag, 1998.

Further Information

International Association for the Properties of Water and Steam
URL: http://www.iapws.org

The association does not answer individual questions, but its website contains useful information along with the answers to some frequently asked questions.

Steam Engine

A steam engine is a device by which heat energy is converted into mechanical energy by means of a piston that moves up and down within a cylinder. A steam engine is an example of an external combustion engine since the fuel that operates the device is burned outside the cylinder in which work is done.

Steps Toward the Steam Engine

Humans have known that steam can do work for many hundreds of years. The Greek engineer Hero (born ca. A.D. 20; date of death not known) invented a glass device consisting of a sphere to which two bent arms were attached. He filled the sphere with water and heated it with a flame. As water inside the sphere began to boil, steam escaped through the bent arms, pushing the device around in a circle. Some modern-day lawn sprinklers operate on a similar principle, using only the force of water, not steam, to drive the arms.

Most early "steam engines" were little more than toys designed to amuse people or to demonstrate scientific principles. The first steam engines intended for actual use were not constructed until the late seventeenth century. An awkward beginning on the research was the work of Dutch physicist Christiaan Huygens (1629–1695), who invented a "gunpowder engine" in 1680. Huygens's device consisted of a metallic cylinder containing a charge of gunpowder and a piston. When the gunpowder was ignited, the force of the explosion moved the piston outward. After a few moments, the cylinder could be recharged with gunpowder and fired again.

There really wasn't any application for Huygens's gunpowder engine, at least partly because it had to be recharged before each use. Also, working with gunpowder was very dangerous. However, Huygens's assistant, Denis Papin (1647–1712), had a glimpse of the way in which the gunpowder engine could be refined and improved. He wrote the following in 1690:

"Since it is a property of water that a small quantity of it turned into vapour by heat has an elastic force like that of air, but upon cold supervening is again resolved into water, so that no trace of the said elastic force remains, I concluded that machines could be constructed wherein water, by the help of no very intense heat, and at little cost, could produce that perfect vacuum which could by no means be obtained by gunpowder." (quoted in Derry and Williams 1993, 315)

Papin proceeded to build a device based on this principle. The device consisted of a tube about 12 centimeters (2.5 in.) in diameter with a little water in the bottom and a piston fitted inside it. When the cylinder was heated, the water converted to steam and pushed the piston upward.

The First Steam Engines

As with all earlier and similar devices, Papin's invention had no practical use. But it established the principle on which all steam engines were later to be based.

The first practical steam engine was built by the English engineer Thomas Savery (ca. 1650–1715) in about 1690. Savery's device did not quite fit the model of the steam engine since it had no piston. However, it was a true steam engine since it used steam (and a vacuum) to perform useful work.

The purpose of Savery's engine, eventually given the name the Miner's Friend, was simply to remove water from coal mines. The problem of flooding in mines was a serious one in England since it posed a limit on how deeply and extensively miners could work. Savery's solution to this problem was to construct a large metallic cylinder with a small amount of water in it. A long tube connected the cylinder to the bottom of the mine shaft. When water in the cylinder was heated, steam was produced, pushing out all of the air in the cylinder. When the cylinder was cooled, air was prevented from returning to the cylinder, leaving a partial vacuum inside it. Normal atmospheric pressure outside the cylinder then pushed water from the bottom of the mine up into the cylinder, from which

it could be emptied. By alternately heating and cooling the cylinder, the operator was able to continue pumping out the mine.

The final stage in the design of the modern steam engine came about twenty years later when the English engineer Thomas Newcomen (1663–1729) reintroduced the piston to what was essentially Savery's device. Newcomen's steam engine occupied a whole section of Dudley Castle in Worcestershire and first began operation in 1712. For its design, a brick boiler was installed on the bottom floor of the building in which the steam engine was housed. Attached to the top of the boiler was a long brass cylinder containing the piston. Attached to the top of the piston was a horizontal wooden beam that could rock up and down. To start the engine, heat was applied to the boiler, where water was converted to steam. The pressure applied by the steam was about the same as atmospheric pressure so that the horizontal beam did not move up or down.

When cold water was sprayed on the cylinder, however, a partial vacuum was produced. Atmospheric pressure pushing down on the top of the piston forced it downward into the cylinder, causing one end of the horizontal beam to move upward. When steam was reintroduced into the cylinder, the piston was forced back upward again, restoring the beam to its original position.

As with Savery's Miner's Friend, the purpose of Newcomen's machine was to remove water from mines, in this case, tin mines. The machine was able to make twelve strokes a minute, lifting about 40 liters (10 gall) of water a height of 45 meters (153 ft) from the mine to the surface. Newcomen's engine was about five times as efficient as the Miner's Friend. Its primary shortcoming was simply that machinery was not available for making cylinders and pistons that fit tightly to each other.

Perhaps the most important break-through in the development of steam engines came from Scottish engineer James Watt (1736–1819). Called upon to repair one of Newcomen's engines, Watt noted a number of improvements that would vastly improve the efficiency of steam engines. One of his most important changes was to use two cylinders instead of one. While steam was being introduced into one cylinder, it was being cooled in the other. This development removed the delay that occurred when a single cylinder was first cooled and then heated. Watt also invented gears that converted the back-and-forth motion of the horizontal beam into rotary motion and a centrifugal governor that kept the engine running at a constant speed.

Applications

The steam engine can really be said to have powered the Industrial Revolution. It quickly replaced many other, more traditional, forms of power. It was far more effective than human or animal labor, and it had the convenience of being portable, capable of being used almost anyplace that water and fire could be produced.

By the middle of the eighteenth century, steam engines had begun to take over almost every known manufacturing task, and was used in textile operations, mining, rolling mills, sawmills, pumping plants, and printing presses. They also brought about a revolution in the field of transportation, where they were used to power boats and ships, trains, and even automobiles. In the 1880s, they were first used to drive turbines, providing both mechanical energy and a way of making electrical energy.

By the early 1900s, however, the role of steam engines in commerce and industry had begun to decline dramatically. The one problem that could never be solved was the very low efficiency of steam engines. In many cases, no more than a quarter of the energy supplied to the engine came out in the form of useful work. As long as no other type of engine was available, the steam engine solved many industrial problems satisfactorily, but with the development of a better option—the internal combustion engine—the day of the steam engine was over. *See also* Boats and Ships; Steam

Further Reading

Derry, T. K., and Tevor I. Williams. *A Short History of Technology from the Earliest Times to A.D. 1900*, chap. 11. New York: Dover Publications, 1993.

Mills, Richard. *Power from Steam: A History of the Stationary Steam Engine*. Cambridge: Cambridge University Press, 1993.

Rutland, Jonathan. *The Age of Steam*. New York: Random House, 1987.

Steam Pressure Sterilizer

A steam pressure sterilizer is a device that can be filled with high-pressure, high-temperature steam for the purpose of sterilizing materials. Such devices are widely used in the medical profession to sterilize instruments, operating gowns, masks, bed linens, and other objects. In similar forms, a steam pressure sterilizer can be used to carry out certain types of chemical reactions that occur only at high temperatures and pressures and to test the properties of certain materials at such conditions.

The earliest form of the steam pressure sterilizer was invented in 1679 by the French physicist Denis Papin (1647–1712). Papin called his invention a *steam digester*, a device that was the forerunner of both the steam pressure sterilizer and the modern pressure cooker. Over time, Papin's device became known as an *autoclave* when it was used for the purpose of sterilizing objects. The term *autoclave* means "self-closing" and arises from the fact that the lid on the device seals automatically when steam pressure inside the container pushes upward on it. The term has somewhat fallen into disuse because many modern types of steam pressure sterilizers are not self-sealing but must be closed and sealed from the outside. The process of sterilizing materials and objects, however, is still generally called *autoclaving*.

The general principle on which the steam pressure sterilizer is based is that most disease-causing organisms are killed by high temperature. However, the lethal temperature for such organisms is much higher in the absence of moisture than in its presence. Hot steam, then, kills such organisms more efficiently than would hot air by itself.

Steam is also an effective sterilizing agent because it gives up so much heat when it changes from a gaseous to a liquid state. When cold equipment is placed into a sterilizer, steam condenses on the equipment, giving up its latent heat of vaporization, about 540 calories per gram. The product of this change of state—water—is a colorless, odorless, safe, nonpolluting substance that can easily be recycled in the sterilizer.

Steam is also an effective sterilizer because it coats and covers all of the surfaces—including small openings, cracks, and other irregularities that may be present—with which it comes into contact.

As with a pressure cooker, *pressurized* steam is used to raise the boiling point of the water inside the device. Early autoclaves used steam under a pressure of about 115 kPa (kilopascals) to attain boiling points of about 121°C (250°F). (One atmosphere of pressure is equal to about 101 kPa.) Today's steam pressure sterilizers use pressures of about 220 kPa to attain boiling points of about 134°C (273°F).

Further Reading

Reichert, Marimargaret, and Jack H. Young, eds. *Sterilization Technology for the Health Care Facility*. Gaithersburg, MD: Aspen Publishers, 1993.

"Steam Sterilizers. How Do They Work?" http://www.medisys.co.nz/art_1.htm.
Webster, John G., ed. *Encyclopedia of Medical Devices and Instrumentation*. New York: John Wiley & Sons, 1988.

Stevinus, Simon (1548–1620)

Simon Stevinus (also Stevin) is often credited with founding the science of hydrostatics—the study of the properties of liquids at rest, especially with regard to the pressure and forces they exert on objects immersed in them.

Personal History

Stevinus was born in Bruges, Flanders (now Belgium), in 1548, and died in The Hague, the Netherlands, in 1620. Stevinus left his position as a bookkeeper in Antwerp in 1583 to attend the Latin school in Leiden, after which he entered the University of Leiden (at the age of 35).

While serving in the Dutch army, Stevinus developed a system of sluices (artificial waterways with gates used to control flooding) that could be opened quickly. This system provided the army with a way of flooding the country should it be invaded by an enemy.

Professional Accomplishments

Stevinus discovered a number of fundamental laws and principles in hydrostatics. Perhaps best known of all was his study of the so-called hydrostatic paradox. This term refers to the fact that the pressure exerted by a liquid on the bottom of a vessel depends only on the height of the liquid column pushing down on the bottom and not on the shape of the vessel itself.

Stevinus also demonstrated that the pressure exerted by a liquid acts upward as well as downward. Although he did not explicitly state this principle (that step was taken by Pascal in the mid-1600s), he seems to have understood that the pressure exerted by a liquid is the same in all directions.

Finally, Stevinus studied the conditions necessary for equilibrium in floating bodies. He found that the center of gravity in such bodies must lie in the same vertical line as that of the displaced liquid.

Further Reading

Dijksterhuis, E. J. *Simon Stevin: Science in the Netherlands around 1600.* The Hague: Martinus Nijhoff, 1970.

Stockholm International Water Institute

The mission of the Stockholm International Water Institute (SIWI) is to contribute scientific, technical, and awareness-building skills to support and encourage international efforts to deal with the world's escalating water-supply crisis. SIWI works toward this goal by educating decision makers from countries around the world about the need to act on and limit the consequences of the escalating water crisis. The institute also offers policies and scientifically sound solutions to water-related problems. SIWI also attempts to build awareness of water issues across a wide range of audiences by disseminating information on water problems, the implications of those problems, and some possible solutions.

One of the institute's most important functions is the annual Stockholm Water Symposium, held each August. The purpose of the symposium is to provide a link among science, practice, policy making, and decision making related to worldwide water resources issues. The symposium offers a multidisciplinary approach, with representatives from the natural sciences, engineering, social science, policy making, and other fields.

SIWI is also involved in the awarding of four major water prizes: the Stockholm Water Prize, the Stockholm Junior Water Prize, the Stockholm Industry Water Award, and the Swedish Baltic Sea Water Award. In addition, the institute publishes a number of books and reports on water-related issues, such as *Water and Development in Developing Countries*, *Water Security for Multi-National Water Systems*, *Our Struggle for Water*, *From the Baltic Sea to Lake Victoria*, *Networking between Parties around the Baltic Sea and Lake Victoria*, and *Water Harvesting*. SIWI also published a quarterly magazine on water issues, *Stockholm Water Front*.

Further Information

Stockholm International Water Institute
Sveavägen 59
SE-113 59 Stockholm
Sweden
Telephone: +46 (0) 8 522 139 60
Fax: +46 (0) 8 522 139 61
URL: http://www.siwi.org
e-mail: siwi@siwi.org

Stommel, Henry (1920–1992)

Stommel gained fame as an oceanographer largely because of his research on ocean currents, especially his studies of the Gulf Stream. Prior to his research, only very general information about this great ocean current had been collected. By the mid-1950s, however, Stommel had worked out a mathematical model that explained most of the physical features of the Gulf Stream.

Personal Life

Stommel was born in Wilmington, Delaware, on 27 September 1920. He earned his bachelor's degree in mathematics from Yale University in 1942, and then stayed on as an instructor in mathematics and astronomy. In 1944, he left Yale to take a post as research associate at Woods Hole Oceanographic Institution, where he remained until 1959. He then accepted an appointment as professor of oceanography at the Massachusetts Institute of Technology. During the period 1960–1963, Stommel also taught at Harvard University.

In addition to his academic posts, Stommel established several field stations where he carried out his research. Perhaps the best known of these was the Bermuda Biweekly Hydrographic Station (PANULIRUS), established in 1954. The station was an important source of information on the physical properties of the Gulf Stream, which flows around Bermuda.

Stommel wrote two important books about the Gulf Stream, one (*The Gulf Stream*, 1965) meant primarily for researchers, and the other (*A View of the Sea*, 1987) intended for the general public. He was elected to the National Academy of Sciences in 1962 and awarded the National Medal of Science in 1989. Stommel died in Boston on 17 January 1992.

Professional Accomplishments

The Gulf Stream was first mentioned in the early sixteenth century by the Spanish navigator Juan Ponce de León. But it was more than two centuries before the first truly scientific description was completed on the Gulf Stream. That study was done by Benjamin Franklin. By the mid–twentieth century, scientists had a relatively good understanding of the path followed by the Gulf Stream, its rate of flow, its temperature, and other physical properties, but virtually no theoretical research had been done on the Gulf Stream or other ocean currents. Oceanographers had essentially no idea as to how the Gulf Stream was formed, the forces that drove it, or how water flowed, if at all, beneath the ocean's surface.

During his tenure at Woods Hole, Stommel developed his mathematical theory describing the flow of water in the Gulf Stream. He showed that the water moving in the current, much like the flow of air in Earth's atmosphere, is caused by Earth's rotation.

Warm water in the equatorial regions is pushed westward against the eastern coast of North America, and then northward as far as the Grand Banks of Newfoundland. It then flows eastward across the North Atlantic to the western coast of Europe. The circular pattern exhibited by the Gulf Stream, Stommel showed, is similar to the movement of large air masses in the atmosphere.

Stommel also demonstrated that the movement of water in the Gulf Stream is more complex than earlier observers had thought. For example, he pointed out that the northerly flow of water on the surface along the coast of North America is matched by a southerly flow of water far below the ocean's surface.

Stommel later generalized his study of the Gulf Stream to ocean currents in other parts of the world. His research is now generally regarded as the pioneering research on the topic of ocean currents.

Further Reading

Warren, Bruce A., and Carl Wunsch, eds. *Evolution of Physical Oceanography: Scientific Surveys in Honor of Henry Stommel*. Cambridge, MA: MIT Press, 1981.

Submarines and Submersibles

The terms *submarine* and *submersible* are used to describe vessels that are designed to be operated underwater. In many instances, no distinction is made between the meaning of these two terms. In official documents on underwater vessels, the U.S. Navy, for example, makes no distinction between the terms, using them interchangeably ("Silent Defense"). Other individuals and agencies suggest that a distinction be made between the two terms. They prefer to use the term *submarine* for naval vessels used in warfare and maintenance of defense. In recent years, vessels meeting this definition have also found a number of other uses, as for tourism and as individual recreational vehicles. By contrast, the term *submersible* can then be reserved for vehicles used primarily for scientific research.

History

The concept of building a submersible chamber that would house humans goes back at least two thousand years. Both the ancient Greeks and the ancient Chinese designed such vessels and put them to use in warfare or for the exploration of near-shore waters. Leonardo da Vinci (1452–1519) also conceived of an underwater vessel capable of sinking other ships. His design consisted of a simple shell large enough to hold one person with a conning tower at the top.

The first known treatise on submarines was written in 1578 by an Englishman named William Bourne, described variously as a former gunner, a naval officer, an innkeeper, and a scientific dilettante. Bourne described the use of ballast tanks to make a vessel rise and sink in the water, the basic principle behind any submarine design. He accompanied the design of his submarine with a description, beginning as follows:

It is possible to make a Ship or Boate that may goe under the water unto the bottome, and so to come up again at your pleasure. [If] Any magnitude of body that is in the water . . . having alwaies but one weight, may be made bigger or lesser, then it Shall swimme when you would and sinke when you list. ("World Submarine History Timeline")

There is no record that Bourne's vessel was ever constructed. Credit for the invention of the first working submarine, therefore, generally goes to the Dutch inventor Cornelius Jacobszoon van Drebbel (1572–1633). Drebbel's vessel was made of a greased leather coating over a wooden framework. The vessel was about 5 meters (15 ft) in length and was propelled by oars sticking out

through leather seals. Fresh air was provided to oarsmen by means of snorkel tubes held in place above the water's surface by means of floats. Drebbel demonstrated his invention for King James I in 1620. He sailed his submarine in the Thames River, just below the surface of the water.

Later designs improved the vessel's diving capabilities, and Drebbel's best models were eventually able to travel at depths of up to 20 meters (60 ft) at speeds of about 10 kilometers per hour (6 mph).

One of the first American submersibles was designed by an engineering student at Yale University, David Bushnell (ca. 1742–1824), in 1775. Bushnell's submersible, called the *Turtle*, held a single person, who maneuvered the craft with a complex system of valves, air vents, and ballast tanks. The *Turtle* was propelled by two hand-cranked screw propellers. It first found use in battle on 6 September 1776 when Sergeant Ezra Lee attempted to use the submersible to attach a mine to the British flagship HMS *Eagle* in New York Harbor. The British discovered the *Turtle* before the mine could be released, however, and the effort failed.

One of America's greatest inventors, Robert Fulton (1765–1815), was also interested in submersibles. While living in France, he designed an underwater vessel along the general lines of the *Turtle*. He eventually built his vessel and called it the *Nautilus*—a ship 6 meters (20 ft) long, powered by a hand-cranked screw propeller with diving planes that improved the craft's vertical mobility. In a demonstration of the submarine's effectiveness, Fulton successfully blew up a target ship with dynamite released from the *Nautilus*. Fulton tried to interest the French government in his invention, but neither the French nor the English, whom Fulton later approached, had any interest in his work.

The Civil War provided an impetus to research on submarines, with the Confeder-

The submersible *Aegir*, named for the Norse god of the sea, was designed for depths up to 180 meters. *OAR/National Undersea Research Program; Makai Range Inc.*

ate Navy eventually building four underwater vessels. One of those ships, the CSS *Hunley*, rammed the USS *Housatonic* in Charleston Harbor, South Carolina, marking the first successful attack by a submarine on another ship. During the attack, one of the *Hunley*'s torpedoes exploded, destroying both ships.

By the early 1890s, a number of nations had begun to envision a role for submersibles in their navies. Germany, France, Spain, and Greece had all added experimental models to their forces. In 1898, the U.S. Navy launched its first submarine, the USS *Holland*. The *Holland* was 16 meters (53 ft) long with a displacement of 75 tons. The *Holland* operated on a gasoline motor and cruised at a speed of 7 knots on the surface.

Submarine technology developed rapidly in the first two decades of the twen-

tieth century. By the middle of World War I, submarines from the German navy were roaming the Atlantic Ocean with essentially unrestricted freedom. They eventually sank more than 11 million tons of Allied shipping, with the subsequent loss of thousands of human lives.

By the start of World War II, most major naval powers had acknowledged the role of submarine warfare and had built up substantial underwater fleets. During the war, the Germans were still enormously successful in the Atlantic, destroying 14 million tons of shipping, while U.S. ships were equally successful in the Pacific, destroying nearly 1,400 Japanese merchant and naval ships.

The most important technological advance in submarine design occurred in the early 1950s with the suggestion by Captain Hyman Rickover (1900–1986) that nuclear power be used to propel submarines. In contrast with fossil-fueled vessels that could remain at sea and underwater for only limited periods of time, nuclear-powered vessels could remain on patrol indefinitely. Their return to base was necessitated not by lack of fuel but by equipment malfunction, completion of mission, or some other factor.

Submarine Design

The principle on which submarines operate is that of buoyancy. An object that displaces a greater weight of water than the weight of the object itself will float. An object that displaces less weight than its own weight will sink. Large ships made of metal are able to float because their interiors are filled with air, which weighs less than the amount of water they displace.

Submarines control their weight by means of ballast tanks. A ballast tank is a large tank that can be filled with air or water. When the ballast tank is filled with air, the total weight of the submarine is reduced. When it is filled with water, its weight is increased.

When submarines operate on the surface, their ballast tanks are filled with air. The vessel's overall weight is less than that of the water it displaces. To dive, water is pumped into the vessel's ballast tanks. As more water is added to the tanks, the submarine's weight increases until it is greater than that of the water displaced. The depth to which the submarine dives can be controlled by the amount of water pumped into (or out of) its ballast tanks.

Vertical movement of the vessel is controlled by short airplane-type wings called *hydroplanes* on the front and back of the submarine. Hydroplanes can be rotated upward and downward to direct the vessel's vertical movement. Turning motion in the submarine is controlled, as it is in airplanes, by means of a rudder attached to the vessel's tail.

Types of Submarines and Submersibles

Submarines and submersibles are used in a number of applications, the best known of which may be military operations. Some other applications of submarines and submersibles include

Luxury submarines: A relatively small number of submarines small enough to fit on large yachts are built each year to satisfy wealthy individuals who want to explore the underwater world in their own personal vessel.

Tourist submarines: A number of amusement parks, resorts, and other vacation facilities now operate submarines to provide guests with interesting undersea expeditions, such as visits to underwater gardens or observations of schools of fish. Such facilities are found at the edge of most major oceans and seas. Tourist submarines operate, for example, in Barbados, Grand Cayman, Aruba, Cozumel, and St. Thomas in the Caribbean Sea.

Recreational submarines: Building and operating one's own submarine has become a popular hobby in many parts of the world. Each year since 1989, the Foundation for Underwater Research and Education has sponsored the International Submarine Races for privately owned and operated submarines. At the 2001 race, a one-person submarine from the University of Quebec set a new world's record of 7.192 knots (about 8.2 miles per hour).

Research submersibles: Submersibles built for research have a different function than those built for any other purpose. Their purpose is to transport researchers to some parts of the ocean not readily accessible by any other means and to provide them an environment in which they can carry out their research. Currently, there are forty-one research submersibles in existence. They are used for surveying and drilling for oil in offshore sites, as well as for basic oceanographic research. The United States owns twelve of these vessels; Russia, Ukraine, and France, six each; Japan, Canada, and Romania, two each; and Switzerland, Germany, Finland, South Korea, and Bulgaria, one each.

See also Oceanography, Oceans and Seas

Further Reading

Bagnasco, Erminio. *Submarines of World War Two*. Herndon, VA: Cassell Academic, 2000.

Friedman, Norman. *Submarine Design and Development*. Annapolis, MD: Naval Institute Press, 1984.

Lawliss, Chuck. *The Submarine Book*. Springfield, NJ: Burford Books, 2000.

Weir, Gary E., "Silent Defense. One Hundred Years of the American Submarine Force," http://www.chinfo.navy.mil/mavpalib/cno/n87/history/fullhist.html.

"World Submarine History Timeline," http://www.submarine-history.com/NOVAone.htm.

Suez Canal

See Canals

Surface Tension

Surface tension is the attractive force exerted on the upper layer of molecules of a liquid by the molecules below it. The force accounts for the fact that some liquids appear to have a "skin" that covers their surface.

Explanation

All molecules exert some force of attraction on each other, a stronger force in some cases than in others. For example, some molecules, known as *polar molecules*, contain an unequal distribution of electrical charge. One or more parts of the molecule are somewhat more negative and one or more other parts are more positive. When two molecules of such a substance are close to each other, the positive part of one molecule exerts a force of attraction on the negative part of the second molecule.

Water is such a molecule. A water molecule contains two regions that are relatively positive (the two hydrogen atoms in the molecule) and one region that is relatively negative (the oxygen atom). When two water molecules are close to each other, the positive part of one water molecule is attracted to the negative part of a second water molecule.

For water molecules below the surface of a liquid, this force of attraction is exerted equally in all directions. That is, a molecule in the center of the liquid is pulled on by molecules surrounding it on all sides. Water molecules at the surface of the liquid are pulled downward and sideways by neighboring water molecules. But no comparable attractive force exerts above the surface molecules. The attractive force experienced by surface water molecules accounts for the property known as *surface tension*.

Magnitude

All liquids display some amount of surface tension, although the amount differs widely from substance to substance. The chart below lists a few common substances and their surface tensions. The surface tension is expressed in terms of the liquid named covered by an atmosphere of air. The presence of some gas other than air produces a somewhat different surface tension depending on the extent, if any, to which the molecules of that gas also pull on surface molecules of the liquid.

Surface tension is a function of temperature. That is, as the temperature of the liquid increases, the surface tension decreases. This trend is predictable. As the temperature of a liquid increases, so does the rate at which its molecules vibrate. An increase in

Surface Tension of Various Liquids

Substance	Temperature	Surface Tension (dyn/cm)
n-butanol (n-butyl alcohol)	20°C	20
ethanol (ethyl alcohol)	20°C	22
acetone	20°C	23
toluene	20°C	29
glycerol	20°C	63
water	0°C	76
water	20°C	73
water	50°C	68
water	100°C	59
mercury	20°C	486

rate of vibration reduces the force of attraction between molecules and, hence, the surface tension of the liquid.

Applications

The surface tension of a liquid (such as water) explains many properties and applications of that liquid. Capillarity is an example. Capillarity is the tendency of a liquid to rise in tubes of small diameter. In such tubes, a liquid sticks to the inside surface of the tube and its surface is then pulled upward as a result of surface tension. In the case of water, capillarity is a property of some significance because water molecules are polar and tend to stick both to the inside surface of a tube (*adhesion*) and to each other (*cohesion*) in the form of surface tension.

Surface tension also explains the fact that both animate and inanimate objects that might be expected to sink in water can actually remain on its surface. The insect known as the water strider (*Gerris paludum*), for example, is able to walk across the surface of water because its weight is distributed across a wide area of water by its long legs and the water's surface tension. A needle, razor blade, or other light piece of metal can also be made to float on water if placed there with care. The skinlike quality produced by water's surface tension supports the weight of the object.

Detergents are added to water used to wash clothing because the detergents reduce the surface tension of the water. They do so because detergent molecules insert themselves between water molecules, reducing the force of attraction and, hence, the surface tension. With reduced surface tension, water is able to soak into the material being washed, increasing the efficiency with which dirt is removed.

Tents used for camping are sometimes marked with a warning label indicating that they should not be touched when wet. The reason for this warning is based on the surface tension of water. When water "beads" on the tent's surface, water droplets cover openings in the tent fabric, preventing water from soaking into the interior. If the wet fabric is touched, surface tension is broken, and this protection is no longer available.

Water droplets are a common example of surface tension. When a small amount of water is separated from a larger body, it tends to take the shape of a sphere because surface tension on its outer surface tends to pull the water inward in all directions equally. In actual practice, a droplet of water does not remain a sphere, however, as the force of gravity tends to pull downward on it, giving it its familiar teardrop shape. *See also* Capillary Action; Hydrogen Bonding; Properties of Water

Surfrider Foundation

The Surfrider Foundation is a grassroots, nonprofit environmental organization founded in 1984 to protect the earth's coastal environments. Surfrider currently has fifty chapters located on the East, West, Gulf, Puerto Rican, and Hawaiian coasts, with affiliates in four foreign countries: Australia, Brazil, France, and Japan. Surfrider currently has more than 25,000 members in the United States and other countries.

Surfrider conducts most of its work through its local chapters. Its projects are designed to educate students, the general public, and coastal-management agencies about local, regional, national, and international environmental issues by providing them with information and tools that they can use in their own local areas.

Some examples of Surfrider programs currently in effect include the following:

- Respect the Beach—a coastal education teaching unit for K–12 students
- Blue Water Task Force—a program of coastal water testing and monitoring
- Beachscape—a program to document and map coastal environmental resources

- Clean Water—the Year 2000 Surfrider Foundation National Campaign
- Snowridge Project—a program dealing with environmental issues in the mountains
- Pratte's Reef—the organization's artificial reef project

Surfrider has established the Thomas Pratte Memorial Scholarship Fund to support academic research on the coastal environment. The organization also publishes *Making Waves*, a bimonthly publication covering coastal issues worldwide, Surfrider chapter news, and the latest advances in scientific research.

Further Information

Surfrider Foundation USA
122 S. El Camino Real, #67
San Clemente, CA 92672
Telephone: (949) 492-8170
Fax: (949) 492-8142
URL: http://www.surfrider.org
e-mail: info@surfrider.org

Swimming

See Aquatic Sports, Swimming

Synchronized Swimming

See Aquatic Sports, Synchronized Swimming

T

Thales (624–546 B.C.)

Thales is probably the first, and certainly one of the most famous, of all the Greek natural philosophers. One of the accomplishments for which he is best known today is his development of the concept of elementary substances, and his suggestion that water is the one most basic substance in the world.

Personal History

Thales was born in the city of Miletus in 624 B.C. Little is known with certainty about his life, although most scholars agree that he traveled widely throughout the ancient world. He is said to have visited Egypt and Babylonia, where he must have learned a great deal about the astronomical, geographic, mathematical, and physical arts of these lands.

When he returned to Greece, he began his own theoretical research about the nature of the world. An indication of the approach he took to this work was his decision to think of points and lines and dimensionless objects, the first time this method had been used in mathematical inquiry.

Professional Accomplishments

For all his accomplishments, the one that most impressed his colleagues was his prediction of a solar eclipse on 28 May 585 B.C. At the time, Greek scholars were largely unaware of the techniques that could be used to make such predictions, although Thales must almost certainly have learned such techniques during his visits to Egypt and Babylonia.

Thales's attack on the problem of *elements* had consequences far beyond what might have been expected from that work. He took the position that the natural world in which we live is not really as complex as it appears to be, but had been formed by transformations in one or more fundamental substances known as elements. Upon reflection, Thales decided that water must be that one elemental substance from which everything else in the universe is made.

His choice of water as the one and only elemental substance makes sense when one recalls the widespread occurrence of water in the world and its crucial role in all living things. But the specific choice he made was much less important than his intellectual decision that an element—or elements—really did exist in the world. *See also* Earth, Air, Fire, and Water

Further Reading

Asimov, Isaac. *Asimov's Biographical Encyclopedia of Science and Technology*. 2d rev. ed., 2–3. Garden City, NY: Doubleday & Company, 1982.

Porter, Roy, consult. ed. *The Biographical Dictionary of Scientists*. 2 ed., 664–65. New York: Oxford University Press, 1994.

Tharp, Marie (1920–)

Marie Tharp is perhaps best know for her research on the Mid-Atlantic Ridge, a mountain chain that runs down the middle of the Atlantic Ocean. In the early 1950s, she and colleague Bruce Heezen constructed a map showing the topographic structure of the bottom of the North Atlantic ocean. They later produced additional maps of underwater features that have proved very useful to oceanographers.

Personal History

Marie Tharp was born in Ypsilanti, Michigan, on 30 July 1920. She received a bachelor's degree from Ohio University in 1943, a master's degree in geology from the University of Michigan in 1944, and a second bachelor's degree in mathematics from the University of Tulsa in 1948. After graduating, she accepted the offer of a position working with famed geologist and oceanographer Maurice Ewing at Columbia University. Two years later, Ewing, Tharp, and the rest of the geophysical laboratory at Columbia moved to the Lamont Geological Observatory in Palisades, New York, where Tharp was to remain until she retired in 1983.

Professional Accomplishments

The Mid-Atlantic Ridge was first discovered during the exploratory voyages of the HMS *Challenger* in the 1870s. Additional information about the ridge was obtained between 1925 and 1927 during an expedition by the German research vessel *Meteor*. Until the 1950s, however, scientists had relatively little detailed information about the structure of the ridge.

In 1952, Tharp began collecting and collating data obtained during oceanic voyages made by the Woods Hole Oceanographic Institution's (WHOI) research ship *Atlantis*. On these voyages, extensive studies were made of seafloor profiles. These profiles led Tharp to the conclusion that the ridge was bisected along its long axis by a valley, now known as a *rift valley*. Soon after, Tharp determined that the Mid-Atlantic Ridge was only one portion of an extensive underwater mid-ocean mountain range that circles the planet.

Tharp's work on sea-bottom profiles suggested to her the need for a detailed topographic map of the ocean bottoms, similar to those already available for nearly every part of the earth's continents. She was joined in the work of preparing such a map by her colleague, Heezen. After publishing their first underwater map in 1952, Tharp and Heezen continued their cartographic studies of ocean floors until 1977, when Heezen died. In that year, they published their now-famous World Ocean Floor map, giving the most complete detail of sea-bottom topography ever produced.

The Tharp-Heezen maps were important for more than their illustrative value, however. The presence of the rift valley running down the middle of ocean floor mountains supported the concept of continental drift. The theory that the upper layer of the earth's crust consisted of enormous "plates" that are constantly in motion had been proposed as early as 1912 by the German meteorologist and geologist Alfred L. Wegener (1880–1930). But Wegener's ideas were met with suspicion because there was little concrete evidence for the existence or movement of such plates. By the 1970s, that evidence had begun to accumulate, and one of the most important pieces in the puzzle was the Tharp discovery of rift valleys in the mid-oceanic mountain ranges. Tharp and Heezen were awarded the Hubbard Medal of the National Geographic Society in 1978. In 1999, Tharp was honored with the Women Pioneers in

Oceanography Award of the Women's Committee of WHOI.

Tidal Power

See Energy from Water

Tides (Ocean)

Tides are periodic rises and falls of the surface level of the ocean and adjacent bodies of water, such as bays and gulfs.

History

Tidal action has been of interest to scholars since the time of the ancient Greeks. The geographer and explorer Pytheas (ca. 330 B.C.–?) is responsible for the first precise description of tides. He also provided an explanation for tidal action—based on the Moon's gravitational attraction—that is remarkably similar to that accepted today. Pytheas's works were largely forgotten until the seventh century A.D., when the English scholar Bede (673–735) resurrected the Greek's tidal theories. Bede's work was itself largely ignored, and a truly scientific explanation of tidal action was not developed until the eighteenth century. At that time, Sir Isaac Newton (1642–1727) proposed a hypothesis, now known as *equilibrium tidal theory*, that adequately explains the occurrence of tides.

Theories of Tidal Action

Earth and the Moon form a gravitational system in which each body exerts a gravitational attraction on the other. One consequence of this system is the formation of tides.

High Tide

The most important factor explaining tidal action is the Moon. The Moon's gravitational force pulls on the ocean's waters, raising their surface level above normal sea level. As Earth rotates on its axis, the borders of the continents slip underneath the raised waters, producing a *high tide* on the shores of the continents.

A comparable high tide occurs at the same time on the opposite side of Earth. This high tide is caused not by the Moon's gravitational attraction but by centrifugal force. Centrifugal force is a force produced when objects tend to move outward away from the center of rotation. In this case, the ocean's waters move outward from the center of Earth, forming a bulge that constitutes a second high tide. The bulge forms because ocean waters on this side of the planet feel no gravitational force from the Moon.

The formation of high tides results simultaneously in the formation of low tides at points on Earth equidistant from the high tides. That is, the Moon's gravitational force does not increase the amount of water in the oceans; it simply moves the water from one place to another. As ocean levels rise on one part of the planet, they fall at other parts.

Based on this theory, Earth should experience two high tides and two low tides every day. One high tide occurs when the Moon is directly over a body of water, and the other, when the Moon is on the opposite side of the planet. In fact, the timing of tides is not as simple as it might seem to be. The reason is that the Moon revolves around Earth and does not appear in the same place in the sky every day. In fact, it takes the Moon about 27 days to revolve around Earth, adding about 50 minutes to the tidal pattern each day. The time between high tides, therefore, is about 24 hours and 50 minutes.

Spring and Neap Tides

The Sun also exerts a gravitational force on the ocean's waters. Although the Sun is a much larger body than the Moon, it is also much farther away. As a result, the Sun's gravitational attraction is only about half of the Moon's.

The overall gravitational effect on the ocean's waters is a result of the relative position of the Moon and the Sun. If the Moon and the Sun are both lined up on the same side of Earth, their gravitational attractions operate in the same direction. The ocean's waters are pulled upward by both the Moon and the Sun. In such cases, the highest (and lowest) tides of the year—known as *spring tides*—occur. Spring tides occur about twice a month, during full and new moons.

If the Moon and Sun are on opposite sides of Earth, their gravitational attractions pull the ocean's waters in opposite directions. The Moon pulls the waters in one direction (away from Earth), while the Sun pulls them in another direction (toward Earth). In such cases, high tides are at their lowest level (and low tides, at their highest level) of the year. Such tides are known as *neap tides* and also occur about twice a month, during the first and last quarter of the Moon.

Types of Tides

Actual tidal patterns are generally more complex than the simple model described above. One of the most important factors that affects actual tidal patterns (in comparison to ideal patterns predicted by mathematical theory) is the shape, volume, and other physical characteristics of Earth's oceans. Two regions on the same latitude (and, hence, subject to comparable gravitational forces) may experience very different tidal patterns because of the orientation of ocean water to shore. In addition, factors such as variations in Earth's axial rotation and variations in the geometry of the Sun-Moon-Earth system alter the ideal tidal patterns one might expect to observe.

As a consequence of these factors, three types of tidal patterns occur. A *diurnal tide* is one in which a single high tide and a single low tide occur each day. Such tides are often found in the Gulf of Mexico and Southeast Asia. By contrast, a *semidiurnal tide* is one in which two high tides and two low tides occur each day. Semidiurnal tides are observed primarily along the Atlantic coasts of North America and Europe.

The most common tidal pattern is that of the *mixed tide*. In a typical mixed-tide area, water levels during a single tidal period may vary from 10 meters to 1 meter to 5 meters to 3 meters, and back again to 10 meters. This pattern is then repeated during the next tidal period of about 24 hours and 50 minutes. Mixed tidal patterns are common along the Pacific coast of North America.

Height of Tides

In general, tides vary in height above and below sea level by about 1 to 3 meters (3 to 10 ft). Unusually high tides occur, however, in some areas where topographical factors tend to funnel the rising water into narrow inlets. Some of the highest tides in the world have been measured in the Bay of Fundy, between Nova Scotia and New Brunswick, where water levels may reach 15 meters (nearly 50 ft) above sea level. In some cases, water rushes into the bay at high tide as if it were a raging river.

By contrast, some areas of the world experience barely distinguishable high tides. In smaller bodies of water, such as the Mediterranean, Baltic, and Caribbean Seas, high tide seldom exceeds a half meter (about a foot).

Tidal Tables

Scientists now understand tidal behavior well enough to make precise mathematical calculations of tidal patterns, given the relative positions of Earth, Moon, and Sun. Such calculations have little practical use, however, because of the disturbing effects of ocean topography and other regional factors. As a result, the prediction of tidal patterns has been almost entirely a trial-and-error effort carried out for hundreds of years. By now, people have been measuring and observing tidal behavior for specific regions for a very long time. Based on these measurements and observations, it is possible to construct a *tidal table* that tells when high and low tide will appear for any given geographical area. Such tables are valid only for those areas, of course, and cannot be used in other locations, even if they are only a few miles away. *See also* Oceans and Seas

Further Reading

Cartwright, David Edgar. *Tides: A Scientific History*. Cambridge: Cambridge University Press, 1999.

Godin, Gabriel. *Tides*. Ensenada, Mexico: Centro de Investigación y Educación Superior de Ensenada, 1988.

Helmut, Wilhelm, Water Zürn, and Hans-Georg Wenzel, eds. *Tidal Phenomena*. Berlin: Springer, 1997.

"Our Restless Tides," http://co-ops.nos.noaa .gove/restles1.htm.

Transpiration

Transpiration is the process by which plants lose water to the atmosphere from their leaves and stems. This water escapes through small openings known as *stomata* (singular: *stoma*). The size of a stoma is controlled by liver-shaped *guard cells* on either side of the opening. When the guard cells are full of water, turgor pressure forces the cells apart and the stoma opens. When the guard cells lose water, they become flaccid and the stoma closes.

Transpiration is an essential function in plants because it is the mechanism by which water is drawn from the ground, through root hairs, and upward into the plant. As water moves upward, it carries nutrients needed for the plant's growth. Some water is also used in the process of photosynthesis, by which water and carbon dioxide are converted to carbohydrates. However, only about 2 percent of all the water that travels through a plant is used for photosynthesis;

the rest is lost by means of transpiration through stomata.

Stomata serve a duel function in plants. In addition to providing for the loss or retention of water, they afford a way for carbon dioxide to enter the plant. The function of stomata is related to the amount of light energy available. During the day, abundant sunlight is available, stomata are open, and carbon dioxide flows readily into the plant. At the same time, water vapor is able to leave the plant by means of transpiration. During the evening, the amount of sunlight decreases, stomata close, the flow of carbon dioxide decreases, and the loss of water is reduced.

Stomata clearly have a critical function in plants because they must be open long enough to allow sufficient carbon dioxide to enter so that photosynthesis may proceed, but not so long as to allow so much water to be lost that the plant begins to wilt.

Plants have developed a number of adaptations that control the rate of transpiration. For example, the rate of plant growth during the winter is usually very slow, primarily because of an inadequate amount of light energy. To avoid losing water during this time of the year, many plants drop their leaves. Without leaves and the stomata they contain, the plants cannot carry out photosynthesis, but they also lose very little water by transpiration.

Trees that retain their leaves all year round have other adaptations. Many conifers have small, narrow leaves (needles) with many fewer stomata than occur in a deciduous plant. The stomata on these leaves are often buried within the epidermis rather than being exposed on its outer surface. Conifers, evergreen rhododendrons, hollies, and some other plants have evolved a thick waxy covering (a *cuticle*) that covers the outer surface of leaves and helps retain water.

Desert plants face the most difficult problem with water loss, and have evolved a number of adaptations to deal with this problem. Some plants, for example, die off and deposit seeds in the soil that germinate only when precipitation once more appears. Other desert plants (the *succulents*) store huge amounts of water in their leaves, stems, or roots. Still other desert dwellers have evolved very small leaves—sometimes no more than a millimeter wide—with few stomata to reduce the rate of transpiration. *See also* Evaporation; Evapotranspiration

Further Reading

Butter, A. J. *Transpiration*. London: Oxford University Press, 1972.

Gates, David Murray, and La Verne E. Papian. *Atlas of Energy Budgets of Plant Leaves*. London: Academic Press, 1971.

Transportation

See Barges; Boats and Ships

Tsunami

The word *tsunami* comes from two Chinese ideographs that mean "harbor" (*tsu*) and "wave" (*nami*). Tsunamis are also sometimes called *seismic sea waves* or *tidal waves*. Both terms are inaccurate, however. Tsunamis may be caused by factors other than earthquakes or other seismic events, and they have nothing to do with tides.

Causes

Tsunamis are caused when large volumes of water are displaced by an earthquake, an underwater landslide, an underwater volcanic eruption, a coastal landslide, the impact of a meteorite on the ocean, or some similar event. In such incidents, the movement of large quantities of solid material displaces large volumes of water, setting in motion a tsunami.

Characteristics

Tsunamis are shallow-water waves, waves whose wavelength is much greater than their amplitude (or depth). The wavelength of any wave is the distance between successive crests, or peaks, while its amplitude is the vertical displacement of the wave from its average height. Tsunamis commonly have an amplitude of no more than a few meters with an amplitude of more than 200 kilometers (120 miles). Tsunami waves are barely noticeable in the middle of the ocean, although they travel at high speeds. It is not unusual for a tsunami wave to move at 200 meters per second (650 ft/s) or more than 700 kilometers per hour (450 mph) on the open sea.

The total energy carried by a tsunami wave is a function of two variables: its speed and its amplitude. As a wave approaches shore, its speed slows down because of friction between water in the wave and the ocean bottom. To conserve the energy carried by the wave, its amplitude must increase as its velocity decreases. Tsunami waves approaching shore thus increase in size from only a meter or so to over 30 meters (100 ft).

Tsunamis can be classified according to the area they affect. For example, some tsunamis are local phenomena, affecting only a few hundred square kilometers or less. As an example, a landslide on one side of a bay may set off tsunami waves that reach to the opposite side of the bay, causing damage there. But waves that escape from the bay move into the open ocean and dissipate their energy before they cause further damage.

At the other extreme, some tsunami waves travel thousands of kilometers and cause extensive damage over widespread areas of coastline. Such waves are often caused by volcanic eruptions or earthquakes in the middle of the ocean. Less than 5 percent of all tsunami waves are produced in this way, but they produce the greatest amount of damage of any tsunami.

Examples

Tsunamis have been common occurrences throughout human history. Some his-

A powerful tsunami takes a boat horizontally. *NOAA News Photo.*

torians believe that the ancient Minoan civilization may have been destroyed largely because of damage caused by a powerful tsunami that struck in 1480 B.C. There are records of more than sixty-five major tsunamis in Japan between A.D. 684 and 1960.

One of the worst tsunamis in modern history occurred on 1 April 1946. An earthquake off the coast of the Aleutian Islands created a surface wave that caused terrible destruction along the islands themselves and as far away as the Hawaiian Islands. On Unimak Island, for example, the tsunami totally destroyed the Scotch Cap Lighthouse, built six years earlier. The lighthouse was 5 stories tall and 40 feet above sea level.

The worst damage from the 1946 tsunami occurred on the island of Hawaii, nearly five hours after it began off the coast of the Aleutians. Waves swept through the city of Hilo, killing ninety-six people and causing $26 million (in 1946 dollars) in damage.

A 1996 earthquake on the bottom of the Pacific Ocean, 130 kilometers (80 mil) off the coast of Peru, set off a tsunami that swept onshore along a 590-kilometer (370 mile) stretch from Lima to the city of Pascasmayo. Tidal gauges measured the amplitude of the original wave at about 60 cm, but waves reached more than 10 meters when they struck the South American coastline.

Research and Prevention

By the early 1960s, coastal nations had begun to think about ways of reducing the damage done by tsunami waves. Some countries began to set up their own national research and warning systems. On the international level, the Intergovernmental Oceanographic Commission of the United Nations Educational, Scientific, and Cultural Organization (UNESCO) established the International Tsunami Warning System (ITWS), modeled on the U.S. Weather Service's Pacific Tsunami Warning System (PTWS). Today, twenty-six nations belong to the ITWS, sharing research and warning facilities with each other.

The PTWS is now an international operation located in Ewa Beach (near Honolulu), Hawaii. Its purpose is to detect, locate, and determine the magnitude of submarine earthquakes with the potential for producing dangerous tsunamis. PTWS serves as the warning center for Hawaii and Pacific coastal nations. A similar tsunami research and warning system, the Alaska Tsunami Warning Center (ATWC), is located in Palmer, Alaska. It serves as the regional tsunami warning center for Alaska, British Columbia, Washington, Oregon, and California.

Some types of research carried out at PTWS include the development of deep-ocean measuring and assessment devices, testing of communication systems for tsunami detection and notification, construction of tsunami-risk maps and information, and development of modeling programs for the prediction of tsunami formation, development, and effects.

Further Reading

Bryant, Edward. *Tsunami: The Underrated Hazard*. Cambridge: Cambridge University Press, 2001.

Keating, Barbara H., Christopher F. Waythomas, and Alastair Dawson. *Landslides and Tsunamis*. Basel: Birkhäuser Verlag, 2000.

Myles, Douglas. *The Great Waves*. New York: McGraw-Hill, 1985.

Tsunami Program, http://www.pmel.noaa.gov.

Tugboat

See Barges

U

Underwater Archaeology

Underwater archaeology is the study of sunken ships, submerged human habitations, and other human artifacts now under water.

Development

The first scientific underwater search for artifacts was probably that commissioned by Cardinal Prospero Colonna in 1446 (or 1450, according to some authorities). Colonna ordered the famous architect Leon Battista Alberti (1404–1472) to search for two Roman ships said to have been sunk in Lake Nemi. Expert divers from Genoa hired by Alberti found the ships, but were unable to raise them with equipment then available.

In 1535, a second effort was made to raise the ships under the direction of the naval architect F. De Marchi. De Marchi used a submersible wooden structure invented by one Guglielmo di Lorena that allowed people to work underwater, the first diving device of its kind to be developed. This effort also failed to lift the ships, as did later attempts in 1827 and 1895. The latter expedition, led by Eliseo Borghi, was successful, however, in collecting bronze statues, marbles, mosaics, and other works of art from the Roman ships.

More sophisticated exploration of underwater artifacts became possible in the 1940s and 1950s with the invention of the Aqua-Lung (scuba), air-filled lifting bags, suction pipes, and other devices for uncovering and removing long-sunken objects.

The first complete excavation of an ancient shipwreck took place in 1960 off the coast of Cape Gelidonya, Turkey. The ship was a merchant vessel thought to have sunk in about 1200 B.C. It was first discovered by a sponge diver in 1954 in about 26–28 meters of water. Six years later, archaeologists from the University of Pennsylvania successfully raised the vessel and much of its cargo, strewn over many meters of ocean floor.

Modern Research

Underwater archaeology is now an active field of research with programs at universities worldwide, including those at Florida State University, Texas A&M University, East Carolina University, Indiana University, the University of Hawaii, and Brown University in the United States; the University of Haifa in Israel; the University of St. Andrews and the University of Bristol in the United Kingdom; James Cook University and Flinders University in Australia; and the Norwegian University of Science and Technology. A number of state, national, and private organizations for underwater archaeology also exist, including

The raising of Henry VIII's *Mary Rose*, which sank in Portsmouth harbor in 1545. Many sixteenth-century artifacts were recovered with the vessel during the 1980s. *Hulton-Deutsch Collection/Corbis.*

those at the Ministere de la Culture in France; the Associazione Italiana degli Archeologi Subacquei in Italy; the Channel Islands Marine Archaeology Resources in the United Kingdom; and the Minnesota Historical Society, the State of Texas Underwater Archaeology Program, and the New York Institute for Marine Archaeology in the United States.

Research in underwater archaeology now has two major foci: the study of shipwrecks and the study of submerged cultural resources. For example, in the United States, the Naval Historical Center includes an Underwater Archaeology Branch (NHC/UAB) that advises the navy in matters relating to the historic preservation of U.S. Navy ship and aircraft wrecks. The NHC/UAB has supervised the study and recovery of two

Confederate ships—the CSS *Alabama* and the CSS *Florida*—the submarine *U-1105*, and the naval ships USS *Alligator*, USS *Cumberland*, USS *Housatonic*, and USS *Tecumseh*, among other shipwrecks. The branch has also studied the wrecks of aircraft such as the Martin PBM Mariner Patrol Bomber BuNo59172, wrecked on 6 May 1949 while being transported across Lake Washington.

The U.S. National Park Service also maintains an active Submerged Cultural Resources program in the areas for which it is responsible. As of late 2001, important submerged cultural resources had been discovered in fifty-nine such areas. For example, more than five hundred shipwrecks remain to be explored along the outer banks of Virginia and North Carolina in Cape Hatteras National Seashore, the oldest of which is the *Tiger*, a British ship sunk in 1585.

Also typical of such underwater shipwreck research is the Dog Island Shipwreck Survey, carried out by the Florida State University Program in Underwater Archaeology from May through August 1999. The waters around Dog Island contain shipwrecks dating back hundreds of years. The oldest remains yet found is that of a 1,200-year-old canoe thought to have been used by residents of the area at the time. The area also contains a number of shipwrecks from Colonial, Revolutionary, Civil War, and more recent eras.

Changes in sea level and land subsidence have resulted in the submersion of many coastal towns and villages over the past centuries. In the United States, for example, land that is now part of the continental shelf along the eastern seaboard, the Gulf of Mexico, and the west coast was once the site of countless human habitations. Underwater studies of these areas provide an intriguing snapshot of life in these areas hundreds or thousands of years ago.

Other parts of the world pose similar challenges. The government of India, for example, has supported the Marine Archeology Centre of the National Institute of Oceanography at Goa since 1981. In recent years, the center has carried out studies of submerged ancient coastal towns near Dwarka, Bet Dwarka, Poompuhar, Vijaydurg, and Sindhudurg, as well as studies of shipwrecks near Poompuhar, Lakshadweep, and Goa. The exploration near Bet Dwarka, for example, has yielded pottery and other artifacts dating to about 1500 B.C., while work near Dwarka has revealed the presence of stone blocks, walls, pillars, and bastions as well as pottery and a variety of art objects.

The significance of underwater archaeology as a means for understanding human history was recognized at a meeting of the United Nations International Council of Monuments and Sites, held in Sofia, Bulgaria, in October 1996. At that meeting, the ICOMOS Charter on the Protection and Management of Underwater Cultural Heritage was adopted, designed to "encourage the protection and management of underwater cultural heritage in inland and inshore waters, in shallow seas and in the deep oceans." *See also* Shipwrecks

Further Reading

Delgado, James P., ed. *Encyclopedia of Underwater and Maritime Archaeology*. New Haven, CT: Yale University Press, 1998.

Gould, Richard A. *Archaeology and the Social History of Ships*. Cambridge: Cambridge University Press, 2000.

Kuppuram, G., and K. Kumudamani, eds. *Marine Archaeology: The Global Perspective*. Delhi: Sundeep, 1996.

Ruppe, Carol, and Jan Barstad, eds. *International Handbook of Underwater Archaeology*. New York: Kluwer Academic/Plenum Publishing, 2001.

UNEP Water Branch

The Water Branch of the United Nations Environment Programme (UNEP) is charged with carrying out a number of water-related activities described in the UNEP Programme of Work for 1997–1998. Its activities fall into four major categories: Freshwater Issues, Regional Seas Programme, International Coral Reef Initiative, and Global Programme of Action for the Protection of the Marine Environment from Land-Based Activities.

Freshwater Issues

The mission of the Freshwater Unit is to:

- promote integrated management and use of the world's freshwaters;

- enhance environmental quality;
- promote socioeconomic development in an environmentally sustainable fashion.

Activities carried out by the unit to achieve these objectives include the organization and conduct of international workshops focused on regional issues; scientific review of major geochemical cycles and their impacts on freshwater resources; and projects with other agencies and national governments to identify, assess, and promote appropriate technologies. Programs for environmental-analysis diagnosis and action planning have been developed for the Zambezi and Nile Rivers and the Lake Chad basins in Africa; the Mekong River, Aral and Caspian Sea basins, and Xinjiang Autonomous Region of China in Asia; and the La Plata and San Juan River and Lake Titicaca basins in South America. Among the services offered by the Freshwater Unit are technical reports and studies on water issues; training courses, workshops, and supporting materials on topics of regional significance; public awareness materials on water-resource management; and aid on water-resources management technology transfer.

Regional Seas Programme

The Regional Seas Programme (RSP) was initiated in 1974 to deal with issues of environmental degradation affecting inland seas. It currently operates in thirteen regions, including the Black Sea, the Caribbean, East Asian Seas, Eastern Africa, Kuwait, the Mediterranean, the Northwest Pacific, the Red Sea and Gulf of Aden, South Asia, the Southeast Pacific, the South Pacific, the Southwest Atlantic, and West and Central Africa.

RSP programs differ somewhat within these thirteen regions, but they tend to have similar structures. The programs all include the following elements:

- Environmental assessment, in which the causes of environmental problems within a region are evaluated and studied. Specific activities include baseline studies, research and monitoring of source levels and effects of marine pollutants, ecosystem studies, studies of coastal and marine activities, and social and economic factors.
- Environmental management, which may involve training in environmental impact assessment; management of coastal lagoons, estuaries, and mangrove ecosystems; and control of industrial, agricultural, and domestic wastes.
- Environmental legislation, under which legal frameworks are provided and/or developed to direct regional and national actions.
- Institutional arrangements, out of which systems are developed through which an agreed-upon work plan can be carried out.
- Financial arrangements, a program through which UNEP often provides seed money to get action programs started and to initiate regional trust funds, through which such actions can be continued.

International Coral Reef Initiative

The International Coral Reef Initiative (ICRI) was created in 1994 by the joint action of eight nations; Australia, France, Japan, Jamaica, the Philippines, Sweden, the United Kingdom, and the United States. The purpose of the organization is to carry out Chapter 17 of UNEP's Agenda 21, adopted at the United Nation's Conference on the Environment and Sustainable Development, held in Rio de Janeiro in 1992. Chapter 12 of the agenda calls on states "to take special care of marine ecosystems, exhibiting high levels of biodiversity and productivity, giving special priority to coral reefs, seagrass beds and mangroves."

The first Global Workshop sponsored by ICRI was held in the Philippines in 1995. Since that time, regional workshops have focused on specific regions of the world, including Tropical Americas (Jamaica, 1995); the Pacific (Fiji, 1995); South Asian Seas (Maldives, 1995); East Asian Seas (Indonesia; 1996); Eastern Africa and Western Indian Ocean (Seychelles, 1996); and Middle Eastern Seas (Jordan, 1997).

Global Programme of Action

The Global Programme of Action for the Protection of the Marine Environment from Land-Based Activities (GPA) was created at the United Nations in 1995 in response to concerns expressed by 108 governments and the European Commission at a conference held in Washington, D.C., from 23 October to 3 November 1995. GPA supports the work of individual member nations in pursuing the following objectives:

- identifying and assessing problems related to food security, poverty alleviation, public health, coastal and marine resources, ecosystem health, impacts of contaminants (such as sewage, organic pollutants, and radioactive substances), physical alteration of the marine environment, and damage to vulnerable areas of concern (such as critical habitats, habitats of endangered species, shorelines, coastal watersheds, and small islands);

- establishing priorities for action among the factors noted above by considering issues such as source categories; the area affected; and the costs, benefits, and feasibility of various actions;

- setting management objectives for priority problems;

- identifying, evaluating, and selecting strategies through which these objectives can be achieved;

- developing criteria for evaluating the effectiveness of the strategies adopted.

Further Information

Water Branch
United Nations Environment Programme
P.O. Box 30552
Nairobi
Kenya
Telephone: (254 2) 621234
Fax: (254 2) 226886/622615
URL: http://www.unep.org/unep/program/natres/
    water
e-mail: cpiinfo@unep.org

United Nations Convention on the Law of the Sea

The United Nations Convention on the Law of the Sea (UNCLOS) is a treaty signed by 130 member nations of the United Nations on 10 December 1982 in Montego Bay, Jamaica. It was agreed upon after a fourteen-year effort by 160 nations to resolve issues concerning sovereignty over ocean waters and the lands beneath them, ocean pollution, marine life, and exploitation of mineral rights on the ocean floor.

Background

For most of human history, the oceans were regarded as a "no-man's land" through which ships from all nations could pass and in which any form of peaceful economic traffic was permitted. All nations did lay claim to a certain stretch of the seas beyond their land-based borders, but these claims varied widely from nation to nation and time to time. At the time negotiations were being conducted on UNCLOS, for example, Australia claimed a 3-mile territorial limit but a 12-mile limit for purposes of economic exploitation. Mexico claimed a 12-mile limit for both territorial and economic purposes, while Costa Rica claimed a 12-mile territorial limit and a 200-mile economic limit. In 1977, the U.S. Congress changed its historic claim of a 3-mile limit for economic utilization to a new 200-mile limit.

The need for an international agreement also became apparent as economic development on the high seas began to increase. In the first place, the growth of commercial shipping, especially of very large oil tankers, meant that the planet's oceans were being more seriously threatened by pollution; runoff from the continents added to that threat.

In addition, interest in the search for oil and minerals on the ocean floor and the technology that made that search possible raised new questions as to who "owned" the ocean floor and where that "ownership" began and ended.

The Treaty

The treaty was divided into 17 major titles, consisting of 320 separate articles. Title 2 of the treaty resolved the issue of territorial sovereignty by establishing a limit of 12 nautical miles for all nations. In Title 5, a limit of 200 nautical miles was set for the exclusive economic zone for every nation. Other parts of the treaty provided extensive definitions and descriptions of ownership and access to the waters in and around straits, archipelagoes, islands, inland seas, continental shelves, and landlocked states. The rights of nations to traverse and make use of the high seas were also defined and described.

Other titles of the treaty outlined provisions for the development and transfer of marine technology, marine scientific research, and protection and preservation of the marine environment.

For the most part, the treaty met with widespread approval among members of the United Nations—only Israel, Turkey, the United States, and Venezuela voted against

the treaty. Seventeen other nations, however—most of them developed countries—abstained from signing the treaty. Of all industrialized nations, only France and Japan signed the treaty.

Title 11

The primary point of contention about the treaty was the portion of Title 11 dealing with deep-seabed mining. Developed nations, such as the United States, felt that the treaty gave undue advantage to less-developed nations at the expense of developed nations. During one debate on ratification of the treaty in the U.S. Senate, John Tower (R-TX) said that the treaty was "drafted so that Third World nations could use multinational institutions and procedures, beyond the control of any individual state, to transfer billions of dollars in technology and material resources from developed to underdeveloped nations" (http://www.senate.gov/~rpe/rva/972/972452.htm). In spite of these objections, President Ronald Reagan announced that the United States would abide by the general principles of UNCLOS while efforts were made to resolve differences of opinion about Title 11.

A long series of meetings were held between 1982 and 1994 to achieve this objective. Modifications were made, and enough nations agreed to them that on 16 November 1994, the treaty entered into force. At that point, the United States had still not ratified the treaty, and was given an extension of four years in which to make that ratification and become a fully functioning member of the UNCLOS secretariat.

In an attempt to meet that deadline, President Bill Clinton submitted to the U.S. Senate both the 1982 UNCLOS treaty and the 1994 amendments to Title 11. However, Senator Jesse Helms (R-NC)—chairman of the Senate Foreign Relations Committee—declined to allow his committee to consider the treaty and amendments. Although Helms had declared his approval of the amended treaty, he apparently decided to continue his general opposition to any treaty that would require the United States to abrogate any of its national sovereignty in any area. As a result, the United States has, as of late 2001, still not signed the UNCLOS treaty and its amendments. The amended treaty has, however, been signed and ratified by 135 nations, including Australia, China, France, Germany, Great Britain, India, Italy, Japan, the Russian Federation, and Spain.

In spite of the failure of the United States and a few other nations to participate in UNCLOS, the agency has been very active for more than seven years in working toward the goals outlined in the original treaty and its amendments. It has established organizations to deal with special issues, such as the International Seabed Authority, the International Tribunal for the Law of the Sea, and the Commission on the Limits of the Continental Shelf. In addition, a dozen regular meetings of the parties to the treaty have been held, and issues about the law of the sea have appeared regularly on the agenda of the United Nations General Assembly.

Further Reading

Bernaerts, Arnd. *Bernaerts' Guide to the 1982 United Nations Convention on the Law of the Sea*. Coulsdon, UK: Fairplay Publications, 1988.

David, Steven R., and Peter Digeser. *The United States and the Law of the Sea Treaty*. Washington, DC: Foreign Policy Institute, 2000.

Leitner, Peter M. *Reforming the Law of the Sea Treaty: Opportunities Missed, Precedents Set, and U.S. Sovereignty Threatened*. Lanham, MD: University Press of America, 1996.

Morell, James B. *The Law of the Sea: An Historical Analysis of the 1982 Treaty and Its Rejection by the United States*. Jefferson, NC: McFarland, 1992.

The full text of the UNCLOS treaty and its amendments is available at http://www.un.org/Depts.los/unclos/closindx.htm.

U.S. Office of Water

The U.S. Office of Water (OW) is a division of the Environmental Protection Agency, established in 1970. The office is responsible for implementing all or part of a number of federal statutes, including the Clean Water and Safe Drinking Water Act; the Coastal Zone Act Reauthorization Amendments of 1990; the Resource Conservation and Recovery Act; the Ocean Dumping Ban Act; the Marine Protection, Research and Sanctuaries Act; the Shore Protection Act; the Marine Plastics Pollution Research and Control Act; the London Dumping Convention; and the International Convention for the Prevention of Pollution from Ships. OW carries out these responsibilities not only through its own staff but also through cooperation with many other federal agencies, state and local governments, Native American tribes, professional and interest groups, land owners and managers, and the public at large.

The Office of Water is organized into five major divisions: the EPA American Indian Environmental Office; the Office of Ground Water and Drinking Water; the Office of Science and Technology; the Office of Wastewater Management; and the Office of Wetlands, Oceans and Watersheds. Some examples of programs initiated and supported by OW include providing multimedia program development grants to Native American tribes, developing drinking-water standards, administering the underground injection control program, establishing human health criteria for drinking water, operating a shellfish protection program, carrying out a clean-water needs survey, monitoring and studying water-quality issues along the U.S.-Mexico border, and administering the American Heritage Rivers program and the National Estuary Program.

Further Information

U.S. Environmental Protection Agency
Office of Water (4101)
1200 Pennsylvania Ave., N.W.
Washington, DC 20640
Telephone: (202) 260-2090
URL: http://www.epa.gov/ow
e-mail: OW-General@wpamail.epa.gov

U.S. Water News

U.S. Water News calls itself "America's premier water news publication." The monthly magazine carries news on a host of water-related issues as well as classified ads, a buyer's guide, a list of meetings and conferences, and a section for buyers and sellers of water rights. The publication also maintains an electronic version of its magazine at http://www.uswaternews.com. The online version of the publication contains a searchable archive for the magazine from 1995 to 2000, classified under major topics such as water supply, policy and legislation, conservation, water quality, litigation and water rights, and global waterfront.

U.S. Water News is an excellent source of information about nearly everything that's going on in the water industry. Its list of conferences, for example, runs for many pages, covering meetings ranging from "IV Inter-American Dialogue on Water Resources" and "OSHA Lab Safety" to "Toxicology" and "Construction Safety for Water Utility Job Sites." The "Action Pack" card deck provides readers with extensive information on products and services available from the magazine's advertisers.

Further Information

U.S. Water News
230 Main Street
Halstead, KS 67056
Telephone: (316) 835-2222
Fax: (316) 835-2223
URL: http://www.uswaternews.com
e-mail: Inquiries@uswaternews.co

V

Vadose Water

See Groundwater

Visual Arts

See Water and the Visual Arts; Watercolor

Vollenweider, Richard (1922–)

Richard Vollenweider is best known for his research on eutrophication—the process by which a body of water gradually becomes richer in nutrients, leading to an increase in primary productivity and a decline in dissolved oxygen. Vollenweider became interested in this issue in the mid-1960s, when he served as a water pollution consultant for the Organization of Economic Cooperation and Development (OECD). At the time, Vollenweider was leading a research effort to learn more about the increased eutrophication of lakes and other bodies of water caused by the release of phosphorus and other minerals by human activities. As a result of his studies, Vollenweider developed a mathematical model that allowed authorities to predict the amount of eutrophication that would occur in a body of water for any given release of phosphorus into that body.

Personal History

Richard Vollenweider was born in Zurich on 27 July 1922. He was awarded his diploma and his Ph.D. in biology from the University of Zurich in 1946 and 1951, respectively. He did postgraduate research at the Italian Hydrobiological Institute (IHI) from 1954 to 1955 and at the Swedish Research Council in Uppsala from 1955 to 1956. In addition to his work with OECD, he has served as field researcher in limnology and fisheries for the United Nations Educational, Scientific, and Cultural Organization (UNESCO) in Egypt from 1957 to 1959; as research associate at IHI from 1959 to 1966; as chief limnologist and head of the Fisheries Research Board for the Canadian Centre for Inland Water (CCIW) from 1968 to 1970; as chief of the National Water Research Institute Lakes Research Division from 1970 to 1973; and as senior scientist with the CCIW from 1973 to 1988.

Professional Accomplishments

The Vollenweider Model has seen wide use around the world in combating problems of water pollution by human activities. For example, the U.S.-Canadian International Joint Commission for the Great Lakes adopted the Vollenweider Model in its efforts to restore Lake Erie to health after its dramatic environmental decline during the mid–twentieth century. Vollenweider also extended his OECD studies on eutrophication to more than two hundred lakes and reservoirs in North America, Western Europe, Japan, and Australia.

Vollenweider was given the International Award, Premio Cervia/Ambiente in 1978, the Tyler International Prize for Global Environmental Achievement in 1986, and the Societas Internationalis Limnologiae Naumann/Thienemann Medal in 1987.

W

WaterAid

WaterAid was established in 1981 as an independent charity by individuals and organizations in the British water industry as an activity of the United Nations Water Decade (1981–1990). The goal of WaterAid is to work with other organizations "to help poor people in developing countries achieve sustainable improvements in their quality of life by improved domestic water supply, sanitation and associated hygiene practices." WaterAid estimates that it has helped more than six million people gain access to improved water supplies and/or better sanitation systems. During the period 2000–2001, WaterAid had programs in Bangladesh, India, Nepal, Ethiopia, Mozambique, Tanzania, Uganda, Zambia, Ghana, Madagascar, Malawi, and Nigeria.

WaterAid has an extensive educational program for schoolchildren of all ages as well as the general public. Some of the resources they offer include a teaching unit, "Water Performance"; a video about two children in Ghana, "Buckets of Water"; a learning resource kit, "WaterLiterate"; a game designed to supplement mathematics lessons, "WaterNumerate"; a grade 5 teaching unit, "Water Counts"; a technology poster; and a book, *The Chimney Swallows*.

Also available are development issue sheets on topics such as "Health and Sanitation," "Women and the Family," "Water and Human Rights," "Water and Poverty," "Dams," "Water and Leisure," "Water and the Environment," and "Water and Conservation in the UK."

WaterAid also produces technical reports, including a series of sixteen "WaterAid Evaluations" outlining changes that have come about in the areas of water, sanitation, and hygiene in certain target countries as a result of the agency's actions. It has also published six major reports on topics such as "Contracts and Partnerships: Working through Local NGO's in Ghana and Nepal," "India: Making Government Funding Work Harder," and "Mega-Slums: The Coming Sanitary Crisis."

Further Information

WaterAid
Prince Consort House
27–29 Albert Embankment
London SE1 7UB
United Kingdom
Telephone: +44 020 7 793 4500
Fax: +44 020 7 793 4545
URL: http://www.wateraid.org.uk
e-mail: information@wateraid.org.uk

Water and the Visual Arts

Painters, photographers, sculptors, and other artists and artisans have found inspiration for their work in all aspects of water, including oceans, rivers, lakes, ice and snow, storms, water plants and animals, and objects developed by humans for the use of water. In the space available here, no more than a brief overview can be provided of the ways in which water themes have been used by visual artists around the world.

Water Gods and Goddesses

Water deities play a major role in the mythology and religion of many cultures dating back to the earliest days of human civilizations. These gods and goddesses have frequently been the subject of work by early painters, sculptors, and other artists. For example, a number of Roman sarcophagi dating to the third century A.D. are covered with representations of river gods sacred to the Romans. In the New World, statues of water gods and goddesses are also to be found. One such statue features that deity Chalchihuitlicue, goddess of water and springs.

Bodies of Water

Cultures that have developed on islands, along a seacoast, or on a land area covered with lakes and rivers might be expected to have an artistic tradition in which water is a common theme. Such is the case, for example, in Canada. Environment Canada (http://www.ec.gc.ca/water/en/culture/art) has put together an extraordinary collection of paintings and other pieces of art showing how water has been depicted over the centuries. Items in this collection include the following:

View of the River St. John near the Poquioq, George Heriot, 1807
The Ice Cone, Montmorency Falls, Robert C. Todd, ca. 1850

The Passing Storm, Saint-Féréol, Cornelius Krieghoff, 1854
Shooting the Rapids, Quebec, France Anne Hopkins, 1879
Evening (color woodcut), Walter J. Phillips, 1921
Stormy Weather, Georgian Bay, Frederick H. Varley, 1921

Similar examples can be found in other cultures. Water-related paintings from the Netherlands, for example, include the following:

Shipwreck off a Rocky Coast, Abraham Willaerts, 1614
Enjoying the Ice, Hendrick Avercamp, ca. 1630–34
Watercourse near 's-Graveland, Pieter Gerardus van Os, 1818
The Shipping Canal at Rijswijk, Hendrik Johannes Weissenbruch, 1868
The Sea, Jan Toorop, 1887
Reedy Lake, by Dorothy Elsie Knowles, 1962
Netsilik River (stonecut), by Pitseolak Ashoona, 1973

Other works of art focusing on lakes and other bodies of water include:

Lake Albano and Castel Gandolfo, Jean-Baptiste-Camille Corot, 1826–27
The Lake of Zug, Joseph Mallord William Turner, 1843
La Grande Vague, Séte, Gustave Le Gray, 1858–59
Lake George, John F. Kensett, 1869
The Meeting on the Lake, Arthur Georg von Ramberg, 1869
At Lake Garda, Hans Thoma, 1907
Boden Sea, Uttwil, Hiroshi Sugimoto, 1993

In the case of islands, artists may focus almost exclusively on water-related topics.

In May and June of 1984, for example, the United Nations sponsored an exhibition of art from Micronesia that included objects such as model houses, canoes, fish traps, paddles, and statues of water gods and goddesses (http://libweb.hawaii.edu/ttp/ttp _htms/3386.html).

Fishing

As an essential means by which many people obtain their food, fishing would naturally be a topic of considerable interest to artists from many cultures. Some of the earliest examples of such art come from Egypt, where wall paintings from many temples and tombs vividly illustrate the fishing practices used as far back as the third millennium B.C. A wall painting from the tomb of Menna in Thebes, for example, shows the pharaoh and his family fishing from boats. The painting dates to the Eighteenth Dynasty, between about 1400 and 1350 B.C. The sport of hippopotamus hunting in rivers is also a common theme in temple paintings. Such a scene is painted on limestone in the tomb of Ti at Saqqâra. Ti was a high official who served under the Pharaoh Kakai (2446–2426 B.C.).

Fishing is a familiar theme among many of the greatest European painters. Examples of this theme include the following:

The Fishing Boat, Jean Désiré Gustave Courbet, 1865
Fishing, Édouard Manet, 1861–63
Fishing Boats on the Beach at Saintes-Maries, Vincent van Gogh, 1888

One of the most famous painters of fishing-related scenes was the American painter Winslow Homer (1836–1910). Homer began his career painting events from the Civil War, but later turned his attention to scenes observed during his trips to the New England coast. Among these paintings are the following:

Sailing the Catboat, 1875
Watching the Tempest, 1881
Fisherwoman, 1882
Mending the Nets, 1882
The Life Line, 1884
Sponge Fishing, Nassau, 1885
Hurricane, Bahamas, 1898–99
The Gulf Stream, 1899
On a Lee Shore, 1900
West Point, Prout's Neck, Maine, 1900
Cape Trinity, Saguenay River, 1904–09

Naval Events

Naval battles and other events taking place on lakes and the ocean are also a common theme among painters and other artists. One of the most famous of these paintings is *Washington Crossing the Delaware*, by the American painter Emanuel Gottlieb Leutze, 1851. The painting shows General George Washington leading a contingent of 2,400 men across the Delaware River in December 1776, where they successfully engaged the enemy in the Battle of Trenton.

Some naval engagements have been the subject of paintings, lithographs, photographs, and other artistic representations many times over. The Battle of Trafalgar, fought on 21 October 1805 off the Cape of Trafalgar on the Spanish coast, is one such engagement. Dozens of artistic works have been based on this battle, including works by Bill Bishop, George Chambers, Graeme Lothian, E. Sartorus, Charles Stanfield, William Stuart, Joseph Mallord William Turner, and Brian Wood. For examples of such works, see http://www.naval-art.com/battle_of_trafalgar.htm.

Another battle that has attracted the interest of artists is that between the USS *Monitor* and the CSS *Merrimac* on 9 March 1862, during the U.S. Civil War. A number of pictorial representations of this battle can be found at http://www.history.navy.mil

Winslow Homer's painting, *The Gulf Stream*, depicts many marine elements, including the boat, the sharks, and the waterspout. *The Metropolitan Museum of Modern Art, Catherine Lorillard Wolfe Collection, Wolfe Fund, 1906.*

/photos/events/civilwar/n-at-cst/hr-james/9mar62.htm.

Other works of art based on famous naval battles include the following:

The Explosion of the Spanish Flagship during the Battle of Gibraltar, 25 April 1607, Cornelis Claesz van Wieringen, 1621
Battle Scene in the English Channel between American Ship "Wasp" and English Brig "Reindeer," Thomas Whitcombe, 1812
Battle of Lissa, Eduard Nezbeda, 1866
Naval Warfare of Preveze (September 27, 1538), Ohannes Umed Beyzad, 1867
Score Another for the Subs, Thomas Hart Benton, 1943
The Battle of Midway, Robert Benny, ca. 1943
Inferno, William F. Draper, 1944

Water-Related Objects

A number of objects are used for drinking, cooking, washing, and using water in other ways. In many cases, these objects are made entirely for utilitarian purposes, with no expectation of producing a "work of art." In other instances, however, the object is designed and made not only to carry out some useful function but also to please the eye. Thus, artists and artisans have been making beautiful bowls, pitchers, steins, urns, and other objects for holding and pouring water that are as much works of art as they are utensils for the kitchen or bathroom.

As an example, the aquamanile (*aqua-* = "water;" *manile* = "hand") was once used to pour water over the hands of wealthy guests at dinner parties. One elaborate form of the device consists of a knight mounted on a horse, all made of pewter. The aquamanile is filled by lifting the knight's helmet. Water is then poured from the container through a forelock on the horse's head.

Additional information about works of art dealing with all aspects of water can be obtained through the website of most major museums. A good beginning point is World Wide Art Resources, http://wwar.com, with connections to dozens of art museums throughout the world, most with searchable collections though which one can look for artwork on storms, lakes, rivers, and other water-related subjects.

Water Animals

See Aquatic Plants and Animals

Water as an Element

See Earth, Air, Fire, and Water

Waterbed

A waterbed is a bed made of a waterproof material, such as rubber or vinyl, and filled with water. Historians report that goatskins filled with water were being used as primitive waterbeds in Persia more than 1,500 years before the Christian era. The first modern waterbed was invented in 1873 by an Englishman, Neil Arnott, and was put to use by Sir James Paget at London's St. Bartholomew's Hospital. The purpose of the bed was to prevent the development of pressure ulcers (or bedsores) in patients who were bedridden for long periods of time. By the end of the nineteenth century, Harrod's Department Store in London was offering waterbeds for sale. The product did not become very popular, however, primarily because an effective waterproof material had not yet been developed.

The concept of a water-filled bed next received wide attention when it was mentioned in a science-fiction novel by Robert A. Heinlein, *Stranger in a Strange Land*, published in 1961. Four years later, two physicians at Tufts University resurrected Arnott and Paget's idea and constructed a waterbed for use by hospital patients. By this time, a variety of waterproof plastic materials had been invented that worked very well as a casing for the waterbed.

Waterbeds are said to have certain advantages over traditional box-spring-mattress designs, and many people find them much more comfortable. A basic drawback of waterbeds has been their tendency to be too cold to sleep on, requiring some type of heater to keep the water warm. Today, solid-state temperature controls are available, which have largely eliminated this problem.

Waterborne Hazards

Waterborne hazards are agents transported in water that pose health risks to humans and other animals. By far, the most common waterborne hazards are disease-causing microorganisms, although chemicals, radioactive materials, and other nonliving materials may also pose health risks. For example, the accidental release of a pesticide into a river or an accident at a nuclear power plant can release hazardous materials into the public water supply.

Types

The Centers for Disease Control and Prevention (CDC) has ranked a number of the most dangerous biological agents that may pose a threat to human life into three major categories. Those agents classified in Category A are regarded as the most serious hazards, as they are known to cause widespread illness and death. Category A agents include anthrax (*Bacillus anthracis*), botulism (*Clostridium botulinum*), plague (*Yersinia pestis*), smallpox (*Variola major*), and tularemia (*Francisella tularensis*).

Organisms in Category B are also capable of causing widespread illness and death, although they are regarded as less of a threat at the present time than those in Category A. Category B organisms include brucellosis (*Brucella sp.*), glanders (*Burkholderia mallei*), psittacosis (*Chlamydia psittaci*), Q fever (*Coxiella burnetii*), typhus (*Rickettsia prowazekii*), and viral encephalitis (alphaviruses).

Finally, Category C consists of agents that are currently considered to be less serious threats than those in Categories A or B, but that are expected to pose more serious threats in the future. Among the organisms in Category C are hantaviruses, the nipah virus, and the yellow fever virus.

A number of less lethal microorganisms are also found in contaminated water. Some of the most common of those organisms are discussed below.

Cryptosporidiosis, often known simply as crypto, is a diarrheal disease caused by the microscopic parasite *Cryptosporidium parvum*. It can be transmitted when it passes from the intestine of one person, by means of a bowel movement, into a public water supply, where it may be ingested during swimming or drinking by a second person. The symptoms of crypto include loose, watery stool; stomach cramps and upset stomach; and low-grade fever. No treatment is available for crypto, and in otherwise healthy people, symptoms tend to disappear in 2–10 days after infection. Very young children, the elderly, and those with weakened immune systems may suffer longer periods of infection and/or relapses at a later date.

Giardiasis is similar to crypto in that it is a diarrheal illness caused by a parasite. The parasite is known as *Giardia intestinalis*, or *Giardia lamblia*. Like crypto, giardiasis is passed from the intestinal tract of an infected person, through a body of water, into the intestinal tract of a second person. Symptoms include loose, watery stool; stomach cramps and upset stomach; and low-grade fever, and last two to six weeks in an otherwise healthy person. A number of drugs are available for the treatment of giardiasis, including metronidazole (Flagyl), quinacrine hydrochloride, and furazolidone.

Legionnaires' disease is a lung infection caused by the bacterium *Legionella pneumophila*. Bacteria are transferred through the air (as when an infected person sneezes or coughs) or, more commonly, through water that has been contaminated with the bacterium. Major sources of *Legionella* infection are large water systems, such as those found in hotels and hospitals, that include humidifiers, spas, and swimming pools.

Symptoms of the disease include high temperature, cough, nausea, vomiting, stomach discomfort, muscle aches, chest pain, and shortness of breath. A variety of antibiotics are effective in treating Legionnaires' disease, including the macrolides (such as azithromycin) and the quinolones (such as ciproflaxacin, levofloxacin, moxiflosacin, and trovofloxacin).

Hepatitis A is the most common form of hepatitis and the seventh most common infectious disease in the United States. It is transmitted through the fecal to oral pathway so common with other waterborne diseases. One of the most common methods of transmission involves the ingestion of shellfish that have been living in water contaminated with the hepatitis A virus.

Common symptoms of hepatitis A include fatigue, nausea, vomiting, fever and chills, jaundice, pain in the liver, dark urine, and light stools. At the present time, there is no treatment for the disease other than extended bed rest and a program of sound nutrition.

Toxoplasmosis is caused by a one-celled parasite, *Toxoplasma gondii*, that occurs widely in waters throughout the world. By some estimates, more than sixty million people in the United States have, at one time or another, been infected with the organism. Such infections rarely produce symptoms in people with healthy immune systems, however. In more serious cases of infection, symptoms of the disease include muscle aches and pains, swollen glands, and other flu-like symptoms. Treatment is usually not necessary and symptoms typically disappear in a matter of weeks.

Helicobacter pylori infections are thought to be the most common chronic infections among humans. In some parts of the world, up to 90 percent of the population has been infected by the bacterium *Helicobacter pylori*, which causes the disease. *H. pylori* is now known to be the cause of most stomach ulcers, which cause severe abdominal pain, nausea, vomiting, and, in some cases, anorexia. Although the complete mechanism by which the bacterium is transmitted is not fully understood, it is clear that the fecal to oral pathway typical of other waterborne diseases is very important. Humans appear to be the only known host for the bacterium. A variety of antibiotics is highly effective in killing the bacterium and eliminating the symptoms of the disease.

Occurrence

The risks associated with waterborne diseases for human health go back to the origins of human life. Disease-causing organisms enter lakes, ponds, rivers, streams, groundwater, and other bodies of water through any number of natural processes. The wastes excreted by wild and domestic animals are a common example of such processes. Any time such water is used for drinking, cooking, swimming, or other purposes, there is a risk that such organisms will enter the body and cause disease and/or death.

It was not until the mid–nineteenth century that humans began to understand the relationship between polluted water and disease. As late as 1872, for example, New York City was building platforms so that citizens could dump their garbage directly into the East and Hudson Rivers, waterways that also served as sources for the city's public water system. Within a few decades, however, most large cities had begun to treat their municipal water supplies with chlorine to kill disease-causing organisms, vastly reducing the risk of waterborne hazards from this source.

The success of chlorination and other water-treatment procedures has been spectacular. Large-scale outbreaks of waterborne diseases have largely disappeared in the developed world. In parts of the world where water-treatment systems are not yet available, waterborne diseases still present widespread and serious health problems.

The World Health Organization (WHO) has estimated that about one-fifth of the world's population does not have access to safe drinking water, resulting in the deaths of more than 5 million people from waterborne diseases annually. In addition, WHO estimates that about 200 million cases of diarrhea and more than 2 million deaths resulting from diarrheal illnesses occur because of agents present in untreated water (http://www.cdc.gov/ncidod/eid/vol7no3;lls upp/hunter.htm).

By contrast, outbreaks of waterborne diseases in the United States are rare. Only 127 such incidents were reported to the CDC between 1990 and 1998, the majority resulting from the use of groundwater by way of wells, rather than from public water systems.

In spite of these statistics, citizens of the United States and other developed countries cannot be totally satisfied that problems of waterborne disease have been completely solved in their parts of the world. One important reason for concern is the pressure on freshwater supplies for irrigation, generation of hydroelectric power, cooling water needs, public water demands, and other uses. As more and more freshwater is withdrawn from natural reservoirs to meet human needs and demands, its overall quality has begun to deteriorate, making water purification a more difficult challenge.

For example, researchers at Harvard University reported in 1999 that about 10 percent of all hospital admissions of children and people over the age of 65 for gastrointestinal illnesses in the city of Philadelphia could be blamed on the presence of turbidity (cloudiness) in the city's water supply. They concluded that the city's water-treatment system was not adequate to deal with the range of disease-causing agents present in the bodies of water on which the city depended for its water supply. They also pointed out that the city's water-treatment facilities met all federal standards, suggesting that the problem of waterborne hazards could well be widespread in the nation.

Waterborne hazards are a source of concern for another, quite different, reason: bioterrorism. In principle, a terrorist might find ways to release a particularly dangerous microorganism, such as the smallpox virus, into the public water supply. In such an instance, disease and death might spread rapidly among the population that uses the water supply.

In actual fact, the dangers posed by bioterrorism are probably less severe than they would at first appear. In the first place, access to supplies of disease-carrying organisms is usually quite difficult. For example, there are only two laboratories in the world that still hold samples of the smallpox virus,

and both are believed to be closely guarded and highly secure. In addition, most disease-causing organisms are killed by water-treatment systems; therefore, they would have to be added to water *after* it is treated, a somewhat more challenging task than simply dumping the agent into a river or lake.

Nonetheless, the events of recent years have shown that terrorists include among their numbers some very intelligent, highly educated individuals with the intellectual and financial resources needed to carry out programs of mass destruction. Bioterrorism that makes use of waterborne hazards is, therefore, a concern that is likely to remain in the public eye for the foreseeable future.

Further Reading

Craun, Gunter F., ed. *Waterborne Disease Outbreaks: Selected Reprints of Articles on Epidemiology, Surveillance, Investigation, and Laboratory Analysis.* Cincinnati, OH: U.S. Environmental Protection Agency, 1990.

———. *Waterborne Diseases in the United States.* Boca Raton, FL: CRC Press, 1986.

"Infectious Disease Information: Diseases Related to Water," http://www.cdc.gov/ncidod/diseases/water/drinking.htm.

Tartakow, I. Jackson, and John H. Vorperian. *Foodborne and Waterborne Diseases: Their Epidemiologic Characteristics.* Westport, CT: AVI Publishing, 1981.

Water Cannon

A water cannon is a large gunlike nozzle mounted on a truck, used to control rioters or demonstrators. The water cannon was developed in the early 1960s and has become popular as an effective and relatively humane method for repulsing, containing, or otherwise regulating the movement of large numbers of people.

In its simplest form, the water cannon is designed to fire streams of water at pressures of up to about 250 pounds per square inch.

The cannon is usually mounted on top of a large truck that contains a tank holding the water used in the cannon. Holding tanks of about 2,000 gallons are not unusual. The truck carrying the water cannon is usually armored to protect driver and crew from artillery fire and other types of attack.

The water cannon can be modified to spray debilitating chemicals, such as tear gas, as well as water. These chemicals can be used to incapacitate as well as to physically move crowds of people against whom they are used. Some water cannons are also able to fire streams of electrically charged liquids that act like stun guns against crowds. Finally, dyes may also be injected in the stream of water fired by the cannon so that protestors and demonstrators can later be identified and, possibly, detained and prosecuted.

Some objections have been raised to the use of water cannons for the purpose of crowd control. They are certainly less likely to cause injury to people on whom they are used than other weapons, such as guns and truncheons. However, the force of a 250-pounds-per-square-inch stream of water can knock people down, break bones, cause bruising, and produce other physical injuries.

Further Information

A number of water cannons are described by manufacturers on their websites. See, for example, http://www.protecharmored.com/manta.htm,http://www.bulldogdirect.com/armored;llwc.html, or http://www.jaycor.com/eme/watcan.htm.

Water Clock

A water clock is a device that keeps time by measuring the flow of water into or out of a container. The first water clocks were probably built in the fifteenth century B.C. A simple water clock was found in the tomb of Amenhotep I, who died in about 1500 B.C. One reason water clocks were needed was that the most common time-measuring device of the time, the sundial, could not be used after dark. When the measurement of time in the evening became important, water clocks were invented.

Water clocks first became relatively common during the Golden Age of Greek civilization, around 325 B.C. One form of the water clock consisted of a metal bowl with a small hole in the bottom. The bowl was floated in a larger container of water. Water seeped upward through the hole in the metal container over time. The passage of time was marked with etchings at various levels on the inside of the bowl. A water clock of this design was called a *clepsydra*, from two Greek words that mean "water thief." Clepsydrae were used, among other purposes, to measure the passage of time during speeches given by lawyers in court.

The floating bowl clepsydra was one of two general designs for the water clock, measuring the amount of time needed for water to run into a container. The second type of water clock operated by measuring the amount of time needed to empty a container. In this design, a bowl or other container is filled with water, which is allowed to drip out of the bowl through a small hole in the bottom. Again, the inside surface of the bowl is marked with horizontal lines to show how much water has left the container—that is, how much time has passed.

Water clocks were also used in other parts of the world, particularly in China and other parts of East Asia. One of the most famous of these devices was built by the Chinese inventor Su Sung in A.D. 1088. Su Sung's water clock was more than 10 meters (30 ft) tall and very elaborate. It contained an escapement that allowed falling water to move the hands on a primitive clock face. The tower housing the clock also contained astronomical devices, mechanized manikins, and bells and gongs that sounded at various times of the day.

Water clocks were used widely until the invention of the pendulum clock in the mid–seventeenth century. In a few isolated places, they continued to be used until much later. There are reports, for example, that they were still in use in North Africa as late as the twentieth century.

Water clocks are, however, inherently inaccurate and of limited value for careful time measurement. For example, as the water in a vertical column falls, the pressure it exerts decreases, meaning that the rate of flow from the column is constantly changing. Factors such as this one are not of much importance when one is measuring time by hours and half hours, but can be a more serious problem when much more precise measurements are needed.

Further Reading

Barnett, Jo Ellen. *Time's Pendulum: The Quest to Capture Time—From Sundials to Atomic Clocks*. New York: Plenum Press, 1998.

Earliest Clocks, http://physics.nist.gov/GenInt/Time/early.html.

Jesperson, James, et al. *From Sundials to Atomic Clocks: Understanding Time and Frequency*. New York: Dover, 1982.

Watercolor

Watercolor is a type of paint that consists basically of three parts: a solid pigment, water, and a binding agent to hold the first two components together. Watercolor paint is a type of dispersion, a homogeneous mixture of two substances that do not dissolve in each other. The most common binder is gum arabic, although pastes made from flour or rice, egg yolk, or the casein obtained from cheese have also been used. The term *watercolor* is also used to describe a work of art produced with this material and the technique used in working with this medium. In

French, the comparable term used to describe watercolor painting is *aquarelle*.

History

The origins of watercolor painting are lost in history. The ancient Egyptians used water-based paints to decorate their temples, tombs, and other buildings and probably created the first true watercolors by painting on papyrus. In China and Japan, early artisans used water-based paints on fine silks and some of the earliest true paper ever made. In the Middle Ages in Europe, watercolor techniques were used by monks to decorate parchments or vellum. The most famous work from this era is probably *The Book of Hours*, produced in about 1415. At about the same time, watercolor-type techniques were used to produce frescoes by applying paint to wet plaster. Michelangelo (1475–1564) and Leonardo da Vinci (1452–1519) were masters in the use of this technique.

The first school of watercolor painting has been attributed to the Flemish master Hans Bol (1534–1593). By this time, watercolor techniques had been developed and mastered by some of the great artists of the period, including Albrecht Dürer (1471–1528), Peter Paul Rubens (1577–1640), and Sir Anthony Van Dyck (1599–1641).

An important step forward in formalizing the art of watercolor painting occurred in about 1630 with the publication by the English artist Edward Norgate of his *Miniatura, or the Art of Limning*, a book that has been reprinted many times, most recently by Yale University Press in 1997. Norgate's book describes methods for making watercolor paints using natural substances. Although a large variety of synthetic materials have since been developed for the manufacture of watercolor paints, the general procedures described by Norgate continue to form the foundation of the technique.

The greatest single obstacle to the development of modern watercolor techniques was the availability of inexpensive paper on which to paint. As long as papermaking remained a slow, difficult, and expensive task, few artists were able to afford the base material on which true watercolor painting was done. By the late eighteenth century, however, methods for the mass production of paper had been developed in England, making it possible for watercolor painting to become a readily accessible technique for almost any artist.

At about the same time, packaged pigments for use in watercolor painting first became available. The Englishman William Reeves made such pigments available in individual cakes, saving artists from having to grind their own pigments and making it possible for them to easily transport their painting materials outside the studio. At least partly because of technological developments like these, British painters dominated the field of watercolor painting through much of the nineteenth century. The skill spread rapidly throughout Europe, however, and then to the United States. Today, watercolor painting is a widely popular technique throughout the world.

Technique

Watercolor painting makes use of a combination of three elements: paint, brush, and paper. In addition to pigment, water, and binder, paints may include a number of other components designed to achieve specific effects. For example, glycerin is often added to keep the paint moist, and a wetting agent may be included to ensure that paint is spread evenly across the paper. The choice of brushes is also important in obtaining the kind of paint distribution needed on the paper.

The surface to which watercolors are applied—the paper—has long been a crucial factor in the technique. In contrast to oil painting, in which opaque paints are laid down on a surface, watercolor painting makes use of transparent paints in which the white of the paper is an essential element. The white shines through the paints and enhances or modifies the color of the pigments used. For example, a bright green pigment may take on a softer pale glow when applied to the white paper. The application of various layers of paint can also produce unexpected and attractive results. The amount of water used in a paint can also have an impact on the color effects produced when laid down on paper. In some cases, nearly dry paint can be used on rough paper to produce an almost crayonlike effect.

Opaque watercolors may also be produced by the addition of lead oxide or, more recently, zinc oxide to the pigment-water-binder mixture. Watercolors of this type are called *gouache*, or *bodycolor*. The techniques used with gouache painting are somewhat different from those of traditional watercolor painting because the whiteness of the paper is no longer a factor. An important characteristic of gouache painting is the brilliant effect produced by the high light-reflection property obtained from the lead or zinc oxide. *See also* Water and the Visual Arts

Further Reading

Appellof, Marian E., ed. *Everything You Ever Wanted to Know about Watercolor*. New York: Watson-Guptill Publications, 1992.

Brett, Bernard. *A History of Watercolor*. New York: Excalibur Books, 1984.

Whitney, Edgar A. *Watercolor: The Hows and Whys: Complete Guide to Watercolor Painting*. New York: Watson-Guptill Publications, 1974.

Water Cycle

See Hydrologic Cycle

Water Dwellers

Water dwellers are people who spend part or all of their lives living on lakes, rivers, marshes, or other bodies of water.

History

The oldest water dwellings yet discovered date to the fourth millennium B.C. They were found in 1853 along the edge of Lake Zurich in Switzerland during an unusually dry summer. As lake levels fell during the drought, Neolithic dwellings were revealed. These dwellings consisted of houses built on large wooden platforms supported by wooden posts. Archaeologists believe that the wooden platforms were originally built over marshy ground.

Similar structures have been uncovered at other locations in Western Europe, including Lake of Neuchâtel, in Switzerland; Upper Swabia, in Germany; the Lagozza area of northern Italy; and the Chassey settlements along the Saône River in France. The largest such structure yet discovered is located at Wangen, Switzerland, and is supported by more than fifty thousand piles. In most cases, archaeologists are still uncertain as to whether these structures were built over marshy ground or over lakes themselves.

Among the most extensively studied lake dwellings are those of Scotland and Ireland. Some of these dwellings, known as *crannogs*, consisted of individual houses built on top of artificial islands. They were occupied over an extended period of time, ranging from the late Bronze Age (about 1600–1100 B.C.) to the end of the seventeenth century.

Archaeological and ethnographic evidence suggests that both the buildings constructed by early lake dwellers and their cultures evolved significantly over time. The earliest water dwellings were probably simple tents or huts made of animal skins and plant materials built on floating or anchored platforms. Over time, more substantial houses, built of clay, mud, stone, and other materials, were constructed. Multistoried structures were designed not only for human habitation but also for storage, for penning domestic animals, and for defense against marauding enemies.

Types of Water Dwellings

Water dwellings take many forms. Crannog structures, for example, consist of simple seawall-type platforms made of stone, mud, wood, and other building materials on the floor of a shallow lake or marsh. Houses and other structures rest on top of this platform.

Stilt construction can also be used to support individual houses or more extended platforms on which multiple buildings can be constructed. Examples of "floating villages" consisting of either type can be found in many locations around the world today. In some places, such as the floating villages of Vietnam and Cambodia, people live permanently in their water dwellings. In other places, such as the floodplains surrounding the Amazon River, people live on land during the dry season and move to their floating houses during the rainy season, when the water level rises and drowns their villages.

In another form of water living, communities have constructed artificial islands on which they build their homes, house their animals, and carry on their lives. Perhaps the most famous of these artificial islands are those found in Lake Titicaca, on the border between Peru and Bolivia. The Uros Indians who inhabit the forty islands on the lake believe that they are descended from Manco Capac and Mama Ocllo, the first Incas and founders of that great empire. The residents' practice of building their floating islands from totora reeds that grow in the lake has, they believe, been handed down from the very beginning of recorded time. As with other water dwellers, the Uros spend the majority—if not all—of their time on their floating islands and fishing boats, maintaining their livestock, operating their schools, entertaining large groups of tourists, and carrying on their everyday lives.

Houseboats are yet another form of water dwelling. In parts of the world where large amounts of still water are available, people have sometimes chosen to construct their homes—and, sometimes, their ancillary buildings—on boats. A favorite tourist destination in Kashmir, for example, is Lake Dal, long noted for its beautiful and sometimes elaborate houseboats and floating gardens. Hong Kong's Aberdeen harbor has for many years been home to tens of thousands of houseboats, some of whose residents are said never to travel to dry land. Many of these houseboats are clustered so closely together in compact communities that they are unable to move, even if their owners should wish to do so.

Reasons for Water Dwelling

Throughout history, people have chosen to live on the water for a variety of reasons. If one's livelihood is based on the water, as is the case for fishermen, it may seem reasonable to make one's residence there. In some floating villages, underwater cages to hold fish are sometimes constructed beneath a house or beneath the village itself. People do not even have to brave the nearest river or lake to take their catch; they just drop a net into the pen beneath their home to collect the day's catch.

In some cases, a water-based village or dwelling may be less expensive than a land-based home, making it the only dwelling people can afford. For example, in the densely populated region of Guangzhon (Canton), China, along the Pearl River, many people simply could not afford to live on land and so made their homes on tens of thousands of houseboats along the river's shores.

In still other instances, a floating or stilt-based house or village may, at one time, have provided greater protection against enemies than would have a similar land-based structure. People who once made that choice may simply have continued to maintain their water dwellings rather than return to land.

Finally, some people have chosen to live on the water simply because they prefer that lifestyle and/or can afford to build and maintain a comfortable home on the water. The floating homes in San Francisco Bay, along the shore of Sausalito, for example, have evolved out of a variety of boats that were home to artists, writers, shipbuilders, and others who chose to abandon more traditional land-based homes in the 1800s. Although some simple and fairly primitive examples of those early homes still exist on the Sausalito shore, most of the members of the city's Floating Homes Association maintain substantial—and, in many cases, spectacular—homes that sell for many hundreds of thousands of dollars.

Threats to Water Dwellers

Water dwellers always face certain risks and dangers unfamiliar to land dwellers. Lake storms, for example, may tear apart reed islands, destroy the stilts on which a house or village is supported, or blow away the relatively flimsy structures themselves. For example, unusually severe flooding in Southeast Asia in the last years of the twentieth century posed a special threat to water dwellers. Lake dwellers at three floating villages in Cambodia's Tonle Sap Lake, for example, raised the stilts that supported their houses to stay above the lake's rising water levels. In so doing, they made the structures top heavy and more unstable, increasing the risk of their being blown over in stormy weather.

Development projects may threaten water dwellers also. For example, the Iraqi government began draining the marshy areas along the Tigris and Euphrates Rivers in 1991 to eliminate hiding places for antigovernment forces. In so doing, however, they destroyed an area that was once home to more than 400,000 Madans. The Madans had traditionally made their living by fishing, raising water buffalo, and weaving reed mats in marshy areas along the rivers. They lived on stilt houses and villages built above the marshes, essentially cut off from any contact with land dwellers except by boat travel to nearby villages.

As a result of the Iraqi government's program of draining the marshes, the Madan have lost both their homes and their traditional ways of making a living. Many have fled to Iran, where they survive only through the support of the United Nations High Commissioner for Refugees. By all accounts, less than 30,000 Madan continue to live in their traditional stilt-based houses above the disappearing marshes of the Fertile Crescent.

Further Reading

Coles, Bryony, and John Coles. *People of the Wetlands: Bogs, Bodies, and Lake-Dwellers*. New York: Thames & Hudson, 1989.

Gabor, Mark. *House Boats: Living on the Water around the World*. New York: Ballantine, 1979.

Horning, Audrey F. "Crannogs in Late Medieval Gaelic Ireland," in *Gaelic Ireland, c. 1250–c. 1650: Land, Lordships, and Settlement*, edited by Patrick J. Duffy, David Edwards, and Elizabeth FitzPatrick. Dublin: Four Courts Press for the Group for the Study of Irish Historic Settlement, 2001.

O'Sullivan, Aidan. *The Archaeology of Lake Settlement in Ireland*. Dublin: Royal Irish Academy, 1998.

Water Environment Federation

The stated goal of the Water Environment Federation (WEF) is to be "the preeminent organization dedicated to the preservation and enhancement of the global water environment." The organization has established the following six strategic initiatives through which it hopes to achieve that goal:

- Be the leading source of information and knowledge of the water environment
- Optimize the use of leading edge information technology
- Develop a strategic marketing and image program
- Expand and build alliances and partnerships
- Enhance our global presence
- Provide technical leadership for global environmental policy formation

WEF sponsors and supports a wide variety of activities. For example, it facilitates online technical discussion groups through which members of the organization and water-quality professionals can exchange ideas, share experiences, and develop solutions to everyday problems of water use and management. Some of the discussions that have been conducted include those on air quality, biological nutrient removal, biosolids and solids, collection systems, industrial water treatment, instrumentation and control, and legislative and regulatory policy form.

The organization also sponsors an annual conference and exposition as well as regional conferences in Latin America, Asia, and other parts of the world. WEF also publishes magazines and newsletters, such as *Industrial Wastewater, Wastewater Technology Showcase, Water Environmental Laboratory Solutions, WEF Reporter, Biosolids Technical Bulletin*, and *Watershed and Wet Weather Technical Bulletin*. Information on a variety of water environment issues is available in the form of WEF fact sheets, such as "Biosolids Recycling: Beneficial Technology for a Better

Environment," "Test the Waters: Careers in Water Quality," "Clean Water: A Bargain at Any Cost," "Guard Your Groundwater," "Nonpoint Source Pollution: You Are the Key to the Cleanup," and "Smoke, Dye, and Television: Ways and Reasons to Fix Sewer Defects on Private Property."

Further Information

Water Environment Federation
601 Wythe Street
Alexandria, VA 22314-1994
Telephone: (703) 684-2452 or
    (800) 666-0206
Fax: (703) 684-2492
URL: http://www.wef.org
e-mail: csc@wef.org

Waterfalls

A waterfall is a portion of a river or stream where water falls nearly vertically. The term *cascade* is also used to describe certain types of waterfalls, especially those of lesser height and steepness, but with large volumes of water.

How Waterfalls Form

The major factors resulting in the formation of waterfalls are (1) differences in rock types over which a river or stream flows, and (2) uplift of land. Some rocks, such as basalt and granite, are more resistant to the erosive effects of running water than are other rocks, such as sandstone and limestone. When a river flows over both hard and soft rocks in different parts of its path, the soft rock will be eroded more quickly than the hard rock. As the soft rock wears away, river water must eventually fall from the less eroded hard rock to the more eroded soft rock at a lower elevation.

Earth movements can also cause waterfalls. If an earthquake or other movement of the ground lifts a portion of land over which a river or stream is flowing, a waterfall may be formed. Earth movements caused by the collision, separation, or sliding of tectonic plates can also cause similar effects that produce a waterfall.

Waterfalls are temporary phenomena in terms of geologic time. Rivers and streams continue to cut downward into their beds, through both hard and soft rock although at different rates. In many cases, the face of the waterfall is gradually cut back and moves upstream over time. The ultimate effect of erosion by running water is to smooth out the vertical profile of the river, even though this process may take many thousands of years.

Some Major Waterfalls

Waterfalls exist almost anywhere that rivers or streams flow across significant changes in elevation. The waterfalls listed here are mentioned because of some special quality, such as height, width, or volume of water flow.

Angel Falls (Salto Angel)

Angel Falls, located in Canaima National Park, Venezuela, are the highest waterfalls in the world. They are located in an upper tributary of the Rio Caroni and fall a distance of 979 meters (3,212 ft). Angel Falls are named for American adventurer James C. Angel, who discovered them in 1935.

Tugela Falls

Tugela Falls are the highest waterfalls in Africa. They occur on the Tugela River in Natal National Park, South Africa. Their height is estimated to be 850 meters (2,800 ft).

Utigordsfoss

Utigordsfoss (Utigord Falls) are located in Norway. They are the highest waterfalls in Europe and are fed by meltwater from glaciers. Utigordsfoss have a height of about 2,625 meters (800 ft).

Yosemite Falls

Yosemite Falls are the highest waterfalls in North America. They are found in Yosemite National Park and consist of three sections: the Upper Falls, the Lower Falls, and an extensive series of cascades that connects them. The total height of the falls is 739 meters (2,425 ft), with the Upper Falls, 436 meters (1,430 ft) high and the Lower Falls, 98 meters (320 ft) high.

Khone Falls

The Khone Falls (Chutes de Khone) are located on the Mekong River in Laos. They are thought to contain the largest volume of falling water of any waterfall in the world. Experts estimate that the average flow over the falls is 11,600 cubic meters (410,000 cu ft) per second. The height of the falls is relatively small, however, only about 70 meters (230 ft).

Victoria Falls

The widest waterfall in the world is Victoria Falls, located on the Zambezi River on the border between Zambia and Zimbabwe. The falls are about 1,700 meters (5,600 ft) wide and fall about 120 meters (400 ft). They were discovered in 1855 by the Scottish explorer David Livingstone in 1855 and are regarded as one of the most beautiful natural phenomena anywhere in the world.

Niagara Falls

Niagara Falls are located on the Niagara River on the border between Canada and the United States. The falls actually consist of two parts, separated by Goat Island. The portions on the American side of the island are called the American Falls, while those on the Canadian side are called the Horseshoe Falls. The American Falls have a straight edge that extends about 320 meters (1,050

ft) and a vertical height of 56 meters (184 ft). The Horseshoe Falls have a curved face 675 meters (2,220 ft) long, with a drop of 54 meters (176 ft). In terms of the volume of water flowing over the dam, Niagara is the third largest waterfall in the world, with an average flow of 6,000 cubic meters (210,000 cu ft) per second. The development of hydroelectric plants at the falls have reduced their natural flow by about half.

Dunn's River Falls

Dunn's River Falls are located on the Dunn's River in Jamaica, about a mile from the town of Ocho Rios on the island's northern coastline. The falls are of special interest because they are the only known waterfall at the mouth of a river. The falls are about 200 meters (660 ft) high. Water falls through a series of cataracts directly into the Caribbean Sea.

Further Information

Waterfalls WebRing, http://www.naturalhighs.net/waterfalls/ring/w;llwebring.htm.

Water Gas

Water gas is a mixture of hydrogen and carbon monoxide formed when hot steam is reacted with coke or a hydrocarbon, such as methane (CH_4). In this reaction, the hydrogen in steam (water) is reduced and the carbon in coke or the hydrocarbons is oxidized. These reactions can be represented by the following equations:

$$H_2O(g) + C(s) \rightarrow CO(g) + H_2(g)$$
$$\text{steam} \quad \text{coke} \quad \text{carbon} \quad \text{hydrogen}$$
$$\text{monoxide}$$

and

$$H_2O(g) + CH_4(g) \rightarrow CO(g) + 3H_2(g)$$
$$\text{steam} \quad \text{methane} \quad \text{carbon} \quad \text{hydrogen}$$
$$\text{monoxide}$$

Since the reaction takes place in air, the final product is primarily a mixture of four gases: carbon monoxide (about 40%), hydrogen (about 50%), carbon dioxide (about 3%), and nitrogen (about 3%). Water gas is one of a group of artificially produced fuel gases known collectively as *synthesis gas*. Because of the flame produced when it is burned, water gas is also known as *blue gas*.

Both of the gases formed in largest volume—carbon monoxide and hydrogen—are flammable, so the resulting mixture can be used as a fuel in industrial operations. It has the advantage of burning with a very hot flame, although the toxicity of carbon monoxide presents some problems in working with the mixture.

Water gas may also be separated into its component gases, each of which then has its own specific uses. For example, carbon monoxide isolated from water gas can be used as a raw material in the preparation of many organic compounds. Methanol (methyl alcohol; CH_3OH) is an example. Carbon monoxide and hydrogen can be combined over a catalyst at temperatures of 300–400°C and pressures of 200–300 atmospheres to produce methanol:

$$CO(g) + 2H_2(g) \rightarrow CH_3OH(l)$$

Water Glass

Water glass is the common name for sodium silicate, Na_2SiO. It is prepared by fusing silica (silicon dioxide; SiO_2) with soda ash (sodium carbonate; Na_2CO_3). The product of this reaction looks very much like ordinary glass, but it has the unusual property of being soluble in very hot water or steam under pressure. When dissolved, water glass is a thick, syrupy liquid with a density of about 1.3. It is then sold in that form or dried and sold as a white powder that can be mixed with water to form a thick emulsion.

Water glass is an important industrial chemical, ranking close to number 50 among the highest-volume chemicals produced in the United States each year. A versatile substance, it's used as a filler in soaps and detergents; in adhesives, especially in the manufacture of wallboard and corrugated paper board; in water-treatment operations; for the bleaching and sizing of paper and textiles; as a flame retardant; in drilling for petroleum; as a binder on abrasive wheels; in the manufacture of casts and artificial stone; in the production of glass foam and certain pigments; and as a catalyst in certain chemical reactions.

Water Gods and Goddesses

See Gods and Goddesses of Water

Water Industry Council

The Water Industry Council (WIC) is an organization of water and wastewater service providers in North America. A primary goal of the council is to help communities make informed decisions on water and wastewater needs. WIC activities include:

- helping communities draft legislation that will provide the best and most appropriate kind of water and wastewater systems;
- working with state and federal governments to obtain greater availability of grant funds and bonds for water and wastewater improvements;
- providing free access to job listings on the council's website;
- sponsoring conferences that bring together government leaders, organized labor, public and private employees, and representatives of the water industry;
- arranging for WIC members to meet with elected and appointed state and local government officials on water-related issues;

A scene from the 1975 film *Jaws*. Universal Studios, Photofest.

- organizing conferences at which members can meet with each other and with representatives of municipalities and public advocacy groups;
- representing the industry to lawmakers and the general public through press releases, testimony, and interviews.

Further Information

Water Industry Council
101 Clark Street
Brooklyn, NY 11201
Telephone: (718) 625-6500
Fax: (718) 624-6135
URL: http://www.waterindustry.org
e-mail: Chertoff@Waterindustry.org

Water in Motion Pictures

Water-related topics have been important themes in motion pictures since the inven-

tion of the medium. The Library of Congress collection of early motion pictures contains many examples of the ways in which early directors used rain, the ocean, rivers, and other water-related subjects as the focus of their work. Examples include:

Surf at Monterey—an 1897 production of the Thomas A. Edison company showing waves breaking over the rocks, produced by James White as part of the Southern Pacific Company series, designed to entice travelers to visit Southern California

U.S. Battleship Indiana—an 1898 Edison picture filmed in the Dry Tortugas during a resupply visit by the battleship

Bathing at Atlantic City—a 1901 Edison film taken from a camera mounted on a tower above the beach, showing people enjoying the sand and water

Fireboat in Action—a 1903 Edison film showing the famous fireboat *New Yorker* in a staged demonstration at the Battery in New York Harbor

The Roosevelt Dam—one of many films produced under the sponsorship of the U.S. government documenting important major federal projects, probably produced in 1928.

A large number of such early films can be viewed and downloaded from the Library of Congress website at http://memory.loc.gov.

Since the earliest days of motion pictures, literally hundreds of films have been made on water-related topics. Many of those pictures can be classified into about a half dozen major categories, including sea creatures: real and imagined sea voyages: real and imagined, waterfront films, storms, river travel, and submarine tales. However, these categories do not cover many other motion pictures in which water-related topics play an important role.

Sea Creatures: Real and Imagined

Animals that live—or that writers may have imagined to have lived—in swamps, the deep oceans, remote lakes, or other bodies of water have intrigued the reading and viewing public for centuries. Motion pictures that have taken advantage of this interest include the following:

The Beast from 20,000 Fathoms (1953) tells of a mythical monster—a carnivorous "rhedosaurus," released from under the Arctic Sea by a nuclear blast—that makes its way to New York City and meets its fate at Coney Island.

Creature from the Black Lagoon (1954) was an early 3-D movie about a prehistoric, amphibious "gill-man," whom researchers suspect to be the last remaining member of a humanoid species.

Jaws (1975) was the most successful of all sea-creature films, generating three follow-up works: *Jaws 2* (1978), *Jaws 3* (1983),

and *Jaws: The Revenge* (1987). The original story tells of the appearance of a massive killer shark off the coast of a fictional seafront community, Amity Island, and the controversy that develops when some community leaders worry about the effect it may have on their busiest weekend of the year. The sequels were variations on the original theme.

The Sea Serpent (1985) tells of another prehistoric sea monster released from its deep ocean hibernation by a nuclear blast.

Terror in the Swamp (1985) is based on a "nutria-man"—a creature part nutria and part human created by scientific research that has run out of control.

Leviathan (1989) reveals another deepwater creature that formed as a result of a genetic experiment gone awry, discovered in a sunken Russian submarine by U.S. researchers.

Deep Rising (1998) is set in the South China Sea, where a luxury liner on its maiden voyage collides with a giant underwater object, which turns out to be a squidlike creature with massive tentacles.

Sea Voyages: Real and Imagined

Filmmakers interested in stories about sea voyages not only have some great literature with which to work (such as Jules Verne's *20,000 Leagues under the Sea* and Herman Melville's *Moby Dick*), but also have the active imagination of science-fiction authors and screenwriters. Examples of films based on sea voyages include the following:

Verne's great science fiction novel has been made into a motion picture twice, the first time a silent version produced in 1916. It is a long, meandering story that incorporates parts of another Verne novel, *Mysterious Island*, but is most important because of the spectacular underwater photography, a breakthrough at the time. The 1954 version was produced by Disney and includes some remarkable special effects, which won an Academy Award.

The Caine Mutiny (1954) was based on Herman Wouk's Pulitzer Prize–winning novel of the same name. The story concerns a mutiny at sea, when ship captain Phillip Queeg panics in the midst of a storm, and the court-martial that follows the ship's return to shore.

Mister Roberts (1955) is another naval story of an eccentric ship captain who loses his crew's respect. The film was followed with a sequel focusing on one of the crew's main characters, *Ensign Pulver* (1964).

One of the most reworked stories on film is the tale of the mutiny on the ill-fated English ship HMS *Bounty* in 1788. The tale was first developed in a 1916 silent film, *The Mutiny on the Bounty*, followed in 1935 and 1964 by films of the same name and in 1984 by yet another production, *The Bounty*.

Melville's *Moby Dick* first appeared as a silent film entitled *The Sea Beast* in 1926. Four years later, it was remade as a "talkie" starring John Barrymore. The film was remade again in 1956 and as a television miniseries in 1998. The story is based on the search for revenge on the part of Captain Ahab, who has lost his leg to the great white whale known as Moby Dick.

Waterfront Films

Life on the waterfront has been an almost endless source of inspiration for filmmakers dating back more than half a century. The best known of those films is almost certainly the 1954 smash hit *On the Waterfront*, directed by Elia Kazan and starring Marlon Brando, Karl Malden, Lee J. Cobb, Rod Steiger, Martin Balsam, and Eva Marie Saint. The story tells of a "coulda-been"

boxing-championship contender (Brando) who helps the authorities gain a conviction involving a gang murder on the docks. The picture won Academy Awards for Best Picture, Best Director, Best Actor (Brando), Best Supporting Actress (Saint), and Best Adapted Screenplay. There are a number of waterfront-related films called simply *Waterfront* (starring Dorothy Mackall, 1928; Dennis Morgan, 1939; John Carradine, 1944; Robert Newton, 1950; and Ray Barrett, 1983), as well as others, such as *Waterfront Wolves*, (1924), *I Cover the Waterfront* (1933), *Frisco Waterfront* (1935), *Gangs of the Waterfront* (1945), *Waterfront at Midnight* (1948), *Murder on the Waterfront* (1943), *Bombay Waterfront* (1952), and *Waterfront Women* (1952).

Storms

The force and drama of storms, both on land and at sea, make them promising subjects for motion pictures. Some of the best documentaries have taken storms as their subject, following the course of tornadoes, hurricanes, and other natural phenomena. The 1999 series *Storm Force*, for example, focused on floods, hurricanes, and tornadoes.

One of the earliest storm-related films was *The Storm*, produced in a silent version in 1916. The film was based on a play by Langdon McCormick entitled *Men without Skirts*. The story centers on the pursuit by the Royal Canadian Mounted Police of a smuggler, and the efforts of the smuggler's daughter to save his life during a terrible storm. The film was remade in 1922 (again as a silent film), with Matt Moore as the hero, and once more in 1930, with Lupe Velez playing the heroine Nanette.

Another *The Storm* appeared in 1938, starring Charles Bickford. In the climactic scene of this film, a ship's crew that has been at odds with each other for much of the story pulls together during a catastrophic storm and save the ship and their lives.

Arguably the most dramatic storm film made to date is the 2000 hit *The Perfect Storm*, starring George Clooney. The story is based on a true-life event that took place in October 1991, when a cold front from the Great Lakes collided with a hurricane moving northward across the Atlantic from Bermuda. The storm produced 100-foot waves and resulted in the loss of a fishing boat and its crew. The film was based on a book of the same name by Sebastian Junger.

River Travel

Just as the oceans have attracted the attention of motion-picture producers, so has life on and around rivers throughout the world. River life was the subject of a number of silent and early talkie movies. For example, the Tom Mix film *Tumbling River* (1927) was filmed on the Merced River, whose raging waters provided an ideal setting for some of the most exciting moments in the film.

The 1936 moving picture *Show Boat* is the second of three films based on the Edna Ferber/Jerome Kern/Oscar Hammerstein II broadway hit (and generally regarded as the best of the three). It starred Irene Dunne, Allan Jones, Paul Robeson, and Helen Morgan. The film tells the complex story of a riverfront gambler (Jones) and the ship's captain who falls in love with him (Dunne). Robeson's "Old Man River" is generally regarded as one of the high points of this film filled with outstanding music.

One of the most famous of all river films was *The African Queen* (1951), starring Humphrey Bogart and Katharine Hepburn. The film relates the adventures of an alcoholic tramp-steamer captain (Bogart) and the sister of a murdered British missionary (Hepburn) in their efforts to escape from German soldiers along East African rivers.

The film was based on a novel by C. S. Forester of the same name.

The Bridge on the River Kwai (1957) was loosely based on an incident that occurred in Burma during World War II. A group of Allied prisoners of war are ordered to build a pair of bridges over the River Kwai to permit the transport of supplies for Japanese troops from Bangkok to Rangoon. The initial focus of the story is on the conflict between the Japanese commander Saito (Sessue Hayakawa) and the British colonel in charge of the prisoners, Colonel Nicholson (Alex Guinness). The story becomes more complex, however, as Nicholson decides to prove the superiority of his troops by building a perfect bridge in spite of the use to which it will be put. The Kwai story is also told in a 1991 book by Peter Davies based on the memoirs of the real Colonel Nicholson, *The Man Behind the Bridge*.

A best-selling novel by James Dickey was also taken as the basis for a river-based film, *Deliverance*, produced in 1972. The film starred Jon Voight, Burt Reynolds, Ronny Cox, and Ned Beatty, and told of the brutal experience of four city-slicker rafters traveling down a backwoods Georgia river. The film was shot on location on the Chattanooga River and was nominated for Best Picture and Best Director.

Submarine Tales

Life on submarines has apparently long held a fascination for readers and filmgoers. Some of the earliest silent movies were based on submarine themes. For example, the 1910 film *Sunken Submarine* told of a young sailor lost at sea, but it focused instead on his mother at home and never actually showed the submarine itself. Five years later, Charlie Chaplin chose a submarine theme as the basis of a four-reel picture entitled *A Submarine Pirate*, in which a waiter (Chaplin) overhears plans to capture a submarine and acts to thwart the plot.

In ensuing years, producers completed films dealing with a range of other submarine-based topics, often dealing with the terrible risks faced by sailors on underwater boats. Some of these films were *Hero of Submarine D-2* (1916); *Trapped in a Submarine* (1931, starring John Batten); *Submarine D-1* (1937, Pat O'Brien), *Submarine Patrol* (1938, Richard Greene); *Phantom Submarine* (1941, Anita Louise); *Submarine Raider* (1942, John Howard); *Submarine Base* (1943, John Litel); *Submarine Alert* (1943, Richard Arlen); *Submarine Command* (1951, William Holden); *Run Silent, Run Deep* (1958, Clark Gable); *Das Boot* (1982, Jurgen Prochinow); *Submarine Warfare* (1987 documentary); *Hunt for Red October* (1990, Sean Connery); *Full Fathom Five* (1990, Michael Moriarty); *Crimson Tide* (1995, Denzel Washington); and *Down Periscope* (Kelsey Grammar, 1996).

Miscellaneous Films

Space does not permit discussion or even a listing of the many motion pictures in which storms, tidal waves, rivers, ice fields, oceans, or other water-related topics are included. Although such events may not represent the main focus of the film, they often play an important role. The "singing in the rain" portion of the 1952 film by the same name, for example, takes up only a few minutes of the 103-minute picture, but it is one of the most memorable scenes in all of moviemaking. The same can be said for *The Rains of Ranchipur*, a 1955 film about the wife of a British officer stationed in India (Lana Turner) who falls in love with an Indian doctor (Richard Burton). The couple consummate their illicit affair just as the annual monsoons strike the region, forcing the doctor to choose his vocation over his new lover.

Watermark

A watermark is a distinctive impression made in paper during its manufacture. The purpose of a watermark is to provide authenticity as to the paper's design or production. Watermarks generally have little or no effect on anything that is later printed on the paper and can be observed only by holding the paper up to the light or conducting some simple physical test on the paper.

The first watermarks were produced in Europe during the thirteenth century, primarily as a means of identifying the artisan or guild member who made the paper. The first watermarks were produced by making a wire pattern that could be laid down on the mesh through which the paper slurry was passed. As water was pressed out of the slurry, the paper formed had different thicknesses according to the pattern woven into the wire design. These differences in thickness produced the watermark in the paper. Over time, the design of watermarks became a sophisticated form of art, with complex designs and actual works of art being used to mark a paper. In some cases, a second watermark was added to authenticate the paper's maker. The second watermark, or *counter-watermark*, usually consisted of the maker's initials.

By the early nineteenth century, the invention of paper-making machines introduced the need for new methods of making watermarks. A device known as the *dandy roll* was developed for this purpose. A dandy roll is a hollow cylinder whose surface is covered with a wire design that corresponds to the watermark desired. As paper slurry is deposited on the roll and water squeezed out, the paper formed has the design woven into the wire template.

Today, a number of computer programs are available for producing watermarks on materials produced by desktop publishing systems. In general, these programs simply add a second image, just slightly darker than the original, on top of the page being printed.

Paper containing watermarks is still popular among individuals and corporations who wish to give their stationery a special quality. Watermarks are also used in the printing of postage stamps to provide authenticity. They have also become an important tool among historians who use watermarks to date very old manuscripts that may otherwise be difficult to authenticate.

Further Reading

Churchill, W. A. *Watermarks in Paper in Holland, England, France, etc., in the XVII and XVIII Centuries and Their Interconnection.* Amsterdam: M. Hertzberger, 1967.

Spector, Stephen, ed. *Essays in Paper Analysis.* Washington, DC: Folger Shakespeare Library, 1987.

Water Needs (Human)

See Biological Functions of Water; Desalination

Water of Hydration

See Hydrate

Water Pipe

See Hookah

Water Plants

See Aquatic Plants and Animals

Water Pollution

The term *pollutant* may be used to describe any substance introduced into the environment that is normally not present and that may pose a health risk to plants or animals, including humans.

Types of Water Pollutants

Water pollutants can be divided into seven major categories: oxygen-demanding wastes, pathogens, synthetic organic compounds, petroleum and petroleum products, heavy metals, sediment, and heat.

Oxygen-demanding wastes are organic materials that react with and, therefore, deplete oxygen dissolved in water. Some examples of oxygen-demanding wastes are dead plants and animals, uneaten food from homes and restaurants, and animal feces. Bacteria in water oxidize these wastes to produce carbon dioxide and water. As this process continues, the supply of oxygen needed by fish, aquatic plants, and other organisms decreases, and they begin to die off.

A common measure of the presence of oxygen-demanding wastes is *biochemical oxygen demand (BOD)*, the amount of oxygen needed to decompose organic wastes in water. A high value of BOD means that large volumes of waste materials are present, and large amounts of oxygen are needed to decompose those wastes. Consequently, lesser amounts of oxygen are available for aquatic organisms, whose survival is threatened. The BOD for pure water is 0 ppm (parts per million); for naturally occurring water, about 2–5 ppm; for municipal sewage after basic treatment steps, about 10–20 ppm; and for runoff from barnyards and food-processing plants, about 100–10,000 ppm. Water is said to be polluted with regard to oxygen-demanding wastes when the BOD exceeds about 5 ppm, at which aquatic organisms may become sick and start to die.

Pathogens are disease-causing organisms. They may be viruses, bacteria, protozoa, parasitic worms, or other microorganisms. The most common source of pathogens in polluted water is wastes from infected humans and other animals. A person infected with cholera, for example, may release cholera bacteria in her or his urine and feces into a river or lake.

Those bacteria may then be passed on to other humans through a community water supply.

At one time, waterborne pathogens were a major health problem throughout the world. The introduction of water-treatment systems in developed nations has largely eliminated the spread of diseases such as typhoid fever, dysentery, and infectious hepatitis through a community water-supply system. At this point, effective water-treatment systems are not yet available in many developing nations, and such diseases continue to pose major health threats. In fact, while access to safe drinking water has reached 100 percent in many developed nations, such as the United States, the United Kingdom, Canada, Denmark, France, and Norway, that percentage may be less than a third in developing countries such as Paraguay (8%), Cambodia (13%), Zaire (25%), and Ethiopia (27%) (World Bank, *World Development Indicators 1997*).

Synthetic organic compounds are chemical compounds containing the element carbon made artificially in the laboratory. Over the past half century, chemists have synthesized millions of such compounds. Many of these compounds have found a variety of useful applications, primarily in the formulation of new pesticides, herbicides, and synthetic detergents (*syndets*). New pesticides and herbicides have revolutionized the fields of agriculture and public health. They make possible the control of pests and weeds that reduce crop yields and cause disease among humans and other animals. Syndets have made all forms of washing and cleaning activities more efficient and, in many cases, less expensive than the use of natural detergents, such as soap.

Unfortunately, most types of synthetic organic compounds (SOCs) also pose serious environmental problems. Pesticides and herbicides are toxic not only to pests but also to many forms of plants and animals. In addition, they tend to be highly stable, remaining in the environment for many years. The threat they pose to plants and animals is, therefore, a long-term problem rather than one that can be solved quickly and easily.

Syndets pose a somewhat different problem for the environment. Many of them contain nitrogen and phosphorus, two essential nutrients for the growth of plants. When released into lakes and rivers, then, syndets contribute to the rapid growth of aquatic plants. Over a period of time, lakes may become so clogged with plants that animal life, such as fish, begin to die out. In the long term, the lake may actually begin to fill up with sediment, turning first into a swamp or a bog, and eventually into a meadow.

The primary sources of pesticides and herbicides is runoff from farms, residential yards and gardens, and dairy facilities where such chemicals are used. The most important source of syndets in the environment are homes, commercial buildings, industrial plants, and other structures where such cleaning products are used. Syndets pass out of such buildings into municipal sewer systems, and then into lakes and rivers.

Since the 1970s, governments have become more sensitive to the threats posed by pesticides, herbicides, syndets, and other SOCs and have passed legislation to deal with possible environmental problems. As a result, the concentration of some of the more dangerous SOCs has been reduced. However, the problem has not been completely solved. The U.S. Geological Survey's 1999 report, reported that 90 percent of all surface water and fish and 50 percent of all wells sampled for the study contained one or more (usually several) pesticides. The report concluded that there has been a significant reduction in the occurrence in water sources of some of the more dangerous SOCs, but that they "continue to occur at levels of concern" (*The Quality of Our Nation's Water: Nutrients and Pesticides*).

An August 1993 photo of an oil-coated beach north of Blind Pass, Treasure, Florida. *NOAA Restoration Center, NOAA.*

Petroleum and petroleum products are used for many different purposes, such as in fuels, lubricants, and the manufacture of plastics. They may escape into the aquatic environment by one of four primary routes: through the dumping of used industrial and motor oil into municipal sewers, into storm sewers, or directly into rivers and lakes; as the result of accidents during normal shipping operations; by means of normal leakage from oil reservoirs in the earth and from storage tanks and supply lines; and from leaks and accidents associated with the offshore production of petroleum.

The presence of petroleum and petroleum products in water can have a variety of effects on organisms. In the first place, some components of petroleum—such as benzene, toluene, naphthalene, xylene, and phenanthrene—are toxic to fish, shellfish, and aquatic invertebrates. These and other components of petroleum may have disabling if not fatal effects on other organisms. For example, sea otters may develop nosebleeds or emphysema or may be blinded by oil and oil products. Other animals are affected indirectly, as when bears and river otters scavenge on seabirds killed by petroleum leaks or spills.

Damage can occur when plants and animals are immersed in oily waters. Bird feathers and mammal fur that have been soaked in oil may lose their natural insulating and weatherproofing properties. As a result, these animals may die from drowning or exposure. Oil slicks can also cover aquatic plants, algae, and lichen, depriving them of carbon dioxide and oxygen and killing them. Films of oil on the water may also decrease the ability of water to absorb atmospheric gases, thereby killing aquatic plants and animals.

In the environmental sciences, the term *heavy metal* refers to a number of chemical elements that are toxic to plants, animals, and/or humans. Some of these metals are arsenic, beryllium, cadmium, chromium, copper, lead, manganese, molybdenum, and selenium. Toxic metals enter the environment through a number of pathways. For example, arsenic, cadmium, and manganese are by-products of mining operations. Cadmium, chromium, and copper are released in the wastewater from electroplating plants. Arsenic and beryllium are found in the wastewater from electrical power generating plants. Lead, manganese, and molybdenum are components of industrial wastes. And selenium is a component of agricultural runoff.

The health effects of heavy metals are extensive. For example, most are thought to be toxic in sufficiently high doses and may also be carcinogenic. Cadmium causes kidney damage and is toxic to many aquatic organisms. Lead can cause anemia and damage to kidneys and the central nervous system. Manganese is toxic to plants in high concentrations. A number of heavy metals (such as chromium, copper, molybdenum, and selenium) are essential to the health of an organism in very low concentrations, but become toxic in higher concentrations.

Sediment is solid material, such as sand and clay, carried off the land by running water and later deposited in a river, lake, or other waterway. When rainwater runs off the land, it usually carries soil with it on its way to a river or lake. The faster water moves, the larger the particles and the greater the quantity of soil it can carry away. When running water comes to rest, as in a lake or pond, the solid materials held in suspension are deposited as sediment.

Erosion and sedimentation are both natural processes and are responsible for some familiar topographic structures, such as the Mississippi River delta. But erosion tends to occur more rapidly on land that is being used for human activities. For example, erosion occurs 5 to 10 times as fast on cultivated land as on uncultivated land, 100 times as fast on land on which construction is taking place, and more than 500 times as fast on mined land.

The presence of sediment in water may reduce the amount of light that reaches aquatic plants, thereby reducing their viability. In large quantities, it can also produce clogged waterways, reducing their usefulness for transportation and recreation. Sediment can also clog the respiratory systems of aquatic plants, shellfish, and other aquatic animals. The presence of sediment also creates a problem for municipal and industrial water-purification systems.

Heat can be regarded as a water pollutant if it reaches a level such that the health of aquatic plants and animals is threatened. The most common source of *thermal pollution* are industrial operations and power-generating plants that withdraw water from lakes and rivers for cooling purposes. When this water is returned to the source from which it was taken, its temperature is typically a few degrees higher. This increase in water temperature poses no threat to humans and, in some cases, may actually be beneficial to certain types of aquatic life. For example, the increase in alligator population in Florida in the last few decades of the twentieth century is thought to be at least partially the result of increased water temperatures in some parts of the state.

But *any* change in temperature may pose a threat to many forms of aquatic life, and larger increases in temperature may put at risk even those organisms for which a small change is beneficial. Many fish live within a relatively narrow comfort zone of temperature, outside of which they rapidly become ill and die. Game fish, such as trout and salmon, are especially sensitive to warm water. In addition, increases in water temperature tend both to affect the migration and spawning patterns of most fish and to damage or kill fish eggs.

Sources of Water Pollution

In general, sources of pollutants entering lakes, rivers, oceans, and other waterways can be classified as *point* or *nonpoint sources*. Point sources are municipal or industrial sites that can be specifically identified as the source from which pollutants are released into a waterway. Such sources might include pipes or canals that flow from a municipal sewage system or industrial plant into a waterway. By contrast, a nonpoint source is a more extended area from which water flows, usually on an irregular basis. Examples of nonpoint sources include agricultural land, developed land, forests, or landfills. The control of point sources in reducing water pollution is usually a much simpler task than controlling pollutants released from nonpoint sources.

The primary source of water pollutants depends on the type of waterway involved. In the United States, for example, the major source of pollutants in rivers, streams, lakes, reservoirs, and ponds is agriculture. According to the most recent data available, farms and dairy operations are responsible for the pollution of about 60 percent of all U.S. rivers and streams and 30 percent of all lakes, reservoirs, and ponds. By contrast, the leading source of pollution along the Great Lakes shorelines is atmospheric deposition (affecting about 20 percent of all shoreline miles), and along ocean shorelines, urban runoff and storm sewers (55 percent of all shoreline miles). The major sources of pollution in estuaries are municipal point sources, urban runoff, and storm sewers (affecting about 28 percent of all estuarine

areas) (*The Quality of Our Nation's Waters*, 7, 9, 11, 13, 15).

Methods of Pollution Control

Efforts to control water pollution have a very long and complex history in most developed nations. As early as 1842, an English lawyer, Edwin Chadwick, published a report entitled "Sanitary Conditions of the Labouring Population of Great Britain," asking for legislation providing more sanitary living conditions for England's working classes. In the United States, the first law dealing specifically with the protection of public waterways was the Rivers and Harbors Act of 1899 (also known as the Refuse Act of 1899). This act was designed to deal with problems of navigation, disease, and oil discharges in navigable waters, but it was drafted broadly enough that it was still being used as late as the 1970s to deal with many kinds of modern water pollution.

The first modern law dealing with water pollution was the Water Pollution Control Act of 1948, which provided funding for new water-pollution-control programs, research, construction loans, and other pollution-control activities. The law was quite weak, however, and was later replaced by more effective pieces of legislation, including the 1956 Water Pollution Control Act; the Water Pollution Control Act Amendments of 1961, 1966, and 1970; and the 1972 Federal Water Pollution Control Act, amended and renamed the Clean Water Act in 1977.

Additional water quality legislation that has been passed in the last few decades are the Marine Protection Research and Sanctuaries Act of 1972, the Water Quality Act of 1987, and the Coastal Zone Act Reauthorization Amendments of 1990. The Clean Water Act is now the cornerstone of efforts in the United States to control all types of water pollution and is mirrored in a number of state laws with a similar objective. Some provisions of the Clean Water Act and comparable legislation are as follows:

- The discharge of pollutants into waterways is discouraged. However, entities that do discharge such pollutants (point sources) are required to obtain a permit from the federal government to operate these facilities. They are also required to make efforts to reduce over time the amount of pollutants discharged. Violators are subject to fines of up to $25,000 per day or more, in the case of repeat offenders.

- Individuals and public interest groups are allowed to seek action against polluters if the federal government has not or will not do so.

- The Environmental Protection Agency (EPA) is authorized to establish standards limiting the kind and amount of pollutants discharged from any given facility. EPA is also provided with funds for the construction of new and improved sewage treatment plants.

- The federal government provides financial, administration, technical, and other types of supports for states and local municipalities who design and introduce new methods of water pollution control.

See also Clean Water Act of 1972; Water Purification

Further Reading

Alley, E. Roberts. *Water Quality Control Handbook*. New York: McGraw-Hill Professional Publishing, 2000.

Laws, Edward A. *Aquatic Pollution: An Introductory Text*. New York: John Wiley & Sons, 2000.

Thornton, J. A., et al., eds. *Assessment and Control of Nonpoint Source Pollution of Aquatic Ecosystems: A Practical Approach*. New York: Parthenon Publishing, 1999.

U.S. Environmental Protection Agency. Office of Water. *The Quality of Our Nation's Waters: A Summary of the National Water Quality Inventory: 1998 Report to Congress*. Publication EPA841-S-00-001. Washington, DC, June 2000.

U.S. Geological Survey. *The Quality of our Nation's Water: Nutrients and Pesticides*. U.S. Geological Survey Circular 1225, May 1999.

World Bank. *World Development Indicators 1997*. Washington, D.C.: The World, 1997. (Available on CD-ROM)

For comprehensive information on water pollution issues, see the U.S. Environmental Protection Agency's Office of Water webpage at http://www.epa.gov/ow.

Water Polo

See Aquatic Sports, Water Polo

Water Pressure

Water pressure is defined as the force exerted by water per square unit of area. The unit of pressure measurement in the SI system is the pascal. The SI system is the modernized version of the metric system to which all scientists in the world have subscribed. The symbol *SI* stands for the French phrase *Système International d'Unités*, or "International System of Units." The SI unit of pressure is named in honor of the French mathematician and physicist Blaise Pascal (1623–1662), who discovered an important law about liquid pressure. One pascal (Pa) in the SI system is equal to a force of one newton acting on a surface one meter square in size, or:

$$1 \text{ Pa} = \frac{1\ N}{1\ m^2}$$

Since the unit of force in the SI system (the newton) is closely related to the unit of mass (the kilogram), the pascal can also be defined in terms of the mass per square area, or:

$$1 \ kg/m^2 = 9.81 \ Pa, \ and$$
$$1 \ Pa = 0.102 \ kg/m^2$$

Calculating Water Pressure

The pressure exerted by any liquid, including water, on the bottom of a container is a function of two variables—the depth of the liquid (h) and its density (ρ):

$$p = h \times \rho$$

For example, suppose a swimming pool 15 meters long by 3.5 meters wide is filled with water to a depth of 2.5 meters. The pressure exerted by the water on the pool bottom is a function only of the depth of the water (2.5 m) and the density of water (1 kg/dm³, or 1,000 kg/m³). The pressure can be calculated by using the above formula:

$$p = 2.5 \ m \times 1,000 \ kg/m^3$$
$$= 2,500 \ kg/m^2$$

or, expressed in pascals:

$$p = 2,500 \ kg/m^2 \times 9.81 \ Pa/kg/m^2$$
$$= 24,500 \ Pa$$

The English system of measurement is still widely used for nonscientific calculations. In the English system, the density of water is 62.4 pounds per cubic foot. Suppose that a swimming pool had dimensions of 50 feet by 10 feet and was filled to a depth of 8 feet. The comparable calculation to determine water pressure in this case would be:

$$p = 8 \ ft \times 62.4 \ lb/ft^3$$
$$= 500 \ lb/ft^2$$

Pascal's Law

Pascal's discovery about water pressure, now known as *Pascal's Law*, is that the pressure of water, and any other liquid, is transmitted equally in all directions. That is, the pressure on some point on the side of the swimming pool described above is equal to the distance from the top of the water to that point times the density of the water.

Imagine, for example, that a hole is bored into the side of the first swimming pool described at a depth of 1.5 meters. The pressure forcing water out of the hole at that point would be:

$$p = 1.5 \ m \times 1,000 \ kg/m^3$$
$$= 1,500 \ kg/m^2$$

or, expressed in pascals:

$$p = 1,500 \ kg/m^2 \times 9.81 \ Pa/kg/m^2$$
$$= 14,700 \ Pa$$

Practical Consequences

Water pressure is a factor of consequence in a number of natural and man-made situations. A number of marine animals, for example, move vertically in the ocean, experiencing a wide variation in water pressure. A sperm whale, for example, may dive to depths of 2,000 meters (6,500 ft), while elephant and Weddell seals dive to depths of 1,500 meters (5,000 ft) and 700 meters (2,000 ft), respectively. Emperor penguins dive to depths of up to 500 meters (1,500 ft). How do such animals survive the enormous changes in water pressure that occur during these experiences?

Marine animals have developed a number of adaptations to deal with changes in pressure. Most whales, for example, have flexible rib cages that allow their lungs to expand or contract as they pass through water at various depths. Many species of fish have gas-filled bladders that "cushion" changes in water pressure around them. Many marine animals also have elevated levels of TMAO (trimethylamine oxide), a compound responsible for maintaining water balance in their bodies. Scientists believe that high levels of TMAO help diving animals adjust to changes in water pressure around them.

Dealing with water pressure is also an important issue in the construction of submarines and other types of submersibles. Such vessels must be constructed of the strongest materials available, such as high-strength steel or titanium alloys, to withstand the crushing pressures to which they are exposed.

Water Properties

See Properties of Water

Water Purification

Water obtained from natural sources, such as rivers, lakes, oceans, and underground sources, is sometimes safe for drinking and other human uses, such as cooking and washing. More commonly, however, that water has to be purified before it is used by humans. Natural waters contain a variety of substances that should or must be removed from water before it is safe and pleasant to drink, acceptable for washing, suitable for industrial processes, and adequate for other human functions. These substances include disease-causing microorganisms, chemical toxins, materials that cause an unpleasant odor or taste, chemicals that leave stains or make cleaning difficult, and solid matter, such as dead plant matter and dirt.

The degree to which water must be purified and the methods through which that purity is reached differs, depending on the uses to which the water will be put. Water

withdrawn from a stream or lake to be used in cooling a nuclear power plant, for example, does not have to meet the same standards of purity as water destined for use as drinking water in homes. Purification methods may also depend on the quantity of water that must be purified. Campers and hikers, for example, can use water purification methods that would be far too expensive to use on a large scale, such as would be required for the public water system of a large city.

Small-Scale Methods

Boy Scouts sometimes learn that water can be purified for drinking purposes by means of a solar still. A solar still is simply a piece of plastic, glass, or other transparent material correctly placed over a basin of impure water. Solar radiation from the sun's rays causes water in the basin to evaporate, after which water vapor condenses on the cool surface of the plastic or glass. As moisture collects on the surface, it runs downward into a collecting vessel. Since it is only water that normally evaporates from the impure sample the water formed and collected in this way is pure and safe to drink. Solar stills can sometimes be constructed on a larger scale for situations in which pure water is in limited supply and sunshine is abundant.

For hikers and campers, the primary concern in purifying water is usually the destruction of disease-causing organisms. The removal of toxic chemicals, which are much less commonly present, is generally not possible by any simple means. Disease-causing organisms can be killed and water made safe to drink by boiling or by adding certain common, inexpensive chemicals, such as solutions or tablets containing compounds of iodine or chlorine.

Public Water Supplies

The purification of water for public use usually involves four primary steps: settling, coagulation, filtration, and disinfection. The sequence in which these steps occur may differ somewhat from facility to facility, but almost all plants include all four steps. The following scheme is only one of many possible schemes.

Settling. The first stage in water purification involves the removal of large materials, such as leaves, branches, fish, stones, clumps of dirt, and the like. One way to remove these materials is to pass incoming water through a large screen, which traps substances with relatively large diameters. Another way is to feed the water into a large settling tank, where heavier materials settle out by gravity over a period of time. The materials extracted by screening or settling are then removed, and partially clarified water passes on to the next stage.

Some chemical treatment may be provided during the first stage of purification. For example, chemicals may be added to adjust the acidity of the water, to correct odor and/or taste, or to carry out a first stage of disinfecting.

Raw water may also be aerated during the first stage of purification. Aeration is the process of forcing streams of water into the air, where oxygen can destroy some bacteria and other disease-causing organisms and improve the taste and odor of water.

Coagulation. The second stage of purification involves the removal of smaller particles of foreign matter in water. These particles are usually too light to settle out in the time available in a settling tank. They also have diameters too small to be trapped by screens.

Coagulation is a chemical process in which two compounds are mixed to produce a product that is insoluble in water. As this precipitate forms, it traps small particles of foreign matter in water and carries them to the floor of the coagulation chamber. A typical reaction is the one that takes place between aluminum sulfate [alum: $Al_2(SO_4)_3$] and calcium hydroxide [slaked lime: $Ca(OH)_2$]. That reaction results in the formation of two insoluble compounds: aluminum hydroxide $[Al(OH)_3]$ and calcium sulfate $(CaSO_4)$. These two compounds occur in the form of a gelatinous, sticky precipitate that settles out of solution, carrying with it many small particles, such as pieces of clay and silt, pollen, spores, and some bacteria.

$$Al_2(SO_4)_3 + 3Ca(OH)_2 \rightarrow 2Al(OH)_3\downarrow + 3CaSO_4\downarrow$$

(The down arrow $[\downarrow]$ represents the formation of a substance that is insoluble in water.)

The muddy mixture of substances formed in the coagulation chamber, known as *sludge*, is removed from the bottom of the chamber and can sometimes be used as fertilizer, as landfill, or for other purposes.

Filtration. Water from the coagulation chamber next passes through a filter to remove any remaining suspended impurities. The simplest filters are those made of a thick layer of sand resting on a layer of gravel. From time to time, the filter is back-flushed (i.e., water is forced upward through the filter) to wash out materials that have been trapped within it. These materials may then be added to the sludge collected from the coagulation chamber.

Activated carbon is sometimes added to the filter bed. *Activated carbon* is a form of carbon that has been broken up into small enough pieces to have a very large total surface area. Carbon has the ability to absorb (form an attachment to) a number of materials that are responsible for unpleasant odors, tastes, and colors in water.

Disinfection. The final stage of purification involves the destruction of disease-causing organisms that may remain in water. The most common chemical used for this purpose is chlorine, which can be added as a gas or in the form of a solid or liquid compound that releases chlorine when dissolved in water. Other methods of disinfecting are also available. For example, *ozonation* is the process by which ozone is added to water to kill organisms. Ozone is a form of oxygen that is produced by passing an electric spark through pure oxygen. Ultraviolet light can also be used to kill bacteria and other disease-causing organisms although this method, like ozonation, is relatively expensive compared to the use of chlorine.

After disinfection, the now-pure water is sent to storage tanks, from which it is pumped throughout a community as needed.

Additional steps may be involved in the purification process of a community water plant depending on the properties of the raw water used and the needs and desires of the community. For example, many communities now add fluoride to their public water supply as a way of reducing the rate of dental caries ("cavities") in their community. Other communities add a water-softening step if the raw water used is unusually "hard." In still other cases, weak acids or weak bases may be added to adjust the pH (acidity) of water produced in the plant. A simple way of increasing the acidity of water, for example, is to force carbon dioxide gas through the water at some step in the purification procedure. *See also* Water Pollution

Further Reading

American Water Works Association. *Water Quality and Treatment: A Handbook of Community Water Supplies.* New York: McGraw-Hill, 1990.

Cheremisinoff, Nicholas P. *Water Treatment and Waste Recovery: Advanced Technology and Applications.* Englewood Cliffs, NJ: PTR Prentice Hall, 1993.

Drinan, Joanne E., and Nancy E. Whiting. *Water and Wastewater Treatment: A Guide for the Nonengineering Professional.* Lancaster, PA: Technomic Publishing, 2001.

Water Rights

The term *water rights* refers to laws, court decisions, policies, and everyday practices that determine who has the right to inland water. In addition to being a major source of dispute between and among nations, water rights are also disputed by individuals, companies, and other groups within a nation.

Policies and Issues in the United States

In the United States, water rights are determined by a large variety of federal, state, tribal, and local laws and regulations. It is difficult to make generalizations about water rights in any part of the country. In general, however, most water laws fall into one of two general categories: riparian rights or appropriated rights.

Riparian and Appropriation Rights

Water laws east of the Mississippi River tend to follow a custom known as *riparian rights*. The term *riparian* refers to the bank of a river or stream or the shoreline of a lake or pond. Early settlers brought this philosophy of water use with them when they emigrated from Europe. According to the custom of riparian rights, people who own property adjacent to a river, stream, lake, or other body of water have the right to use that water. They may remove as much water as they like from the river or lake as long as they do not disturb the natural flow or quality of the water. In many cases, state and local governments have instituted additional conditions to the use of water under riparian rights. For example, water use may be restricted in times of drought.

Riparian rights belong with land ownership. If a person sells a piece of land, any and all riparian rights associated with the land are transferred to the new owner. Riparian rights usually fall equally to everyone within a watershed. No one person has more right to water use than another. As with many other aspects of water law, however, there may be many exceptions to this general principle. For example, a state or municipality may allocate water use during a drought for some purposes (such as a drinking-water system) but not others (such as irrigation).

West of the Mississippi, water rights are controlled by a different doctrine, known as *appropriated rights*, or *prior appropriation*. Under the principle of prior appropriation, the first person to use water from a stream, river, or lake for some beneficial purpose (such as watering crops) receives priority for use of the water from that resource.

The prior appropriation doctrine allows for the possibility that two or more individuals may own rights to water use at the same time. However, a priority exists among these water owners. That is, ten landowners adjacent to a stream may all have earned the right to take and use water from the stream, but some landowners have a prior right to that of others. Priority is determined by length of ownership. The person who first received rights to water use has first priority. In time of drought, the landowner with the oldest claim to water rights gets the first opportunity to take as much water as needed from the resource. Any water left is then passed to the second landowner in order of priority, then to the third landowner, and so on down the list. In contrast to riparian rights, where every owner usually gets to draw some water from a source, the prior appropriation system may result in some users *not* having access to their water.

In another difference from riparian rights, appropriated rights can be sold independently of land. A person who has decided to give up farming may continue to live on a piece of land, but sell to another individual the rights to the water that he or she had previously been using.

Water Rights and Development of the American West

Water rights tend to be a more contentious issue in the western states than in the states east of the Mississippi. The reason for this fact is precipitation. Eastern states receive a considerably larger annual rainfall than do western states, generally more than they need for most purposes. Serious droughts are rare in the East, while they are common in the West.

Development of the western states in the late nineteenth century could never have occurred without the development of a system for finding and using water. For agriculture, that meant building dams to collect and store water, from which it was then transferred through large and complex irrigation systems to farms. The development of such systems turned naturally arid areas into blooming gardens. The Central Valley of California may be one of the best examples of the way an abundant supply of water, taken from hundreds of miles away, has turned a desert into one of the richest agricultural regions in the world. The complex water systems needed for this kind of development would not have been possible unless farmers could be assured that they would always have ready access to an abundant supply of water, even if it came from many miles away. It was this scenario, then, that led to the development of the doctrine of prior appropriation.

Today, water rights have become a major topic of debate in the West. There are an increasing number of individuals, municipalities, farm groups, industries, Native American tribes, and others vying for the region's limited water resources. Perhaps the most important component in this debate has been the burgeoning population of the U.S. West. In the last decade of the twentieth century, the fastest growing region of the country was the mountain states, and the five fastest growing states (Arizona, Colorado, Idaho, Nevada, and Utah) were all in the West.

In the early 1990s, the U.S. Congress recognized the growing need to determine how the West's water resources could best be distributed. It commissioned a large-scale study, the Western Water Policy Review Commission, to investigate this problem. In 1998, the commission recommended that significant changes in water policy be made and that farms and ranches give up more water to growing cities and urban areas. In view of the myriad web of water laws in the West, however, achieving that objective is likely to be a daunting task.

Water Rights Issues around the World

Debates about ownership of water are common almost everywhere in the world. In most cases, those debates are *transboundary* disputes—that is, they involve the use of a river, stream, lake, or other body of water on which two or more nations lie. The United Nations has identified more than two hundred river basins that are shared by two or more nations. For example, nine nations border on the Zaire River; at least eight nations on the Danube River; seven on the Amazon, Nile, Rhine, and Niger Rivers; and six on the Zambezi River. The same pattern holds with large lakes. The Caspian Sea is bordered by six nations, while four of the five Great Lakes are bordered by the United States and Canada.

Problems arise when two or more nations bordering on a waterway want to use that water for some purpose, such as irrigation, hydroelectric power, or domestic or industrial needs. Of course, if there is more than enough water for all users, then no problem arises. But it is common that water supplies are insufficient to meet all demands by all nations. In such cases, disputes may arise among neighbors as to who actually controls the water in that waterway.

Pollution of waterways can also create disputes among neighboring nations. Questions arise as to which nation or nations were responsible for the pollution, how the pollution will be cleaned up, and who will pay the cost of the cleanup.

Controversies over water rights can become serious enough to result in armed conflict. In February 1951, for example, Israel began construction on a project to remove water from the Huleh Basin of the Jordan River. Objections by the Syrian government to this project led to an outbreak of gunfire between the two nations. The dispute was eventually resolved when Israel moved its water intake project to the Sea of Galilee, which lies entirely within Israeli borders.

Military action over water disputes is relatively rare. The United Nations Food and Agriculture Organization (FAO) has discovered that nations are far more likely to negotiate a treaty to resolve water disputes than to go to war over such issues. In a 1984 study, the FAO found that more than 3,600 treaties relating to transboundary water resources had been signed between A.D. 805 and 1984. Since 1814, nations had negotiated about 300 treaties dealing with non-navigational issues, such as flood control, hydroelectric projects, and allocation of water for domestic and industrial use.

One of the most dramatic examples of such treaties is the 1960 Indus Water Agreement between India and Pakistan. Before the partition of India in 1948, the Indus River lay entirely within the boundaries of that country. After partition, the Indus formed part of the boundary between India and newly created Pakistan. Tensions began to develop almost immediately as to which nation had what kinds of rights to the Indus water. Instead of becoming a source of war

Oregon's dry Klamath Basin during the summer of 2001. *Rebecca Clarren,* **High Country News.**

between the two nations (although the question was certainly a source of ongoing disputes between them), the issue was resolved by the 1960 agreement in which each nation acknowledged certain rights for the other nation.

Another impressive example of agreements among nations over water rights is the Mekong Committee, established in 1957. The four nations bordering on the Mekong River—Cambodia, Laos, Thailand, and Vietnam—signed an agreement regulating the use of river water by the four signatories. The original agreement was revised and readopted in 1975 and again in 1994. Remarkably, the organizations and working mechanisms established by the 1957 committee continued to function throughout the turbulent 1960s and 1970s, when at least three of the four nations were at war with

each other or with other nations. *See also* Water-Use Issues

Further Reading

Burns, Bryan Randolph, and Ruth S. Meinzen-Dick. *Negotiating Water Rights*. London: Intermediate Technology, 2000.

Faure, Guy Olivier, and Jeffrey Z. Rubin, eds. *Culture and Negotiation: The Resolution of Water Disputes*. Newbury Park, CA: Sage Publications, 1993.

Lowi, Miriam R. "Political and Institutional Responses to Transboundary Water Disputes in the Middle East," http://wwics.si.edu/PROGRAMS/DIS/ECS/report2/lowi.htm.

Miller, Char, ed. *Fluid Arguments: Water in the American West*. Tucson: University of Arizona Press, 2001.

Wright, Kenneth R., ed. *Water Rights of the Eastern United States*. Lanham, MD: American Water Works Association, 1998.

Water Signs

The water signs are the three astrological symbols and concepts related to water: Cancer, Scorpio, and Pisces. A fourth water-related astrological symbol is Aquarius, the water carrier. In spite of the fact that Aquarius is a water carrier, the sign is one of the air signs, not a water sign.

Many ancient Greek philosophers believed that all aspects of the universe were made up of four fundamental properties: earth, air, fire, and water. Although that belief died out long ago, it has had an ongoing impact in a number of fields, one of which is astrology.

Astrology is based on the belief that events that take place in the world, including those that occur in human lives, are determined by celestial bodies. Astrologers believe that by studying the positions of the stars and planets at specific times, patterns of human behavior, choices that people will make, and events that are likely to occur or not likely to occur can be divined.

An important element in astrological study is the zodiac—the band of stars that circles the heavens along a path roughly traced out by the Sun's orbit in the sky. Astrologers divide that band into twelve parts, or "houses," to each of which is assigned a name derived from some distinctive constellation found within that zodiacal section. Those zodiacal houses are associated with each of the ancient Greek elements: earth, air, fire, and water. The three water houses are associated with the constellations Cancer (the crab), Scorpio (the scorpion), and Pisces (the fish).

According to astrological thought, a person's personality and fate in life is determined by and can largely be predicted on the basis of the zodiacal house that was dominant at the time of his or her birth and that is ascendant at any given time for which one wishes to predict events. Making a specific astrological forecast is a somewhat complex procedure, involving the study of the location of planets as well as the position of zodiacal houses at any given time.

Nonetheless, astrologers have assigned certain personality traits to those who are born under each of the zodiacal signs. Those traits are thought to be associated in some fundamental way with the organism represented by the sign. Those born under the sign of Cancer, for example, are thought, like crabs, to have a soft, tender internal self around which they have built a tough outer shell for protection.

The complete personality profile associated with any one zodiacal sign however, is not so easily connected with the organism represented. People born under the sign of Pisces (a pair of fish swimming in opposite directions), for example, are thought to be highly imaginative with a tendency toward the mystical, though it is not immediately obvious how those traits relate to fish. Similarly, those born under the sign of Scorpio are said to have a strong tendency toward jealousy and self-protection, and become easily angered.

Further Reading

Editors of Time-Life Books. *Cosmic Connections*. Alexandria, VA: Time-Life Books, 1988.

Goodman, Frederick. *Zodiac Signs*. London: Brian Trodd, 1990.

Parker, Derek, and Julia Parker. *The Compleat Astrologer: The Practical Encyclopaedia of Astrological Science*. Toronto: Bantam Books, 1975.

Waterskiing

See Aquatic Sports, Waterskiing

Water Softening

See Hardness of Water

Water Sports

See Aquatic Sports

Waterspouts

A waterspout is a funnel-shaped column of rotating air that extends from a storm cloud (usually a cumulus, cumulonimbus, or cumulus congestus cloud) to the surface of a body of water. Scientists do not yet fully understand the mechanism by which waterspouts form.

Types of Waterspouts

Waterspouts are sometimes described as tornadoes that travel over water, but that description is incomplete. Some waterspouts do begin as land-based tornadoes that later sweep outward over water. Waterspouts of this type have the same characteristics as land-based tornadoes except that they tend to dissipate more rapidly over water than over land.

But other waterspouts form only over open water. Such waterspouts are known as *fair-weather waterspouts*. Fair-weather waterspouts tend to be smaller and less destructive than tornadic waterspouts, which are often as devastating as land-based tornadoes. Fair-weather waterspouts typically have central cores of about 50 meters (150 ft) in diameter, although some grow to twice that size. By comparison, an average tornado can be more than 500 (1,500 ft) meters in diameter. Wind speeds in fair-weather waterspouts are also moderate compared to those in land-based tornadoes. The average speed for the former is usually less than 100 kilometers per hour (60 mph), while the highest speeds reached by winds in land-based tornadoes are thought to reach 500 kilometers per hour (300 mph) in some cases. (No method has yet been found to calculate accurately the wind speed in most tornadoes.) Fair-weather and tornadic waterspouts tend to travel at similar speeds (an average of about 12 kilometers per hour [20 mph]) and for about the same length of time (usually no more than about 10 minutes).

Tornadic waterspouts are produced as a result of weather conditions similar to those that generate other tornadoes. They form during unstable atmospheric conditions, when a cold air mass moves across a warmer pocket of surface air. As warm air just above ground begins to move upward rapidly, it begins to rotate. Pressures inside the rotating column, or *vortex*, are significantly less than those outside the column. Air pulled inside the vortex expands rapidly and cools below the dew point. Tiny droplets of water formed by this process create a visible funnel-shaped cloud characteristic of a waterspout.

Formation of Waterspouts

Fair-weather waterspouts form by a process similar to that for tornadic waterspouts, but usually under less severe conditions. The clouds from which they form, for example, usually extend no more than 3,000 meters above sea level, while those from which land-based tornadoes form may extend five times as high. The shorter vertical distance traveled by winds in a fair-weather waterspout partially account for the lower energy and reduced risk posed by such storms.

Waterspouts occur along coastal regions in many parts of the world, including the Gulf of Mexico, the Mediterranean Sea, and the Indian Ocean. The largest concentration of waterspouts in the world is thought to occur off the lower Florida Keys, where up to 500 such events a year may occur. Fair-weather waterspouts are not very destructive, although they can pose a threat to small sailing craft. The National Weather Service now provides waterspout forecasts when atmospheric conditions are favorable for the formation of such storms.

Multiple waterspouts off the Bahamas. *Dr. Joseph Golden, NOAA.*

Further Reading

Golden, Joseph H., "Wild Waterspouts," *Weath-erWise*, September/October 1999, 14–19.

Grotjahn, Richard, "Multiple Waterspouts at Lake Tahoe," *Bulletin of the American Meteorological Society*, April 2000, 695–702.

Lane, Frank W. *The Elements Rage: The Extremes of Natural Violence*. Newton Abbot, UK: David & Charles, 1966.

Water Table

See Groundwater

Water Toys

Water toys are of two general kinds: those that use water to operate and those that can be used on water. One of the oldest and most familiar toys of the first type is the *water pistol*, or *water gun*. The water pistol is a device that is filled with water and then fired by pulling a trigger. The trigger operates a plunger that forces water out the nose of the pistol.

More forceful water pistols can be made by adding a second chamber filled with air. When the trigger on this type of water pistol is pulled, it activates a plunger that compresses air in the auxiliary chamber. Compressed air then forces water out the nose of the gun.

Modifying the size of the air and/or water chambers can significantly increase the amount of water released from a water pistol and the force with which it is delivered. The Super Soaker, invented by aerospace engineer Lonnie Johnson in 1988, is one of the most powerful versions of the traditional water pistol. Other variations on the basic design of the water pistol have resulted in devices that have uses other than those of a toy. A high-powered water pistol capable of delivering a mist of water droplets at speeds of up to 250 miles per hour, for example, has been used both as a fire extinguisher and as a method of crowd control.

A *water rocket* is a device that consists of a bottle containing water and compressed air, sealed at the mouth with a cap containing a small hole in its middle. The simplest water rocket can be made out of a plastic soda bottle into which has been inserted a tightly fitted cork or plastic adapter. A hole large enough to fit a small plastic or rubber tube is inserted into the cork or adaptor. Finally, some predetermined amount of water is added to the bottle, and the cork or plastic adapter is sealed to the mouth of the bottle.

To operate the water rocket, air is pumped into the bottle through the hole in the cork or adapter. The easiest way to add air is with a bicycle pump. When the required amount of air has been added to the bottle, the plastic or rubber tube from the bicycle pump to the bottle is removed. At that point, the compressed air within the bottle pushes water out through the hole in the cork or adapter, propelling the bottle forward or upward into the air.

The height and distance to which a water rocket travels as well as its velocity can be increased with a variety of modifications. For example, the size of the bottle used and the amount of water and air added can be adjusted to provide greater travel distance and velocity. Fins can also be added to the bottle to improve the stability of the rocket's flight and to reduce its drag. The shape of the bottle used can also be modified to improve both of these characteristics.

Two examples of toys designed to be used on and in the water are the AquaSkimmer and the AquaDisc, both of which are frisbee-like objects. The former can be bounded off the water's surface—similar to skipping stones—and the latter can be tossed back and forth underwater at depths of up to 30 feet, as one does with a Frisbee on land.

A young inventor with a number of water-toy inventions to his credit is Richie Stachowski, a California teenager. In the late 1990s, Stachowski invented a device with which two people can talk to each other underwater. Stachowski followed up his Water Talkie with other water-toy inventions, including the Scuba Scope, which allows people to see each other underwater and above water at the same time; the Bumper Jumper Water Pumper, a sit-down float whose paddle can also be used as a water gun; an underwater hockey game; an underwater pogo stick; and novelty diving goggles.

Further Reading

"Rockets," http://ourworld.compuserve.com/homepages/pagrosse/h2orocketsi.htm.

Water Treatment

See Water Purification

Water Use

The uses to which water is put by human communities can be divided into three major categories: domestic, industrial, and agricultural. Data on water use for these three categories is generally available for most countries in the world, although those data are more complete for some nations than for others. In the United States, for example, the U.S. Geological Survey has conducted extensive surveys of the way water is used in this country every five years since 1950. For some less developed nations, water data are less complete and less reliable.

Water Use Worldwide

Dramatic differences exist worldwide in the uses to which water is put in various regions and nations. In most nations of the

Two boys playing with Super Soaker water guns. *Dirk Douglass/Corbis.*

world, the vast majority of water consumed is for agricultural purposes, often for irrigation. The percentage of water used for such purposes exceeds 90 percent of all water consumed in most parts of Africa, South and Central America, and Asia. Agricultural uses account for more than 90 percent of all water consumed in nations such as Madagascar (99%), Guyana (99%), Afghanistan (99%), Turkmenistan (98%), Mali (97%), Somalia (97%), Pakistan (97%), Swaziland (96%), Sri Lanka (96%), Nepal (95%), Bangladesh (94%), Rwanda (94%), Sudan (94%), Oman (94%), Syria (94%), Uzbekistan (94%), and the Kirgiz Republic (94%) (The World's Water).

By contrast, some countries in the world use very little water for agricultural purposes. Such countries include the United Kingdom (3%), Finland (3%), Lithuania (3%), Singapore (4%), Belgium (4%), Switzerland (4%), Estonia (5%), Equatorial Guinea (6%), Iceland (6%), Canada (8%), Austria (8%), and Norway (8%).

The allocation of water for industrial purposes worldwide tends to range between 10 and 20 percent, although it is much higher in most European nations and a few other countries, such as Belgium (85%), Finland (85%), Canada (80%), the United Kingdom (77%), Ireland (74%), Switzerland (73%), Norway (72%), and Austria (73%).

The proportion of water used for domestic purposes varies widely throughout the world. In some parts of Africa, for example, the percent of water used for domestic purposes can be less than 10 percent, as in Madagascar (1%), Mali (2%), Swaziland (2%), and Somalia (3%), while in other nations, it can be very high, as in Equatorial Guinea (81%), Gabon (72%), Congo (62%), Togo (62%), and Guinea-Bissau (60%).

Domestic water-use trends follow no discernible pattern in other parts of the world either. In Europe, for example, some nations use only small fractions of their water for domestic purposes, such as the Netherlands (5%), Greece (8%), Romania (8%), and Hungary (9%), while other nations allocated much larger proportions, such as Malta (87%), Luxembourg (42%), Sweden (36%), Iceland (31%), and Denmark (30%).

In Asia and Oceania, similar patterns prevail. Domestic use accounts for a very small amount of all water consumed in Afghanistan (1%), Pakistan (2%), Sri Lanka (2%), India (3%), Bangladesh (3%), and Iraq (3%), but for a relatively large amount in Australia (65%), New Zealand (46%), Singapore (45%), Bahrain (39%), Kuwait (37%), and Bhutan (36%).

Water Use in the United States

According to the U.S. Geological Survey (USGS), water use is divided into eight major categories: public water supplies, domestic, commercial, irrigation, livestock, industrial, mining, and thermoelectric. In 1995 (the most recent year for which data are available), USGS estimated that an average of 341,000,000,000 gallons of water were withdrawn daily from both surface and underground reservoirs. Of this amount, the majority was used for irrigation and thermoelectric production (39% each), with public water supplies (12%) and industrial uses (6%) being the next most important consumers. Livestock (2%), domestic (1%), commercial (less than 1%), and mining (less than 1%) accounted for the remaining water used in the country (Solley, Pierce, and Perlman 1998).

For some purposes, saline water can be utilized. About 7 percent of all water used for industrial purposes, 32 percent of all water used in mining operations, and 31 percent of all water used in the production of thermoelectric energy comes from saline sources. The total amount of saline water used for these purposes each day is estimated to be about 60,000,000,000 gallons.

Historical Trends in the United States

During the period over which the USGS has been conducting its water-use surveys, the U.S. population has nearly doubled, going from 150.7 million in 1950 to 267.1 million in 1995. During that same period, the amount of water withdrawn for all uses in the nation has slightly more than doubled, from 180 billion gallons per day in 1950 to 402 billion gallons per day in 1995. Water use increased quite dramatically during the first three decades of the surveys, with increases of 90 bg/d (billion gallons per day) in the first decade (1950–1960), another increase of 100 bg/d in the second decade (1960–1970), and an additional increase of 70 bg/d in the third decade (1970–1980). Over the next decade and a half, however, water consumption leveled off and remained close to about 400 bg/d in the 1985, 1990, and 1995 surveys.

Patterns of water consumption among the eight major groups did not change to any great extent during the forty-five years of the water-use surveys. Public and rural water supplies have shown a modest increase over the period, although the proportion of consumption in these categories has remained relatively small (about 10% in the former case and about 1% in the latter). The relative amounts of water allocated for irrigation, thermoelectric production, and industrial uses have followed the general pattern of water use, with significant increases from 1950 to 1980 and a leveling off in the fifteen-year period that followed. *See also* Desalination; Water-Use Issues

Further Reading

Rijsberman, Frank R., ed. *World Water Scenarios: Analysing Global Water Resources and Use.* London: Earthscan Publications, 2001.

Solley, Wayne B., Robert R. Pierce, and Howard A. Perlman. *Estimated Use of Water in the United States in 1995.* U.S. Geological Survey Circular 1200. Washington, DC: Government Printing Office, 1998.

Speidel, David H., Lon C. Ruedisili, and Allen F. Agnew. *Perspectives on Water: Uses and Abuses.* New York: Oxford University Press, 1988.

The world's water, http://www.worldwater.org.

Young, G. J., James C. I. Dooge, and John C. Rodda. *Global Water Resource Issues.* Cambridge: Cambridge University Press, 1994.

Water-Use Issues

The allocation of water resources is one of the most contentious issues faced by national, state, and local governments; industries; farmers; and others for whom water is an essential raw material. In many instances, disputes over water use have their origin in debates as to which governmental body has ownership rights to a body of water. The Middle East, for example, has long been the site of acrimonious arguments as to who "owns" or has rights to use the limited water resources available in the area.

Natural Water Supplies

Disputes over rights to water use tend to be more common and more intense in regions where water resources are limited and/or during periods of drought, when normal water flow is reduced. However, such disputes arise for reasons other than scarcity. In Louisiana, for example—the U.S. state with the most extensive water resources per square mile—debates over water use have become more common in recent years.

Louisiana is the only state in the union with no regulations of any kind on the use of water. Any person or corporation can take any amount of water from any resource at any time for any use. This philosophy developed over time at least partly because of the abundance of water in and under the state.

However, residents of the state have begun to see the effects of such a policy in a period of rapid population growth and increasing industrialization and expanding agricultural systems. In the early 2000s, the state's aquifers began to drop anywhere from a half-foot to eight feet per year, prompting state officials to begin considering new water regulations for Louisiana ("Watery Louisiana").

Water-Use Issues in the American West

In the United States, conflicts over water use tend to be more common and more intense in areas west of the Mississippi River than those east of the river. The reason for this is that most of the eastern United States receives relatively large amounts of precipitation, more than 20 inches annually. The western states, by contrast, often receive less than—and sometimes much less than—20 inches per year.

Other crucial factors that lead to disputes over water use are also common in the western states. One factor is increasing population. Some of the fastest growing regions of the country are located in areas where water resources are very limited. While populations in areas like Las Vegas, Los Angeles, Phoenix, and Salt Lake City continue to rise, the amount of water available for domestic use remains essentially constant.

Another factor contributing to disagreements over water use in the West is the need for irrigation. Some of the most productive farms in the nation, if not the world, are located in areas that are so arid as to be classified as semi-deserts. Yet these areas have become agriculturally productive because of vast irrigation systems developed before the middle of the twentieth century. As long as annual precipitation remains in the normal range and other demands are not placed on water supplies, these irrigation systems continue to operate efficiently. In times of drought or as other demands on water supplies increase, sufficient water supplies may no longer be available for farming and dairy activities.

The development of hydroelectric power in the West is also a factor in determining who gets how much water. Although hydroelectric power plants normally use only *instream* water (water that does not leave a river or stream), their operation can affect the allocation of water in other ways, such as in reducing the flow of water needed by fish populations.

Examples

In spite of dramatic differences in precipitation patterns across the United States, disputes over water use are now common in almost every part of the country. The examples below highlight the kinds of situations that can develop when different groups of users compete over a relatively constant supply of water.

The Pacific Northwest's Columbia River has long been one of the world's greatest natural water reservoirs. Over time, humans have tapped the river for the operation of huge irrigation systems on both sides of the river and for large hydroelectric power-generating systems. In addition, the river has traditionally been a major source of sports and commercial fishing, with various types of salmon high in popularity.

During periods of drought, however, the Columbia is not able to meet all of these needs, and disagreements develop among power producers, farmers, Native American tribes, local communities, and other groups as to how the river's waters are to be used. In the late 1990s and early 2000s, concern for the survival of salmon and other species of fish prompted the U.S. Fish and Wildlife

Service to place the chinook, chum, and coho salmon as well as certain varieties of the cutthroat trout on the endangered species list. As a result, sufficient amounts of Columbia River water had to be provided to maintain the ecosystem shared by these fish.

One result of this decision is that farmers using Columbia water for irrigation suffered cutbacks in the amount they were allowed to draw down each year. In 1999, for example, for the first time in one hundred years, farmers in Washington State's Okanogan County were told that they would not be permitted to draw water from the Columbia for use in their irrigation systems.

Operators of the giant Bonneville Power Plant on the Columbia were also required to increase the amount of water spilled over the top of the dam and decrease the amount run through its turbines. This change resulted in a decrease in the amount of electricity produced for the region. (In spite of the new regulations, Bonneville operators decided on a number of occasions that they could not maintain an adequate supply of water *over* the dam for fish and *through* the dam for power generation and, in almost every case, decided in favor of power generation.)

The Detroit Lake in northern Oregon was the site of another debate over water use arising out of the need to provide water for endangered fish. In 2001, the U.S. Army Corps of Engineers decided that it would have to start draining the lake to provide enough water to protect endangered spring chinook and winter steelhead populations in the Willamette River. But in so doing, lake levels fell so severely that boaters in the lake were left literally high and dry. Owners of recreational facilities located on the lake formed a Save Our Lake committee to demand that the Corps of Engineers give more attention to the recreational opportunities provided by the lake.

Residents of Colorado's San Luis Valley have been involved in a water-use dispute that pits farmer and ranchers against developers. During the 1990s, American Water Development Inc. was created to make water available from the valley's aquifers to communities in the region who wanted to promote development. When the corporation later failed, its water rights were purchased by rancher Gary Boyce, who preferred to offer the resource to farmers and ranchers. In the early 2000s, the debate over water use between developers and farmers and ranchers continues.

The dispute over the Basin Mills Dam Project in Maine highlights a common controversy between Native American tribes and other groups over the use of water resources traditionally controlled by Native Americans. In the case of the Basin Mills project, the Bangor Hydro-Electric Company wanted to build a dam for hydroelectric power production on the Penobscot River. The Penobscot Nation objected, however, pointing out that the tribe had used the river for fishing for hundreds of years. Eventually, the tribe agreed to accept a multi-million-dollar compensation for relinquishing its objections to the dam.

Competing demands for water resources in central Nebraska in 1998 involved nearly a dozen different governmental and nongovernmental agencies concerned about hydroelectric power, irrigation, and endangered species. The crux of the issue was the relicensing of two hydroelectric generation plants in the area. After extended discussions, all interested parties signed an agreement that carried in perpetuity sufficient water resources for continued operation of the dam, maintenance of the Platte River Whooping Crane National Habitat, and supply of existing irrigation systems in the area.

Possible Solutions

The occurrence of drought conditions highlights the problems that can develop when various users compete for limited water supplies. In the past, such disputes have usually come down to acrimonious arguments among special-interest groups. Many experts now believe that a new framework must be developed within which such disputes can be handled more amicably. An example of this approach is the report issued in early 1998 by the Western Policy Review Advisory Committee, a federal advisory board chartered by the Department of the Interior in 1995.

The committee recommended a number of steps to alleviate the crisis atmosphere surrounding debates over water-use issues. These steps include the following:

- Increasing the cooperation of governmental activities located within a given watershed
- Taking additional steps to maintain and improve water quality
- Developing methods for increasing water supplies in environmentally sensitive ways
- Promoting methods for better management of floodplains
- Finding ways to restore and improve the nation's river systems
- Recognizing and dealing with issues of special concern to Native Americans.

See also Water Rights

Further Information

"Watery Louisiana Faces Declining Water Supply," *U.S. Water News Online*, http://www.uswaternews.com/archives/arcsupply/1watlous5.html.

Young, G. J., James C. I. Dooge, and John C. Rodda. *Global Water Resource Issues*. Cambridge: Cambridge University Press, 1994.

A useful source of information about disputes over water use is the periodical *U.S. Water*

News, whose online version is available at http://www.uswaternews.com.

Water Waves

A water wave is a mass of water characterized by two types of motion: that of the water particles of which it is made, and that of the overall mass of water itself, the *waveform*.

Characteristics

Like other types of wave motion (light and sound, for example), water waves can be described by three characteristics: wavelength, period, and amplitude. The wavelength (λ) of a wave is defined as the horizontal distance between two successive wave crests, wave troughs, or other comparable parts of the wave. The crest of a wave is the highest point water reaches in the wave; the trough is the lowest point reached by water in the wave.

The period of a wave (T) is defined as the time required for any given part of a waveform to pass a given point in space. Amplitude (A) is defined as the distance from the undisturbed water surface to the wave crest or trough. The depth of water (h) is also an important factor in determining certain characteristics of a wave. The rate at which a wave is propagated—its velocity (v)—is a product of its wavelength times its period, or

$$v = \lambda \cdot T$$

Waves can be classified according to the process by which they are generated, namely those produced as a result of astronomical factors (tides), those produced by earth movements (tsunamis), and those produced by movement of air (wind). The first two types of waves are discussed elsewhere in this book; this entry deals only with wind-generated waves.

Wave Generation

Waves are generated when the energy of moving air (wind) is transferred to a body of water. This transfer occurs when wind travels over a water surface that is not perfectly smooth. Differences in adjacent water heights result in differences in air pressure on the water's surface. This difference in air pressure forces water particles to begin moving upward and downward.

This pattern of particle movement is easily observable if a cork is placed on the surface of a wave. The cork appears to bob up and down with no noticeable movement in a forward or reverse direction. In fact, this observation is somewhat deceptive, since the cork's movement is actually more complex. The cork, like the water particles on which it rides, actually follows a circular path. As the wave front passes, water particles travel along the crest of the wave, then backward in a direction opposite that of the wind, then downward to the trough of the wave, then forward in the direction of the wind, and finally back upward to the wave crest—a completely circular path.

During this process, the normal tendency within the wave is for surface tension to restore water particles to a smooth surface, a surface with the smallest possible surface area. At some point, however, the interaction between air and water movements tends to become synchronous, increasing the magnitude with which water particles are displaced.

The first and simplest manifestation of this interaction is the formation of very small waves, known as *ripples*, or *capillary waves*. Patches of water containing ripples are easily distinguishable on the water's surface because they do not reflect light very well and, therefore, tend to appear as dark patches on the water's surface, patches sometimes known as *cat's-paws*. Cat's-paws form, disappear, and move across the water's surface quickly.

Wave Development

The way in which a wave develops depends primarily on three factors: how strongly the wind blows, how long it acts on the water, and the distance over which it acts on the water, or its *fetch*. An increase in either of the first two factors results in an increase in the wavelength and the amplitude of a wave, while an increase in fetch produces an increase in wavelength. On the open seas, where the fetch is very great, wavelengths may reach a few hundreds of meters. On lakes and other smaller bodies of water, the fetch is much smaller and water wavelengths seldom exceed a few meters. The greatest amplitude of waves is not known because such waves occur during the most severe storms, when measurements are difficult or impossible. However, scientists believe that the maximum amplitude for an open-ocean wave is about 25 meters (80 ft).

The types of waves observed on open water vary, depending primarily on two factors: wind and water depth. In areas where moving air is still disturbing surface water, *sea waves* are formed. Sea waves consist of many different kinds of waves with different wavelengths, periods, and amplitudes. The wild, irregular scene of turmoil at the center of an ocean storm is characteristic of a sea wave.

As waves move outward and away from the winds that produced them, they undergo a process known as *sorting*, or *dispersion*. In this process, waveforms of different amplitude and period are separated from each other. Those with longer wavelengths and shorter periods (greater velocity) begin to outdistance those with shorter wavelengths and longer periods (lesser velocity). They form a distinct, more clearly defined pattern of water movement known as *swell waves*. Swell waves lose their energy very slowly and tend to move over very long stretches of water. Indeed, observations indicate that swell waves formed as far south as New Zealand or the southern tip of South America in the Pacific Ocean may not dissipate until they reach the coasts of Alaska.

As a wave approaches shallow water, its characteristics begin to change. When water depth is about half that of a wave's wavelength, water particles begin to come into contact with the ocean bottom. Friction between water particles and the bottom changes the orbit in which particles travel from circular to elliptical. The shallower water gets, the more elliptical the orbit.

At a point where the water depth is about 1/20 that of the wavelength, water particles are essentially moving back and forth rather than elliptically. At this point, the wave's amplitude and velocity have increased significantly, creating the *surface waves* observable along most beaches. The waves actually begin to break when the horizontal velocity of water particles is greater than that of the waveform, a condition that may occur when the wave height is less than 10 percent that of the water depth.

Internal Waves

Waves can also form beneath the water's surface, where they are observable only indirectly. Such waves are generated by differences in water density rather than by interaction with air movements, as is the case with surface waves. Differences in water density are usually produced by variations in water temperature at various depths. One of the few observable ways in which internal waves can be identified is by the regular patterns of oil slicks, plankton, or other floating material that often form above their troughs.

Wave-Shore Interactions

When waves come into contact with land, changes occur both in wave patterns and in the land itself. In the simplest but probably least common instance, water waves break on a perfectly straight beach, cliff, or other land form that is at right angles to the direction of the wave front. In such a case, the wave is reflected back on itself, much as a beam of light is reflected off a mirror.

More commonly, a wave front strikes the land surface at an angle. In such a case, one part of the wave front comes into contact with the land first. That part of the wave is slowed down, while the rest of the wave continues to move at its characteristic speed. As the whole wave front gradually comes into contact with land, the direction in which it is moving changes, heading in a direction more at right angles to the land than initially. This change in the direction of a wave is called *refraction* and is similar to the change that takes place when light passes from one medium (such as air) into a second medium of different density (such as glass or water).

The refraction patterns that form along a shoreline are many and complex, depending on the topography of the shore. For example, the refraction that occurs around a peninsula of land jutting out into the water results in water waves bending in such a way that they appear to "attack" the peninsula from every direction along its shores.

An opposite pattern may develop when waves pass through a narrow opening in the shoreline that leads into a bay. In such a case, the waves are *diffracted*, or spread out, as they pass through the opening. Wave fronts reaching the shorelines inside the bay are fanned out after passing through the opening to the bay, reaching the shore with much diminished magnitude.

Erosional Effects

Water waves erode the shoreline in two primary ways. First, the force with which waves strike the shore is often large enough to tear apart rocks and break them into smaller pieces. A wave 10 feet high, for example, strikes the shore with a pressure of 1,675 pounds per square foot. Because of the way in which waves are refracted onto landforms that project into the ocean, such landforms receive the greatest force and, therefore, are eroded to the greatest degree. For this reason, the ultimate effect of wave action on a shoreline is to straighten out the shoreline and remove projections, such as promontories, that jut out into the water.

Second, water waves tend to move previously eroded materials on a beach from place to place. When a wave strikes a beach at right angles, the movement of sand and finer particles is away from the beach, into the water, and then back again onto the beach. When waves strike the beach at an angle, however, sand is carried in one direction or another. Because of such action, waves can dramatically change the shape of a beach, essentially removing all or much of the sand from one area and depositing it in another area.

Further Reading

Crapper, G. D. *Introduction to Water Waves.* Chichester, UK: Ellis Horwood, 1984.

Mytton, Craig. "Waves," http://www.brookes.ac.uk/geology/8361/1998/craig/craig.html.

Wright, John, Angela Colling, and Dave Park, eds. *Waves, Tides, and Shallow-Water Processes.* London: Butterworth-Heinemann, 2000.

Waterwheels

A waterwheel is a large wheel capable of doing work when it is turned by flowing water. It consists primarily of two parts: the wheel itself, and an axle that passes through the center of the wheel. As water causes the wheel to turn, the axle rotates. The movement of the axle can be used to grind corn or wheat or to power some type of machinery, such as a mechanical saw for the sawing of logs.

The first waterwheel is thought to have been invented by the early Greeks or Norsemen. It differed from later versions in that the wheel was horizontal and the axle, vertical. Cups or vanes were attached to the outer rim of the wheel to catch water flowing past the wheel. The force of the running water pushing against the cups or vanes turned the wheel and its axle. Water wheels of this design were not able to generate much power and were used primarily for small operations, such as the grinding of corn.

As early as the first century B.C., a variation on this design was suggested by the Roman architect Vitruvius (ca. 70–25 B.C.). He suggested using the modern form of the waterwheel, in which the wheel is vertical and the axle, horizontal. The wheel is suspended in a fast-flowing river, whose waters push against slats in the wheel to make it rotate.

Scholars believe that Vitruvius may have gotten his idea for the vertical waterwheel from an older device known as the *Persian wheel*, or *saqiya*. In this form of the device, cups were attached to the outer perimeter of the wheel. When placed into a river, the cups captured water, which was then transferred to a holding area next to the wheel. The Persian wheel was thus used to collect water, not as a source of power. Vitruvius saw that the function of the wheel could be changed simply by reversing the position of the cups, having the flowing water strike their convex side rather than their concave side.

The earliest form of this wheel is now known as the *undershot* wheel, since the wheel was positioned above the river and flowing water pushed against the lower part of the wheel. It was later found that the wheel operated more efficiently if it was positioned below a source of flowing water. In this orientation, the wheel was moved not only by the motion of the water itself but also by the weight of the water as it falls downward along the outside circumference of the wheel. This form of the waterwheel is known as the *overshot* wheel.

Although the overshot wheel is more efficient than the undershot wheel, it is more difficult to design and build. Running water must be diverted from its normal path in a river into a channel arranged so that the water can be directed at the top of the wheel. Under the best operating circumstances, early waterwheels were able to develop a power of about 3 horsepower. With this power, they were able to grind up to 400 pounds of corn an hour, more than 40 times the capacity of wheels operated by humans or animals.

Although waterwheels were an important source of energy in specific parts of the world over a rather extended period of time, they could never be used efficiently in regions where swift rivers were not available, and they were not developed during the many years when human workers (slaves) and animals were cheap and abundant. In Europe and North America, where these conditions no longer held true after about the fourth century A.D., waterwheels became important sources of power. They were used not only to grind corn, wheat, and other grains, but also to operate sawmills, ore-crushing plants, metal-working operations, and other industrial plants.

Further Reading

Derry, T. K., and Trevor I. Williams. *A Short History of Technology from the Earliest Times to A.D. 1900*, chap. 8. New York: Dover Publications, 1993.

McGuigan, Dermot. *Small Scale Water Power*. Dorchester, UK: Prism Press, 1978.

Reynolds, Terry S. *Stronger than a Hundred Men: A History of the Vertical Water Wheel*. Baltimore: Johns Hopkins University Press, 1983.

Wave Power

See Energy from Water

Weather

See Clouds; Drought

Weather Modification

The term *weather modification* refers to any change in normal weather patterns as a result of human activity, unintentional or intentional. Some examples of unintentional weather modification include the development of urban areas, which changes surface temperatures that may alter air movement and precipitation patterns; agricultural activities, which may change surface albedo (reflectivity), humidity, and surface temperature; and industrialization, which may result in the introduction of gases not normally present in the atmosphere, such as the oxides of sulfur and nitrogen.

Weather-modification activities tend to fall into two major categories: the reduction of undesirable forms of precipitation, such as hail, frost, lightning, fog, and hurricanes; and the enhancement of desirable forms of precipitation, primarily rain.

History

Humans have attempted to influence weather patterns for thousands of years. For most of this time, the techniques used for weather modification were crude and based on faith and mythology rather than science. For example, the Mayans practiced human sacrifice in the form of throwing women into wells to appease the rain gods. Also, most Native American tribes had (and have) rain dances that could be performed when precipitation was needed. In eighteenth-century England, some communities rang their church bells to shake moisture out of the clouds. After the Civil War, the firing off of cannons and other explosive materials became popular as a way of increasing rainfall.

The founder of the modern science of weather modification, however, was Vincent Schaefer (1891–1993). While working at the General Electric Research Laboratory in Schenectady, New York, Schaefer became interested in the problem of cloud seeding to induce precipitation. In the early 1940s, he accidentally discovered that a piece of dry ice introduced into a moist atmosphere caused condensation of the moisture. He decided to see if this principle could be used in the atmosphere by spraying dry ice over the tops of clouds. He experienced his first success in November 1946 when he induced a rain shower by seeding clouds with dry ice from an airplane. The military quickly became interested in Schaefer's work and appointed him director of its weather-modification program, Project Cirrus.

Enthusiasm for weather-modification programs reached a peak during the 1960s and 1970s. A number of states established agencies responsible for developing and implementing programs for the enhancement of rainfall and the reduction in undesirable forms of precipitation. By the early 1980s, however, the initial enthusiasm for weather modification had begun to abate. A number of reasons account for this change.

First, the scientific evidence for the effectiveness of weather-modification programs was difficult to produce. If one seeded a cloud and rain fell, one could not be sure whether the seeding had caused the rain or whether the rain might have fallen anyway. Second, weather-modification programs often create animosity among geographically related areas. An effective precipitation program in one region might deprive another region of the moisture it might have had without the modification program. Third, some people believe that humans should not interfere with natural processes (or, at least, some natural processes) and should not try to "play God" with the weather. These factors ultimately combined to result in a reduction of funding for research on weather modification, further reducing the scientific basis for such programs.

Types

The major forms of weather modification include cloud seeding, fog and cloud dispersal, hail suppression, and storm modification.

Cloud Seeding

The theoretical basis of cloud seeding is that precipitation will not form, even if large amounts of water vapor are present in clouds, if particles are not present on which the water vapor can condense. Normally, there are sufficient amounts of dust, ice crystals, and other particles to allow the formation of ice crystals and water droplets from which rain, snow, and other forms of precipitation are formed.

Schaefer discovered that fine particles of dry ice are effective nuclei for the formation of ice crystals and water droplets in clouds. His first experiments involved the scattering of pulverized dry ice (solid carbon dioxide) over clouds. A number of other materials have also been tried in the seeding of clouds, the most popular of which is silver iodide.

Cloud seeding can be carried out in a variety of ways. Seeding from the ground can be accomplished by igniting solid silver iodide in large burners. The smoke formed then rises into overlying clouds to provide the nuclei needed for ice-crystal and water-droplet formation. Alternatively, canisters of powdered silver iodide or dry ice can be fired into the atmosphere inside rockets that explode when they reach cloud level.

The most common method of cloud seeding, however, is by airplane. The plane is flown through or above a cloud and powdered silver iodide or dry ice is sprayed into or onto the cloud.

One of the difficult research issues involved in cloud seeding is deciding which clouds are likely to respond to seeding operations. It is not unusual for seeding to work in some cases and to fail in others with clouds that appear to have similar physical characteristics.

Fog and Cloud Dispersal

In principle, the same methods used to produce precipitation with silver iodide or dry ice should be applicable to the removal of fog and low-lying clouds. When such particles are sprayed into fog (which is nothing other than a low-lying cloud), they may induce condensation, converting water vapor to water droplets, which then precipitate out of the fog. This method is now used routinely at a number of airports to clear runways and allow for normal operation of aircraft.

Hail Suppression

Hail suppression is an issue of great significance to many groups. Each year, for example, farmers lose millions of dollars worth of crops during hailstorms. Some scientists have argued that the methods used for cloud seeding might also be effective in reducing the formation of hail—the argument being that the more nuclei there are on which moisture can condense, the more likely it will be that many small ice crystals and water droplets will form, and the less likely it will be that larger particles will develop. While that argument is interesting and attractive, hail-suppression programs have thus far been relatively ineffective, and most people interested in such programs understand that much more research is necessary before they can have widespread economic value.

Storm Modification

Hurricanes, tornadoes, cyclones, and typhoons constitute another major threat to human life and property around the world. Some scientists have suggested that seeding the center of such storms might lead to a reduction in their severity. They suggest that the formation of water droplets that occurs as a result of seeding will drain off some of the energy stored in a storm.

From 1961 to 1983, the United States government sponsored a research program called Project Stormfury to test this hypothesis. During early stages of the program, seeding was carried out in the eyes of Hurricane Ester (1961), Hurricane Beulah (1963), and Hurricane Debbie (1969). At its most effective level, seeding appears to have reduced the force of hurricane winds by about 30 percent. The results of Project Stormfury, however, were not impressive enough for the federal government to continue funding of this project.

Military Applications

Many people are familiar with the existence and nature of civilian weather-modification programs. Similar programs operated by the U.S. military are, however, probably less well known to the general public, yet the Army, Navy, and Air Force have been interested in the potential of weather modification as a military weapon since the earliest days of Vincent Schaefer's research.

Perhaps the first case in which military use of weather modification became a subject of discussion was during the Vietnam War. Advisors to the U.S. military suggested that weather-modification techniques be developed and used against the North Vietnamese, while withholding information about this program from the general public.

The major weather-modification program developed during this time was known as Project Popeye. The program was described in a 1967 memorandum to Secretary of State Dean Rusk from Deputy Undersecretary of State for Political Affairs Foy D. Kohler Foreign Relations at the United States. Kohler explained that the purpose of the project was

to produce sufficient rainfall along these lines of communication to interdict or at least interfere with truck traffic between North and South Vietnam. Recently improved cloud seeding techniques would be applied on a sustained basis, in a non-publicized effort to induce continued rainfall through the months of the normal dry season.

Kohler also pointed out the importance of keeping the project secret. He said that

The DOD proposes to conduct the project under conditions of strictest secrecy, noting that publicity would create vulnerability to communist charges of US manipulation of weather, in the affected areas or elsewhere. (Foreign Relations of the United States, 1964–1968)

Kohler noted that a test phase of Popeye had been carried out in October 1966, producing results that were "outstandingly successful."

Weather modification as a military tool continues to be of interest in the United States and some other nations. In a position paper published in June 1996, members of the Air War College described some ways in which weather modification could be used in future military conflicts. Techniques could be used to degrade enemy forces, the paper said, by enhancing precipitation (thereby flooding lines of communication and decreasing the comfort level and morale of enemy troops); increasing the frequency and severity of storms; reducing or eliminating precipitation (to deny freshwater to troops and induce drought conditions); and removing fog and clouds (to deny opportunities for concealment and increase vulnerability for friendly reconnaissance).

Weather modification can also be used as an adjunct to friendly forces by reducing or eliminating precipitation (to improve visibility and troop comfort and morale); ameliorating the intensity of storms; increasing fog and clouds (to improve concealment of troops); and reducing or eliminating fog and clouds (to maintain airfield operations and improve reconnaissance activities) ("Weather as a Force Multiplier: Owning the Weather in 2025: Military Applications of Weather Modification," http://www.abovetopsecret.com/pages/af2025a.html). *See also* Clouds; Precipitation

Further Reading

Davis, Ray Jay, and Lewis O. Grant. *Weather Modification: Technology and Law*. Boulder, CO: Westview Press, 1978.

Foreign Relations of the United States, 1964–1968. "Memorandum from the Deputy Under Secretary of State for Political Affairs (Kohler) to Secretary of State Risk." 13 January 1967. Vol, 28, 274.

Jiusto, James E. *The Physics of Weather Modification*. College Park, MD: American Association of Physics Teachers, 1983.

Kahan, Archie M., et al. *Guidelines for Cloud Seeding to Augment Precipitation*. New York: American Society of Civil Engineers, 1995.

Mörth, Hans. *Weather Modification: Prospects and Problems*. Translated by Geog Breuer. Cambridge: Cambridge University Press, 1979.

"Weather as a Force Multiplier: Owning the Weather in 2025: Military Applications of Weather Modification," http://www.above-topsecret.com/pages/af2025a.html.

Wells

A well is a hole dug or bored into the ground to recover some resource, such as water.

Wells in Ancient History

Humans have been digging wells since the dawn of civilization. Archaeological surveys often find remnants of wells built thousands of years ago to provide communities with a reliable water supply. Excavations in 1968 in the Lake Van area near Cappadocia, Turkey, for example, uncovered a complex network of water wells, water tanks, and air shafts dating to about the thirteenth century B.C.

Water wells are mentioned frequently in the Bible also. For example, the success in finding freshwater for the Jews who were fleeing Egypt under Moses is described in the book of Exodus. In chapter 15, the people begin to "murmur against Moses, saying, What shall we drink?" In response, Moses brings them to the town of Elim, "where were twelve wells of water, . . . and they encamped there by the waters" (Exod. 15:27). In 2 Chronicles, the construction work of Uzziah is described, including the building of towers and the digging of many wells, "for he had much cattle" (2 Chron. 26:10).

One of the most famous wells of ancient times is the so-called Jacob's well, where Jesus sat and talked with the Samaritan woman (John 4:5–26). That well survives today and is located in the vicinity of Shechem, 65 kilometers (40 mi) north of Jerusalem. Scholars believe that the well was dug by Jacob himself. It is about 3 meters (9 ft) in diameter and 23 meters (75 ft) deep.

Wells Today

Wells are still an important and widespread source of water. There are an estimated 15 million working water wells in the United States today, more than in any other nation. India has the second largest number of water wells with 12.3 million, followed by Germany (500,000), and South Africa (500,000).

In the United States, there are just over 280,000 public water systems that obtain their water from wells, and an additional 15.1 million households that depend on wells for their water supplies. About 800,000 new water wells are drilled each year. Michigan has the largest number of households who rely on well water (1,121,075), followed by Pennsylvania (978,202), North Carolina (912,113), and New York (824,342) (all data as of late 2001 from WaterBank at http://www.waterbank.com/faq1.html).

Types of Wells

Wells can be constructed by one of three general methods: digging, driving, or drilling. For most of human history, wells were dug by hand, with a pick and shovel. This method of well construction is limited by the depth to which a person can dig with simple tools and the kinds of materials through which the well is to be built. In general, the deepest wells dug by hand extend no more than about 10 meters (30 ft) in depth. Such wells are possible only when the water table is relatively close to the earth's surface. If the water table is very deep, hand digging is not sufficient to penetrate to the depths at which water is found. Hand-dug wells can be made deeper with the use of a power auger (drilling tool). With a handheld power auger, a well can be extended to about 25 meters (80 ft).

Wells are usually enclosed in some kind of casing to prevent their walls from collapsing. The type of casing used depends primarily on the type of material through which the well is dug. If the material is hard and solid, such as hard rock, the type of casing is somewhat less important. If the material is soft and friable, such as sandstone, a strong casing is essential. Hand-dug wells are usually lined with sheet metal, concrete, steel, or wrought-iron material.

A second method of well drilling is driving. Driven wells are constructed by hammering a hollow pipe into the ground until it reaches the water table. The end of the pipe is usually covered with a metal screen to filter out sand and other small particles. Most wells made in this way cannot go very far into the earth, and typically have a maximum depth of about 15 meters (50 ft). Driven wells don't require any casing because the pipe itself acts as a protective barrier against the surrounding earth.

The most popular way of constructing wells today is with a drill. Four types of drill bits are used in making such wells:

- a *rotary drill*, which has a sharp point that turns at a high speed to cut into soil and rock;
- a *rotary-percussion drill*, which smashes rocky material while it cuts into the material with its sharp point;
- a *reverse-circulation drill*, which has a point that continuously reverses direction as it works its way downward;
- an *auger drill*, which has a large, strong auger point attached to its end.

Each type of drill has special uses and certain limitations. For example, the reverse-circulation drill is most commonly used for wells with a large diameter. The typical maximum depth for such wells is about 60 meters (200 ft). By contrast, the deepest wells are dug with rotary and rotary-percussion drills, which are effective to depths of 500 meters (1,600 ft) or more. The most common casing material used in all kinds of drilled wells is steel or wrought-iron pipe. *See also* Groundwater

Further Reading

Brassington, Rick. *Finding Water: A Guide to the Construction and Maintenance of Private Water Supplies*. Chichester, UK: J. Wiley, 1995.

Driscoll, Fletcher, G. *Groundwater and Wells*. St. Paul, MN: Johnson Division, 1986.

Lehr, Jay, et al. *Design and Construction of Water Wells: A Guide for Engineers*. New York: Van Nostrand Reinhold, 1988.

Rowles, Raymond. *Drilling for Water: A Practical Manual*. 2d ed. Aldershot, UK: Avebury, 1995.

Further Information

National Ground Water Association
(formerly the National Water Well Association)
601 Dempsey Road
Westerville, OH 43081
Telephone: (800) 332-2104
Fax: (614) 898-7786
URL: http://www.ngwa.org
e-mail: info@ngwa.org
publishes *Water Well Journal*

Wetlands

A wetland (or wetlands) is an area that is covered by water for at least some part of the year. Sufficient moisture is present so that soil is wet for a significant period of time every year. Plants, humans, and other animals have adapted special techniques for surviving in the special environmental conditions present in wetlands. Although they are among the most biologically productive ecosystems in the natural world, wetlands, for the most part are not highly valued by humans and have been destroyed at a rapid rate over the past century.

Types

Wetlands can be divided into categories depending on their source of water, the length of time they are covered with water, the types of plants and animals they support, and other factors. The U.S. Environmental Protection Agency's Office of Water (EPA-OW) lists about a dozen major types of wetlands, as follows:

Marshes

Marshes can be divided into tidal and nontidal marshes. *Tidal marshes* occur along coastlines in middle and high latitudes almost everywhere in the world. In the United States, they are most commonly found along the eastern coast and the Gulf of Mexico. Tidal marshes obtain their water from the ocean, from freshwater sources (such as a river), or from both. They can be classified as *saline, brackish*, or *freshwater* marshes depending on the concentration of salts present in the water.

Nontidal marshes occur inland, adjacent to or near rivers, streams, lakes, ponds, and other bodies of water. They are usually filled with freshwater, although they may sometimes be classified as brackish or alkaline depending on the types of salts they contain. Nontidal marshes are the most common type of wetlands in the United States.

Nontidal marshes can be further categorized as *wet meadows, prairie potholes, vernal pools*, and *playa lakes*. Wet meadows tend to be drier than other forms of wetlands, with no standing water for much of the year. However, the water table is high enough in such areas for the soil to remain damp all or most of the year. Prairie potholes consist of large depressions in the ground, formed after the last North American glacier receded. These depressions fill with water in spring and after heavy rainfalls. They may be filled with water for part of the year, although they tend to dry up for at least a few months every year. Some potholes eventually dry up completely and never fill with water again.

Vernal pools are usually found along the West Coast in regions with a Mediterranean climate. They range in size from small puddles to small lakes and are filled with water for part of the year, although they tend to dry up during the summer and fall. Playa lakes are forms of wetlands common to the Southern High Plains that cover West Texas, Oklahoma, New Mexico, Colorado, and Kansas. Some fill with water after spring rains and remain wet for at least a few months. Others apparently receive their water from under-

Interior wetlands on Ashepoo-Combahee-Edisto (ACE) Basin Island, South Carolina. *National Estuarine Research Reserve Collection, NOAA.*

ground sources. Playa lakes are often quite salty because the dissolved minerals they hold remain behind when the lake water evaporates during the summer heat.

Swamps

Swamps are wetlands covered with water for most or all of the year and characterized primarily by the presence of woody plants. Swamps often occur close or adjacent to rivers, streams, lakes, and other bodies of water. They are widely found in the eastern part of the United States, although vegetation patterns differ to some extent between northern and southern swamps.

The EPA-OW recognizes two major types of swamps—*forested swamps* and *shrub swamps*—the basis of distinction being the primary form of plant life found in each.

One of the most common forms of forested swamp is the *bottomland hardwoods*, found primarily in the southern United States. Bottomland hardwoods usually occur on extensive, flat floodplains adjacent to rivers. River flooding is responsible for the formation of these wetlands, which often remain submerged for the greater part of the year.

Shrub swamps may occur in connection with forested swamps, but are characterized by plants such as dogwood, swamp rose, willow, and buttonbush, rather than the cypress, cedar, tupelo, and other large trees found in forested swamps. A familiar example of a shrub swamp is the *mangrove swamp*, found in coastal areas in tropical and subtropical regions. Mangrove trees are the dominant form of plant life in mangrove swamps, although other forms of plant life have evolved that are able to survive in the salty environment of this type of wetland.

Bogs and Fens

Bogs and *fens* are primarily found in the Northern Hemisphere. Both are covered with spongy soil that is saturated most of the year, and both support lush communities of plant life, including grasses, sedges, rushes, wild-flowers, and, especially in the case of bogs, sphagnum moss. Bogs receive all of their water from precipitation, while fens may also receive water from the runoff from higher areas that surround them. The water in both bogs and fens is quite acidic, although more so in the former than in the latter. Because of the high level of saturation found in bogs and fens, plants that die do not decompose nor-mally but change into *peat*, a sooty and not very efficient type of fuel once widely used by humans who lived close to these wetlands.

Bogs are sometimes further classified into *northern bogs* and *pocosins*. The former tend to occur in the north-central and northeastern regions of the United States, where there is sufficient precipitation to keep bogs at least partially filled with water all year round. Pocosins occur in states along the southeast-ern coastline and seldom contain open bodies of water, although their subsoil remains satu-rated for all or most of the year. In contrast to other types of bogs, pocosins tend to be dom-inated by evergreen trees and shrubs, such as loblolly pine, sweet bay, fetterbush, and zeno-bia. The largest concentration of pocosin wet-lands is found in North Carolina.

Benefits

Humans have long been tempted to drain or fill wetlands because they were unable to perceive any practical value for them. Over the past half century, researchers have identified a number of functions served by wetlands, many of them not immediately obvious to the onlooker. These functions include the following:

Enhancement of water quality. Waters that flow into a wetland may contain pollu-tants, sediment, and other materials that can potentially harm plant and animal life down-stream of the wetland. As these waters flow into the wetland, their rate of flow decreases and some of these pollutants and sediments are deposited in the wetland. In this way, some wetlands act as a natural water-purifi-cation plant built and operated at no cost to humans. In one widely quoted study, the Congaree Bottomland Hardwood Swamp in South Carolina was found to serve a purifi-cation function equivalent to that of a $5 million wastewater treatment plant (Wet-lands).

Flood control. Wetlands are nature's method for controlling the potential devas-tation of flooding. As swollen rivers and streams overflow their banks, they often empty into marshes, swamps, and other types of wetland. There, excess water is held in storage and absorbed by the roots of trees, shrubs, grasses, and other vegetation found in wetlands—vegetation that has evolved to survive in an aquatic environment. One of the major factors leading to increased flood losses along the Mississippi River over the past century has been the loss of bottomland hardwood swamps to human development.

Shoreline protection. Coastal wetlands act as buffers against heavy storms that would otherwise wash away plants and erode soil along the shoreline.

Fish and wildlife preservation. Cursory overviews of wetlands give little or no hint of their enormous biological productivity. Indeed, swamps and marshes are among the most biologically productive ecosystems in the natural world. According to one author-ity, only algal beds and reefs and tropical rain forests produce more biomass annually than do swamps and marshes. The compar-ative values are 2,500 g/m^2/y (grams per square meter per year) for algal beds and coral reefs, 2,200 g/m^2/y for tropical rain forests, and 2,000 g/m²/y for swamps and marshes (Whittaker 1975).

The list of plants and animals that breed, are born, grow to maturity, and/or spend their lives in wetlands is virtually endless and includes ferns, grasses, sedges, wildflowers, many types of shrubs and trees, mosses, aquatic birds, otters, muskrats, frogs and other types of amphibians, salamanders, bald eagles, jackrabbits, raccoons, and many kinds of fish and shellfish. Of special concern is the list of endangered species who depend on wetland areas for all or part of their lives. According to one estimate, more than a third of all endangered and threatened species in the United States live only in wetlands, while nearly half spend at least part of their lives in a wetlands area (Wetlands).

Commercial value. Many of the trout, bass, menhaden, flounder, shrimp, clams, oysters, crabs, and other fish and shellfish that end up on American dinner tables began their lives—or lived their entire lives—in coastal wetlands. The EPA-OW has estimated that about three-quarters of the commercial fish crop produced by the American fishing industry spent all or part of their lives in coastal wetlands. The agency says that marshes on the Louisiana coastline accounted for an annual harvest of 1.2 billion pounds, worth $244 million in 1991 (the last year for which data are available).

Recreation. The U.S. Office of Technology Assessment (OTA) has estimated that more than half of all adults in the United States take part in some type of recreational activity that involves wetlands. These activities include hunting, fishing, birdwatching, and photography. According to the OTA, these wetland-related activities brought in $59.5 million to the U.S. economy in 1991 (Office of Technology Assessment, chap. 4).

Other functions. Some wetlands have special value for the regions in which they are located. The prairie potholes found throughout the Upper Midwest in the United States provide breeding and feeding stops for more than half of all migratory birds that traverse the North American flyway from Canada to Mexico and south. In the southwestern United States, playa lakes are an important source of water in a region that commonly receives less than 20 inches of rain annually. By one estimate, the lakes provide a winter home to more than two million waterfowl every year (Wetlands).

Status of Wetlands

In the 1600s, before the North American continent was colonized by Europeans, over 220 million acres of wetlands are thought to have existed in the lower forty-eight states of the United States. By the mid-1980s, that number had been reduced to about half. The largest concentrations of wetlands were found in Florida (11 million acres), Louisiana (8.8 million acres), Minnesota (8.7 million acres), and Texas (7.6 million acres). In addition, the state of Alaska contained an estimated 170–200 million acres of wetlands in the mid-1980s.

A number of factors were responsible for the dramatic decrease in wetland acreage between the 1600s and the mid-1980s. These factors included drainage and filling of wetlands for construction of new buildings and structures and for new agricultural land; dredging and steam channelization; diking and damming; construction of levees; logging; and mining.

In 1986, the U.S. Congress passed the Emergency Wetlands Resources Act of 1986 that required, among other things, that the U.S. Fish and Wildlife Services (FWS) conduct regular status and trend studies of U.S. wetlands. The 1997 study found that the amount of wetlands in the forty-eight coterminous states had been reduced by 644,000 acres during the period from 1986 to 1997, leaving 105.5 million acres of wetland still in existence. The vast majority of this area (100.5 million acres, or 95%) was freshwater wetland, while the remaining 5 million acres (5 percent) were saltwater wetland.

Of special significance was the finding that the rate of wetland loss had dropped by 80 percent over the previous decade and that, in 1997, no more than 58,500 acres of wetland were being destroyed each year. That number was still greater than the "no net loss" goal established by President George Bush in 1990, but it was regarded as a step in the right direction by FWS researchers.

The FWS study also identified the major forces contributing to the destruction of wetlands. Urban development was responsible for 30 percent of all wetland loss, agriculture for 26 percent, silviculture for 23 percent, and rural development for 21 percent.

Wetlands Protection

For nearly two decades, the protection and restoration of wetlands in the United States has been a major issue of concern to federal, state, and tribal governments, all of which have developed laws and administrative rules to reduce and repair the damage done to the nation's wetlands. One of the earliest of these regulations was promulgated in Executive Order 11990 by President Jimmy Carter. Carter ordered all federal agencies to "provide leadership and . . . take action to minimize the destruction, loss or degradation of wetlands, and to preserve and enhance the natural and beneficial values of wetlands in carrying out the agency's [various] responsibilities."

Arguably the strongest weapon provided for the preservation of wetlands was contained in Section 404 of the Clean Water Act of 1994. This section contained detailed specifications for the filling and dredging of certain types of waterways, wetlands among them. Governments have also introduced incentive programs to encourage farmers, developers, and others to reduce their destruction of wetlands. The Wetlands

Reserve Program of the U.S. Department of Agriculture, for example, pays farmers 75 percent of the cost of restoring a wetland area previously converted for agricultural purposes.

Further Reading

Keddy, Paul A. *Wetland Ecology: Principles and Conservation*. Cambridge: Cambridge University Press, 2000.

Mitsch, W. J., and J. G. Gosselink. *Wetlands*. 2d ed. New York: John Wiley & Sons, 2000.

Niering, W. A. *Wetlands*. New York: Alfred Knopf, 1998.

Office of Technology Assessment. *Preparing for an Uncertain Climate*. Vol. 2 (October 1993). OTA-0-568.

Wetlands, http://www.epa.gov/owow/wetlands.

Whigham, Dennis, Dagmar Dykyjová, and Slvaomil Hejný, eds. *Wetlands of the World: Inventory, Ecology, and Management*. Dordrecht: Kluwer Academic Publishers, 1993.

Whittaker, R. H. *Communities and Ecosystems*. 2d ed. New York: MacMillan, 1975.

Further Information

U.S. Environmental Protection Agency
Office of Wetlands, Oceans and Watersheds
Wetlands Division (4502F)
401 M Street S.W.
Washington, DC 20460
Telephone: (800) 832-7828
URL: http://www.epa.gov/owow/wetlands

Whales and Whaling

Whales are members of the order of aquatic mammals *Cetacea*. The Cetacea consists of three suborders: the *Mysticeti*, commonly known as *baleen whales*; the *Odontoceti*, or *toothed whales*; and the *Archaeoceti*, a group of extinct mammals known only through their fossil remains.

Characteristics

The baleen whales consist of about a dozen species, including the right (*Balaena*), gray (*Eschrichtius robustus*), blue (*Balaenoptera musculus*), fin (*Balaenoptera physalus*), humpback (*Megaptera novaeangliae*), and sei (*Balaenoptera borealis*) whales. They are the largest members of the whale family and among the largest animals that have ever lived on earth. The largest blue whale ever taken was 34.6 meters (113.5 ft) in length and weighed about 170 metric tons (375,000 lb).

Baleen whales feed by swimming slowly along the ocean bottom with their mouths open. Water streams into their mouths and is then forced out through keratinous (hornlike material) strainers called *baleen* that trap plankton (small aquatic organisms). From time to time, the whale wipes the baleen clean of plankton with its tongue and forces it down its throat. The main diet of baleen whales is a shrimplike plankton known as *krill*.

The toothed whales include the pilot (*Globicephala*), beaked (*Mesoplodon*), beluga or white (*Delphinapterus leucas*), killer (*Orcinus orca*), sperm (*Physeter catodon*), and pygmy sperm (*Kogia breviceps*) whales, as well as all of the dolphin and porpoise species. Members of this suborder are active predators, with fish, squid, and other larger aquatic organisms as their main prey.

Distribution

At one time, whales could be found in large numbers in all of the earth's oceans and seas. As the result of extensive hunting, their numbers have declined dramatically over the past 200 years, although they continue to survive almost everywhere in the ocean. Some species follow migration paths that take them more than 22,000 kilometers (14,000 mi) along the coasts of eastern Asia and western North America. Their movement appears to be dictated to some extent by ocean currents, changes in water temperature, and availability of plankton and other food sources.

Commercial Value

Whales have had significant economic value to peoples for whom other natural plant and animal resources were limited. For example, when western explorers first encountered Inuit (Eskimo) civilizations in North America, they found that virtually every part of the whale was used for some purpose or another. Whale blubber (fatty tissue) was used for food and as a source of oil for light and heating; sinews were used as ropes; and skeletal parts were converted to tools and structural components.

In more recent times, various types of whale oil have been used for lamp fuel and in the production of soaps, margarines, paint oils, machine oils, lubricants, candles, crayons, cosmetics, and shoe polish. The material known as ambergris is used as a stabilizer in the manufacture of perfumes and is found only in the sperm whale. Whale meat continues to be a popular part of the diet among the Inuit and among the Japanese, among whom it is a popular delicacy. In some parts of the world, where whale meat is not very popular as a food for humans, it is used in the production of foods for domestic pets and commercial animal farms. Baleen and other bony parts of the animal have been used in the manufacture of corsets, bustles, hooped skirts, and umbrellas.

Whaling

Archaeological evidence suggests that whaling was common along the North Atlantic and North Pacific coasts at least as far back as 3000 B.C. This evidence includes rafters made of whale bones, as well as scraps of bones found in kitchen trash heaps. Methods of hunting differed from place to place, with the Japanese employing nets to snare beached whales or whales trapped in shallow waters, the Aleuts using poisoned spears, and the Inuit making use of primitive harpoons thrown by hunters in skin boats.

The first large-scale commercial whaling was conducted by the Basques along the coasts of France and Spain as early as the twelfth century. The Basques' primary target was the black right whale, which came to breed in the Bay of Biscay. The animals were docile and slow moving—easy prey for hunters who pursued them with harpoons. After being stabbed to death, the whales were dragged to shore and dismembered. Over the centuries, the Basques expanded their whale-hunting efforts until, by the sixteenth century, they were regularly "fishing" off the coasts of southern Labrador.

By the end of the seventeenth century, the Dutch, English, French, and German had also become involved in whaling. Their primary target was the right whale, which they pursued along the northern rim of the Atlantic Ocean. Whaling in the United States received an important boost in 1712 when the first sperm whale was caught, providing important new substances (spermaceti and a clean, waxy oil) not generally available from other whales.

The face of whaling changed dramatically in the 1760s with the invention of *tryworks*. Tryworks were brick ovens used to render blubber. Vessels carrying tryworks could remain at sea for extended periods of time—sometimes as long as four years—catching and processing whale blubber without the necessity of returning to port. As a result, it became possible for whaling vessels to make long trips to the Pacific Ocean and other more remote seas in search of their prey.

Whaling experienced a significant downturn in about 1860 because of two factors. First, whalers had given little thought to the survival of the species over the long term, slaughtering juveniles and nursing calves along with mature whales. In many parts of the world, some species of whales had been hunted virtually to extinction. Second, the discovery of oil in Pennsylvania in 1859 made available new products, such as kerosene and candle wax, that could be used in place of whale oil at much less cost.

The decline in whaling was slowed to some extent in the 1860s by the inventions of Svend Foyn, a Norwegian whaler. Foyn developed a steam-powered boat capable of traveling at speeds of 7 knots or better. The boat was equipped with a bow-mounted gun, from which was fired a harpoon carrying a time-delay grenade. It also carried a large crane that could be used to lift whales out of the water and tow them behind the boat to its home harbor. Using boats of this type, whalers were able to kill and return whales to land quickly enough to prevent spoilage of the animal. Before long, shore stations were being built along the coasts of Norway, Scotland, and Newfoundland for the processing of whales killed and transported by Foyn's boats.

Shortly after the turn of the century, a technological breakthrough once again made whaling an economically viable business. In 1904, the French chemist Paul Sabatier (1854–1941) invented a method for converting liquid oils into solid fats by means of *hydrogenation*. Almost overnight, interest in whaling as a source of oil exploded. The number of whales taken worldwide jumped from about 2,000 per year in 1900 to more than 20,000 annually a decade later. By 1930, six shore stations had been constructed along the Antarctic coast alone, serviced by 41 factory ships and 232 catcher ships.

Whaling received yet another stimulus in the 1950s as demands for whale meat for human consumption increased in Europe. Whaling fleets from Great Britain, Norway, the Netherlands, the former Soviet Union, and Japan extended their search for whales to every part of the world's oceans. So aggressive and successful were these expeditions that one whale species after another soon became threatened or endangered: the blue whale in the 1940s, the fin and hump-

A slaughtered whale in Iceland. *Wolfgang Kaehler/Corbis.*

back whales in the 1960s, and the sei after 1970. Today, the Convention on International Trade in Endangered Species of Wild Fauna and Flora lists all species and subspecies of whales as "vulnerable," "threatened," or "endangered."

Conservation Efforts

The first international effort to monitor whaling was sponsored by the League of Nations in 1931. The League's International Whaling Conference (IWC) experienced moderate success in raising awareness of the threat posed by whaling to whales around the world, and some reduction in whaling catches were reported as a result of its efforts. But, like the league itself, the IWC had disappeared from sight by the end of the 1930s.

More successful has been the work of the International Whaling Commission (IWC), established in 1946 for the purpose of regulating whaling, helping to restore threatened and endangered species, and increasing the population of all whale species around the world. The IWC banned the killing of gray and white whales entirely and established limits on the annual take of other whale species in the Antarctic according to a formula based on the number of "blue whale units."

An initial effort by the United Nations to ban whaling completely in 1972 was unsuccessful, although a similar attempt in 1982 was approved by the IWC. In that action, commercial whaling was to be brought to an end in 1986. In the United States, a ban on all whale products was adopted in 1970, and two years later, the Marine Mammal Protection Act of 1972 banned the killing of all whales within U.S. national waters.

Prohibitions on whaling have had less-than-complete success for two reasons. First, neither the IWC nor any other concerned group has the resources to monitor effectively the world's oceans to ascertain that bans are being observed. Second, a few nations where whale meat and whale products are still in great demand have continually insisted on their right to carry out their traditional whaling activities. In 1992, for example, Norway unilaterally decided that it would resume hunting of minke whales. And the Japanese have, on occasion, continued their whaling activities, sometimes on the pretext of carrying out scientific studies of the animals. *See also* Aquatic Plants and Animals; Oceanography; Oceans and Seas

Further Reading

Clapham, Phil. *Whales of the World*. Stillwater, MN: Voyageur Press, 1997.

Ellis, Richard. *Men and Whales*. New York: Lyons Press, 1999.

Friedheim, Robert L., ed. *Toward a Sustainable Whaling Regime*. Seattle: University of Washington Press, 2001.

Stoett, Peter J. *The International Politics of Whaling*. Vancouver, BC: University of British Columbia, 1997.

Further Information

International Whaling Commission
The Red House
135 Station Road, Impington, Cambridge
United Kingdom CB4 9NP
Telephone: +44 (0) 1223 233971
Fax: +44 (0) 1223 23287
URL: http://ourworld.compuserve.com
e-mail: iwc@iwcoffice.org

Wild and Scenic Rivers Act of 1968

The Wild and Scenic Rivers Act was passed by the United States Congress and signed into law by President Lyndon Johnson in October 1968. The purpose of the act was to preserve the character of rivers in the United States with special scenic, recreational, geologic, fish and wildlife, historic, cultural, or other kinds of value. In the original act, 135 rivers and sections of rivers and the land adjoining them were set aside for future development as part of the Wild and Scenic Rivers system.

At first, responsibility for the administration, study, and development of these rivers was divided among various agencies within the Departments of Agriculture and Interior. In 1995, responsibility for management of the program was turned over to an interagency council composed of representatives from the Bureau of Land Management, National Park Service, U.S. Fish and Wildlife Service, and U.S. Forest Service.

As of 2001, 156 rivers have been designated as wild, scenic, or recreational. Oregon contains the largest number of these rivers, 48, with Alaska holding the second largest number, 25. Of all rivers in the system, those designated as "wild" have a total length of 5,345 miles; those designated as "scenic," 2,446 miles; and those designated as "recreational," 3,501 miles.

Designation as a wild, scenic, or recreational river does not necessarily prevent development on or around the river. The goal of the program is to restrict such development, however, so as not to damage the particular quality for which the river was selected. *See also* Oceans and Seas; Rivers and Streams

Further Reading

Bonham, Charlton H. *The Wild and Scenic Rivers Act and the Oregon Trilogy*. Portland, OR: Northwest Water Law & Policy Project, 2000.

Wild and Scenic Rivers, http://www.nps.gov/rivers.

Wild and Scenic Rivers Act of 1968, http://www.nps.gov/rivers/wsract.html.

Windsurfing

See Aquatic Sports, Windsurfing

Wishing Wells

A wishing well is a small body of water thought to contain special powers, such as the ability to grant a person his or her wish if a token (such as a coin) is dropped into the water.

Water Worship

The belief that certain bodies of water possess supernatural powers goes back many centuries and appears in nearly every culture that has been studied. Such bodies of water are often regarded as the home of gods and goddesses and/or the entryway to some unknown world. The casting of gifts into such bodies of water is traditionally seen as a votive offering to the water's residents, a request for protection or succor.

In his classic study "Religion in Primitive Culture," Sir Edward Burnett Tylor relates a number of examples of water worship. Some Native American tribes, for example, believed that rivers, streams, and other bodies of water were home to a variety of spirits who had to be appeased with gifts from time to time. Tylor cites one observer who saw a Native American man throw a bundle containing a knife, some tobacco, and other small objects into a river in hopes of achieving a cure for his wife's illness. In Peru, people sometimes threw maize into a stream or drank a cup of its water to receive permission to cross the water or to improve their success at fishing.

Similar practices have been observed in Africa, where people sometimes throw stones, grain, slaughtered animals, or other objects into a river in order to ask for a safe crossing, in hopes of preventing a drought, or as a way of curing illness. Further to the north, communities located along the Ob River in Russia have traditionally tried to improve their fish catch by tying a stone around the neck of a live reindeer and throwing it into a river.

Animism and Christianity

In Western Europe, many of these animistic beliefs and practices were gradually absorbed and transformed by Christian theology and philosophy. At first, leaders of the Christian church attempted to wipe out such beliefs and replace them with practices from their own religion. Tylor tells, for example, of efforts by the Duke of Břetislav in Bohemia and by England's King Ecgberht to ban the offering of gifts to water deities. Both failed, as did most Christian leaders who made similar prohibitions. As Tylor observed,

the old veneration [to water deities] was too strong to be put down, and with a varnish of Christianity and sometimes the substitution of a saint's name, water-worship has held its own to our day (299).

Some of the most thoroughly studied connections between pagan and Christian wishing-well practices come from the British Isles. Some of these studies have produced problematic but suggestive results. One researcher, for example, reports on the discovery of six small stone figures dating to the late Iron Age at the bottom of a dry ditch in Yorkshire. Five of the six figures had been decapitated, suggesting to the researcher that they may have been sacrificed to a water deity (Lovegrove).

The evidence from less ancient digs seems more substantial. For example, a spring dating to the Roman period in Bath, England, contains about sixteen thousand coins and other offerings, including fifty curses written on lead tablets. The typical format for such a curse was the request that some person be made ill until he or she repaired some misdeed. Curses were a special type of wish found in other places as well, such as at St. Elian's cursing well, into which stones containing the victim's initials were dropped.

With the rise of Christianity, the thousands of holy wells in the British Isles were given new names (usually those of saints) and incorporated into Christian theology. For example, St. Cormac's well in County Offaly, Ireland, is reputedly the place where the saint was attacked by snails. That story almost certainly grew out of a much older pagan myth based on mollusks from "the other world," which lived in the well and were capable of transforming themselves into wolves—certainly a more serious threat to humans than were Cormac's snails.

Many wells throughout the British Isles carry some variation of the name Lady Well. Modern observers believe that the name may have been given in honor of the Virgin Mary, of some particularly pious Christian woman of an earlier century, or, during the process of evolution from pagan to Christianity, of a pagan goddess, such as the daughter of Dagda—mythical Scottish goddess of fire and poetry.

Wells of Special Powers

Most wishing wells today are "general purpose" wells, in which, for the price of a penny, one can ask to win the lottery, earn the love of a beautiful woman or handsome man, or regain lost good health. Historically, however, many wishing wells were "special purpose" bodies of water, offering succor related to a mythical deity's or Christian saint's special area of expertise.

In Ireland, for example, a number of wells are named in honor of St. Brigid, thought by some to have been descended historically from the Druid fire goddess Brighid. Most of St. Brigid's wells are devoted to the cure of eye disorders, although some are reputed to offer fertility to childless couples. The water of another Irish well—dedicated to St. Brendan, the Navigator—is reputed to offer cures for back problems. And some wells, such as those along the Irish coast, are believed to control weather patterns.

Modern Wishing Wells

Most people who drop a coin into a wishing well today probably know little or nothing about the history of the custom, and probably have only a limited hope that their request will be honored by the well, its water, or its governing deity.

But beliefs in the curative powers of holy wells have by no means disappeared. For example, many thousands of people travel to supposedly curative spas each year, although medical evidence for the effectiveness of such cures is limited. In addition, many of the religious faithful in some parts of the world still travel to holy wells with the firm belief that their requests will be answered as long as they perform the correct rituals. Each year, for example, many Christians in countries with a Celtic heritage make pilgrimages to wells dedicated to saints, hoping to achieve some desired wish or to pay homage to the saint. During such pilgrimages, highly formal rituals are observed. For example, a pilgrim may be required to walk or crawl three times around the well before dropping an offering into the well. Or the pilgrim may dip a piece of cloth into the well, wash a part of his or her body with the holy water, and then hang the cloth on a nearby tree. As water drips from the cloth drop by drop, the pilgrim's wish is (supposedly) granted. Interestingly enough, rituals of these kinds, common in today's world, are very similar to those described in pagan times for similar bodies of water.

Further Reading

Harvey, Steenie, "Wishing Wells," http://www.twoh.com/autumn98/wells.cfm.

Hope, Robert Charles. *The Legendary Lore of the Holy Wells of England: Including Rivers,*

Lakes, Fountains, and Springs. Felinfach: Llanerch, 2000.

Jones, Francis. *The Holy Wells of Wales.* Cardiff: University of Wales Press, 1992.

Logan, Patrick. *The Holy Wells of Ireland.* Gerrards Cross, Buckinghamshire: Smythe, 1980.

Lovegrove, Chris, "Wishing Wells and Votive Offerings," http://www.bath.ac.uk/lispring/sourcearchive/fs8/fs8cl1.html.

Tylor, Sir Edward Burnett. *Religion in Primitive Culture.* New York: Harper Torchbook, 1958.

Woods Hole Oceanographic Institution

The Woods Hole Oceanographic Institution (WHOI) was established in 1930 following a recommendation by the National Academy of Sciences that the government create an independent research laboratory on the East Coast "to prosecute oceanography in all its branches." A $3 million grant from the Rockefeller Foundation made possible the construction of the first laboratory building and the 142-foot research vessel, *Atlantis*, whose profile now forms the institution's logo. World War II provided a major impetus to the growth of the institution, which now employs about 650 staff members housed in 58 buildings and laboratories on 219 acres of land and waterfront property. Each year, about 130 graduate students pursue their research at WHOI, and 35 more students are awarded summer fellowships.

Research work at WHOI is divided into the following eight major categories:

Applied Ocean Physics and Engineering focuses on the interaction between atmosphere and the oceans, sediment transport on the ocean bottoms, and internal water movements within the oceans

Biological Oceanography concerns itself with all types of marine life, ranging from viruses and bacteria to whales and birds; special attention is paid to the effects of pollutants on marine animals

Geology and Geophysics studies the structure and evolution of the earth's oceanic crust, earthquakes and volcanoes in the mid-ocean ridges, and techniques for mapping and sampling the ocean floors

Marine Chemistry and Geochemistry investigates global climate change, ocean circulation patterns, radioactive and chemical pollutants in the ocean, seafloor hydrothermal systems, and the history of the earth and its marine life

Physical Oceanography surveys circulation patterns in the oceans, the mixing of water masses offshore and along coasts, and other kinds of physical processes

Marine Policy involves the work of economists and social scientists who study current national and international oceanic issues and make recommendations about future public policy decisions

Coastal Science works to develop new approaches to better understand and manage coastal resources

Climate Research functions in connection with the National Oceanic and Atmospheric Administration to learn more about the role of oceans in shaping and modifying climate patterns and variability

The institution publishes two publications of general interest. *Oceanus* is a 32-page semiannual report written by WHOI scientists for the general public. Each issue is devoted to a special topic, such as ocean circulation or climate change. *Woods Hole Currents* is a 16-page quarterly publication describing oceanographic and laboratory research, profiling WHOI staffers, and reporting on institution news and developments.

Further Information

Woods Hole Oceanographic Institution Information Office

93 Water St., MS#16
Woods Hole, MA 02543-1050
Telephone: (508) 289-2252
Fax: (508) 457-2034
URL: http://www.whoi.edu
e-mail: information@whoi.edu

World Water Council

The World Water Council (WWC) was established in 1996 as a result of recommendations made at the United Nations Conference on Environment and Development, held in 1992 in Rio de Janeiro. The council consists of 294 members from 40 countries. Members come from public institutions, private sector firms, United Nations organizations, and nongovernmental organizations. The mission of WWC is "to promote awareness of critical water issues at all levels, including the highest decision-making level, to facilitate efficient conservation, protection, development, planning, management and use of water in all its dimensions on an environmentally sustainable basis for the benefit of all life on the earth."

The council sponsors a number of meetings and events, such as a conference on Water Development for Poverty and Alleviation in Ahmadabad, India (2001), a symposium on Frontiers in Urban Water Management in Marseille, France (2001), the Third World Water Council Forum in 2003, and an annual World Water Day (March 22 in 2001). The council also publishes a quarterly newsletter, featuring reports on activities, announcements of forthcoming water meetings, and interviews with and articles from members. WWC also publishes a number of reports, such as *Changing Course*, a report of the technical sessions held at the 2nd General Assembly of the WWC; *A Water Secure World*; and *Making Water Everybody's Business*. The council also publishes a bimonthly research journal, *Water Policy*.

Further Information

World Water Council
Les Docks de la joliette
13002 Marseille
France
Telephone: +33 (4) 91 99 41 00
Fax: +33 (4) 91 99 41 01
URL: http://www.worldwatercouncil.org
e-mail: wwc@worldwatercouncil.org

Y–Z

Yachting

See Aquatic Sports, Yachting

Ymir

Ymir was the earliest parent of the Jotuns, a race of giants in Norse mythology. According to legend, the world originally consisted of two regions: Niflheim, a land of cold, darkness, frost, and mist; and Muspellsheim, a sea of turbulent fire and flame. The space between them, Ginnungagap, was completely empty. Over time, however, the heat from Muspellsheim began to melt the snow of Niflheim, and the first creature, Ymir, was formed from the meltwater. Shortly thereafter, a giant cow named Audhumia was also formed from meltwater.

Audhumia survived by licking frost and salt from the ice that remained in Niflheim. She produced milk on which Ymir was able to survive. Eventually, Ymir began to produce a race of giants, some from his legs and others from his armpits. These giants became the founders of the race of Jotuns.

Norse mythology tells of the battles that later developed both between Ymir and the other Jotuns and between Ymir and Odin and the other gods. In one of these battles, Ymir is slain. As he lays dying, he releases so much blood that all of the other Jotuns, save Bergelmir and his wife, are drowned. The two remaining Jotuns then begin to replenish the race of Jotuns, who go on to do battle with both gods and humans for many years.

Meanwhile, Odin and the other gods dismember Ymir's body and use its parts to create a new world. They use his blood to make the oceans and lakes, his bones to build the mountains, his brains to form the clouds, and his hair to make grass and trees. *See also* Gods and Goddesses of Water; Jotuns

Zooplankton

See Plankton

Selected Bibliography

Ackefors, Hans. *Introduction to the General Principles of Aquaculture*. New York: Food Products Press, 1994.

Ahrens, C. Donald. *Meteorology Today: An Introduction to Weather, Climate, and the Environment*. 6 ed. Pacific Grove, CA: Brooks/Cole, 2000.

Alley, E. Roberts. *Water Quality Control Handbook*. New York: McGraw-Hill Professional Publishing, 2000.

American Water Works Association. *Water Quality and Treatment: A Handbook of Community Water Supplies*. New York: McGraw-Hill, 1990.

Andrews, Tamra. *Legends of the Earth, Sea, and Sky*. Santa Barbara, CA: ABC-CLIO, 1998.

Appellof, Marian E., ed. *Everything You Ever Wanted to Know about Watercolor*. New York: Watson-Guptill Publications, 1992.

Asimov, Isaac, Martin H. Greenberg, and Charles G. Waugh, eds. *Atlantis*. New York: New American Library, 1988.

Avault, James W. *Fundamentals of Aquaculture: A Step-by-Step Guide to Commercial Aquaculture*. Baton Rouge, LA: AVA Publishing Company, 1996.

Avery, William H., Chih Wu, and John P. Craven. *Renewable Energy from the Ocean: A Guide to OTEC*. New York: Oxford University Press, 1994.

Bagnasco, Erminio. *Submarines of World War Two*. Herndon, VA: Cassell Academic, 2000.

Baker, D. James, and Vicky Cullen. *Careers in Oceanography and Marine-Related Fields*. Virginia Beach, VA: Oceanography Society, 1995.

Balaban, Miriam, ed. *Desalination and Water Re-Use: Proceedings of the Twelfth International Symposium*. Rugby, UK: Institution of Chemical Engineers, 1991.

Barnett, Jo Ellen. *Time's Pendulum: The Quest to Capture Time—From Sundials to Atomic Clocks*. New York: Plenum Press, 1998.

Barnhart, Robert K., ed. *The Barnhart Dictionary of Etymology*. New York: H.W. Wilson Company, 1988.

Battan, Louis J. *Harvesting the Clouds: Advances in Weather Modification*. Garden City, NY: Doubleday, 1969.

Baum, Joseph L. *The Beginner's Handbook of Dowsing*. New York: Crown Publishers, 1974.

Beaglehole, J. C. *The Life of Captain James Cook*. Stanford, CA: Stanford University Press, 1992.

———, ed. *The Journals of Captain Cook* New York: Penguin, Penguin Classics, 2000.

Becher, Paul, ed. *Encyclopedia of Emulsion Technology*. New York: M. Dekker, 1983.

Bennett, Matthew. *Glacial Geology: Ice Sheets and Landforms*. Chichester: Wiley, 1996.

Bernaerts, Arnd. *Bernaerts' Guide to the 1982 United Nations Convention on the Law of the Sea*. Coulsdon, UK: Fairplay Publications, 1988.

Berner, Elizabeth Kay, and Robert A. Berner. *Global Environment: Water, Air, and Geochemical Cycles*. Upper Saddle River, NJ: Prentice Hall, 1996.

Berry, Arthur John. *Henry Cavendish: His Life and Scientific Work*. London: Hutchinson, 1960.

Biswas, Asit K. *History of Hydrology*. Amsterdam: North-Holland, 1970.

Blackburn, Graham. *The Illustrated Encyclopedia of Ships, Boats, Vessels, and Other Water-Borne Craft*. Woodstock, VT: Overlook Press, 1999.

Bonham, Charlton H. *The Wild and Scenic Rivers Act and the Oregon Trilogy*. Portland, OR: Northwest Water Law & Policy Project, 2000.

Boon, P. J., Bryan Davies, and Geoffrey E. Petts, eds. *Global Perspectives in River Conservation*. New York: John Wiley & Sons, 2000.

Bowen, Robert. *Geothermal Resources*. London: Elsevier Science Publications, 1989.

Brassington, Rick. *Finding Water: A Guide to the Construction and Maintenance of Private Water Supplies*. Chichester, UK: J. Wiley, 1995.

Brater, Ernest F., et al., eds. *Handbook of Hydraulics for the Solution of Hydraulic Engineering Problems*. New York: McGraw-Hill, 1996.

Brett, Bernard. *A History of Watercolor*. New York: Excalibur Books, 1984.

Brewer, E. Cobham. *Dictionary of Phrase and Fable*. Philadelphia: Henry Altemus, 1898.

Brönmark, Christer, and Lars-Anders Hasson. *The Biology of Lakes and Ponds*. Oxford: Oxford University Press, 1998.

Brooks, Paul. *The House of Life: Rachel Carson at Work*. New York: Houghton Mifflin, 1972.

Brown, William Elgar. *Hydraulics for Operators*. Boston: Butterworth, 1985.

Brutsaert, Wilfried. *Evaporation into the Atmosphere: Theory, History, and Applications*. Dordrecht: Reidel, 1982.

Bryan, T. Scott. *The Geysers of Yellowstone*. Niwot, CO: University Press of Colorado, 1991.

Bryant, Edward. *Tsunami: The Underrated Hazard*. Cambridge: Cambridge University Press, 2001.

Burns, Bryan Randolph, and Ruth S. Meinzen-Dick. *Negotiating Water Rights*. London: Intermediate Technology, 2000.

Buros, O.K. *The ABCs of Desalting*. 2d ed. Topsfield, MA: International Desalination Association, 1999.

Buros, O. K. et al. *The USAID Desalination Manual*. Englewood Cliffs, NJ: IDEA Publications, 1982.

Butler, Trent C., ed. *Holman Bible Dictionary*. Nashville: Holman Bible Publishing, 1991.

Butter, A. J. *Transpiration*. London: Oxford University Press, 1972.

Bynum, W. F., E. J. Browne, and Roy Porter. *Dictionary of the History of Science*. Princeton, NJ: Princeton University Press, 1981.

Campbell, Steuart. *The Loch Ness Monster: The Evidence*. Amehurst, NY: Prometheus Books, 1997.

Carr, Michael H. *Water on Mars*. New York: Oxford University Press, 1996.

Cartwright, David Edgar. *Tides: A Scientific History*. Cambridge: Cambridge University Press, 1999.

Casson, Lionel. *Ships and Seafaring in Ancient Times*. Austin, TX: University of Texas Press, 1994.

Cheremisinoff, Nicholas P. *Water Treatment and Waste Recovery: Advanced Technology and Applications*. Englewood Cliffs, NJ: PTR Prentice Hall, 1993.

Chorlton, Windsor, and the editors of Time-Life Books. *Ice Ages*. Alexandria, VA: Time-Life Books, 1983.

Choudhuri, Usha. *Indra and Varuna in Indian Mythology*. Delhi: Nag Publishers, 1981.

Churchill, W. A. *Watermarks in Paper in Holland, England, France, etc., in the XVII and XVIII Centuries and Their Interconnection*. Amsterdam: M. Hertzberger, 1967.

Clapham, Phil. *Whales of the World*. Stillwater, MN: Voyageur Press, 1997.

Clark, Champ. *Planet Earth: Flood*. Alexandria, VA: Time-Life Books, 1982.

Cohn, Norman Rufus Colin. *Noah's Flood: The Genesis Story in Western Thought*. New Haven, CT: Yale University Press, 1996.

Colbeck, Samuel C., ed. *Dynamics of Snow and Ice Masses*. New York: Academic Press, 1980.

Colcord, Joanna C. *Songs of American Sailormen*. 1938. Reprint, New York: W. W. Norton, 1967.

Cole, Gerald A. *Textbook of Limnology*. Prospect Heights, IL: Waveland Press, 1994.

The Complete Guide to Home Plumbing. Minnetonka, MN: Creative Publishing International, 2001.

Complete Home Plumbing. Menlo Park, CA: Sunset Books, 2001.

Cone, Joseph. *Fire under the Sea: The Discovery of the Most Extraordinary Environment on Earth*. New York: Morrow, 1991.

Cousteau, Jacques, with James Dugan. *The Living Sea*. New York: Harper, 1963.

———.*The Silent World*. New York: Harper, 1953.

Dahl, Thomas E. *Status and Trends of Wetlands in the Conterminous United States 1986 to 1997*. Washington, DC: U.S. Fish and Wildlife Service, 2000.

Dailey, Stephanie, ed. *Myths from Mesopotamia: Creation, the Flood, Gilgamesh, and Others*. Oxford: Oxford University Press, 1998.

David, Steven R., and Peter Digeser. *The United States and the Law of the Sea Treaty*. Washington, DC: Foreign Policy Institute, 2000.

Davis, Ray Jay, and Lewis O. Grant. *Weather Modification: Technology and Law*. Boulder, CO: Westview Press, 1978.

Dawson, E. Yale. *Marine Botany*, 2d ed. New York: John Wiley, 1998.

Day, David. *The Whale War*. San Francisco: Sierra Club Books, 1987.

Deacon, Margaret. *Scientists and the Sea 1650–1900: A Study of Marine Science*. Aldershot, UK: Ashgate Publishing, 1997.

Delgado, James P. *Lost Warships: Great Shipwrecks of Naval History*. New York: Facts on File, 2001.

————, ed. *Encyclopedia of Underwater and Maritime Archaeology*. New Haven, CT: Yale University Press, 1998.

Derry, T. K., and Trevor I. Williams. *A Short History of Technology from the Earliest Times to A.D. 1900*. New York: Dover Publications, 1993.

Dickson, Mary H., and Mario Fanelli. *Geothermal Energy*. New York: Wiley, 1995.

Dijksterhuis, E. J. *Archimedes*. Translated by C. Dikshoorn. Princeton, NJ: Princeton University Press, 1987.

————. *Simon Stevin: Science in the Netherlands around 1600*. The Hague: Martinus Nijhoff, 1970.

Dinsdale, Tim. *Loch Ness Monster*. 4th ed. London: Routledge & Kegan Paul, 1982.

Doneen, L. D. *Irrigation Practice and Water Management*. 2d ed. Rome: Food and Agriculture Organization, 1984.

Donnelly, Ignatius. *Atlantis: The Antediluvian World*. New York: Dover Publications, 1985.

Dorcey, Anthony H. J., ed. *Large Dams: Learning from the Past, Looking at the Future*. New York: World Bank, 1997.

Dorfman, Robert, and Peter P. Rogers, eds. *Science with a Human Face: In Honor of Roger Revelle*. Boston: Harvard School of Public Health, 1997.

Dregne, H. E., and W. O. Willis, eds. *Dryland Agriculture*. Madison, WI: American Society of Agronomy, 1983.

Drinan, Joanne E., and Nancy E. Whiting. *Water and Wastewater Treatment: A Guide for the Nonengineering Professional*. Lancaster, PA: Technomic Publishing, 2001.

Driscoll, Fletcher, G. *Groundwater and Wells*. St. Paul, MN: Johnson Division, 1986.

Duxbury, Alison, Alyn C. Duxbury, and Keith A. Sverdrup. *Fundamentals of Oceanography*. New York: McGraw-Hill, 2001.

Dyer, K. R. *Estuaries: A Physical Introduction*. 2d ed. New York: John Wiley, 1997.

Dyson, James Lindsay. *The World of Ice*. New York: Knopf, 1962.

Dzurik, Andrew Albert. *Water Resources Planning*. Lanham, MD: Rowman & Littlefield Publishers, 1996.

EAO Scientific Systems. *Introduction to Earth's Oceans*. Halifax, NS: EOA Scientific Systems, Inc. 2001.

Edwards, L. M., ed. *Handbook of Geothermal Energy*. Houston: Gulf Publishing Company, 1982.

Ellis, Richard. *Imagining Atlantis*. New York: Knopf, 1998.

————. *Men and Whales*. New York: Lyons Press, 1999.

————. *Monsters of the Sea*. New York: Knopf, 1994.

Everett, Lorne G., and Igor S. Zektser. *Groundwater and the Environment: Applications for the Global Community*. Boca Raton, FL: CRC Press, 2000.

Ewbank, Thomas. *A Descriptive and Historical Account of Hydraulic and Other Machines for Raising Water*. New York: Arno Press, 1972.

Exploring the Rivers of North America. Washington, DC: National Geographic Society, 2000.

Fairchild, D. M. *Ground Water Quality and Agricultural Practices*. Chelsea, MI: Lewis Publishers, 1988.

Farner, Henry George. *The Organ of the Ancients*. London: William Reeves, 1931.

Faure, Guy Olivier, and Jeffrey Z. Rubin, eds. *Culture and Negotiation: The Resolution of Water Disputes*. Newbury Park, CA: Sage Publications, 1993.

Feder, Kenneth L. *Frauds, Mysteries and Myths*. Mountain View, CA: Mayfield Publishing Company, 1990.

Fingerman, Milton, and Rachakonda Nagabhushanam, eds. *Aquaculture*. Enfield, NH: Science Publishers, 2000.

Fletcher, N. H. *The Chemical Physics of Ice*. London: Cambridge University Press, 1970.

Food Safety Issues Associated with Products from Aquaculture: Report of a Joint FAO/NACA/WHO Study Group. Geneva: World Health Organization, 1999.

Franks, Felix. *Water*. 2d ed. London: Royal Society of Chemistry, 2000.

Freeze, R. A., and J. A. Cherry. *Ground Water*. Englewood Cliffs, NJ: Prentice Hall, 1979.

French, Peter W. *Coastal and Estuarine Management*. London: Routledge, 1997.

French, Richard H. *Salinity in Watercourses and Reservoirs: Proceedings of the 1983 International Symposium on State-of-the-Art Control of Salinity*. Boston: Butterworth, 1984.

Friedheim, Robert L., ed. *Toward a Sustainable Whaling Regime*. Seattle: University of Washington Press, 2001.

Friedman, Norman. *Submarine Design and Development*. Annapolis, MD: Naval Institute Press, 1984.

Galdorisi, George. *Beyond the Law of the Sea: New Directions for U.S. Oceans Policy*. Westport, CT: Praeger, 1997.

Garbrecht, Günther, ed. *Hydraulics and Hydraulic Research: A Historical Review*. Rotterdam: Balkema, 1987.

Gardiner, Robert. *The Earliest Ships: The Evolution of Boats into Ships*. Annapolis, MD: United States Naval Institute, 2001.

Garrison, Tom. *Oceanography: Invitation to Marine Science*. Belmont, CA: Brooks/Cole, 2001.

Gartner, Carol B. *Rachel Carson*. New York: Ungar, 1983.

Gaskell, T. F. *The Gulf Stream*. New York: John Day Company, 1973.

Gates, David Murray, and La Verne E. Papian. *Atlas of Energy Budgets of Plant Leaves*. London: Academic Press, 1971.

Geothermal Resources Council. *Geothermal Energy: Bet on It*. Davis, CA: Geothermal Resources Council, 1985.

Gericke, W. F. *The Complete Guide to Soilless Gardening*. Englewood Cliffs, NJ: Prentice Hall, 1940.

Giorgetti, Franco. *The Great Sailing Ships*. New York: Metro Books, 2001.

Gladwell, John S., and Calvin C. Warnick. *Low-Head Hydro: An Examination of an Alternative Energy Source*. Moscow, ID: Idaho Water Resources Research Institute, 1978.

Godin, Gabriel. *Tides*. Ensenada, Mexico: Centro de Investigación y Educación Superior de Ensenada, 1988.

Gonda, Jan. *The Indra Hymns of the Rgveda*. Leiden: Brill, 1989.

Gould, Richard A. *Archaeology and the Social History of Ships*. Cambridge: Cambridge University Press, 2000.

Graver, Dennis. *Scuba Diving*. 2d ed. Champaign, IL: Human Kinetics Publishers, 1999.

Griffiths, Tom. *Sport Scuba Diving in Depth: An Introduction to Basic Scuba Instruction and Beyond*. Princeton, NJ: Princeton Book Company, 1991.

Grimal, Pierre, ed. *Larousse World Mythology*. New York: G. P. Putnam's Sons, 1965.

Guillemard. F. H. H. *The Life of Ferdinand Magellan and the First Circumnavigation of the Globe, 1480–1521*. New York: AMS Press, 1971.

Gupta, Ram S. *Hydrology and Hydraulic Systems*. 2d ed. Prospect Heights, IL: Waveland Press, 2001.

Haber, Louis. *Black Pioneers of Science and Invention*. New York: Harcourt Brace, 1970.

Hadfield, Charles. *World Canals: Inland Navigation Past and Present*. New York: Facts on File, 1986.

Hall, W. David, ed. *Waterpower 1993: Proceedings of the International Conference on Hydropower*. New York: American Society of Civil Engineers, 1993.

Hallegraeff, Gustaaf M. *Plankton: A Microscopic World*. Leiden: E. J. Brill, 1988.

Hamblyn, Richard. *The Invention of Clouds: How an Unknown Meteorologist Forged the Language of the Skies*. New York: Farrar, Strauss & Giroux, 2001.

Hambrey, Michael, and Jürg Alean. *Glaciers*. Cambridge: Cambridge University Press, 1992.

Harris, Robert. *Canals and Their Architecture*. New York: Praeger, 1969.

Haslam, S. M. *The River Scene: Ecology and Cultural Heritage*. New York: Cambridge University Press, 1997.

Hawthorne, Daniel. *Ferdinand Magellan*. Garden City, NY: Doubleday, 1964.

Heath, Sir Thomas Little. *Archimedes*. New York: Macmillan, 1920.

Heiel, Alexander. *The Gilgamesh Epic and Old Testament Parallels*. 2d ed. Chicago: University of Chicago Press, 1970.

Heitzman, William Ray. *Opportunities in Marine and Maritime Careers*. New York: McGraw-Hill, 1999.

Helmut, Wilhelm, Water Zürn, and Hans-Georg Wenzel, eds. *Tidal Phenomena*. Berlin: Springer, 1997.

Henderson-Sellers, Brian. *Reservoirs*. London: Macmillan, 1979.

Hendrickson, Robert. *The Facts on File Encyclopedia of Word and Phrase Origins*. Rev. and exp. ed. New York: Facts on File, 1997.

Hickok, Ralph. *New Encyclopedia of Sports*. New York: McGraw-Hill, 1977.

Hillier, Chris, and Owen Hill. *The Devil and the Deep: A Guide to Nautical Myths and Superstitions*. Dobbs Ferry, NY: Sheridan House, 1997.

Horne, Alexander J., and Charles R. Goldman. *Limnology*. 2d ed. New York: McGraw-Hill, 1994.

Houck, Oliver A. *The Clean Water Act TMDL Program: Law, Policy and Implementation*. Washington, DC: Environmental Law Institute, 1999.

Hough, Richard. *Captain James Cook*. New York: W. W. Norton, 1997.

Hugill, Stan. *Shanties and Sailors' Songs*. New York: F. A. Praeger, 1969.

Hwang, Ned H. C., and Robert J. Houghtalen. *Fundamentals of Hydraulic Engineering Systems*. Upper Saddle River, NJ: Prentice Hall, 1996.

Images Publishing Group. *Water Spaces of the World: A Pictorial Review of Water Spaces*. Mulgrave, Victoria, Australia: Images Publishing Group, 1997.

International Water Tribunal. *Dams*. Utrecht: International Books, 1994.

Iselin, Columbus O'Donnell. *Matthew Fontaine Maury, 1806–1873, Pathfinder of the Seas; The Development of Oceanography*. New York: Newcomen Society in North America, 1957.

Jackson, Donald C., and David G. McCullough. *Great American Bridges and Dams: A National Trust*

Guide. Washington, DC: Preservation Press, 1988.

Jellicoe, Susan. *Water: The Use of Water in Landscape Architecture*. London: A. and C. Black, 1971.

Jesperson, James, et al. *From Sundials to Atomic Clocks: Understanding Time and Frequency*. New York: Dover, 1982.

Jiusto, James E. *The Physics of Weather Modification*. College Park, MD: American Association of Physics Teachers, 1983.

Jones, Frank E. *Evaporation of Water: With Emphasis on Applications and Measurements*. Chelsea, MI: Lewis Publishers, 1992.

Jones, H. G., ed. *Snow and Ice Covers: Interactions with the Atmosphere and Ecosystems*. Wallingford, Oxfordshire, UK: Office of the Treasurer, IAHS, 1994.

Jungnickel, Christa, and Russell McCormmach. *Cavendish*. Philadelphia: American Philosophical Society, 1996.

Kahan, Archie M., et al. *Guidelines for Cloud Seeding to Augment Precipitation*. New York: American Society of Civil Engineers, 1995.

Kalff, Jacob. *Limnology*. Upper Saddle River, NJ: Prentice Hall, 2001.

Kardon, Redwood, Michael Casey, and Douglas Hansen. *Code Check Plumbing: A Field Guide to the Plumbing Codes*. Newtown, CT: Taunton Press, 2000.

Keating, Barbara H., Christopher F. Waythomas, and Alastair Dawson. *Landslides and Tsunamis*. Basel: Birkhäuser Verlag, 2000.

Keddy, Paul A. *Wetland Ecology: Principles and Conservation*. Cambridge: Cambridge University Press, 2000.

Keenan, Joseph H., et al. *Steam Tables: Thermodynamic Properties of Water including Vapor, Liquid, and Solid Phases*. Malabar, FL: Krieger, 1992.

Keston, Joseph, et al., eds. *Sourcebook on the Production of Electricity from Geothermal Energy*. Washington, DC: Government Printing Office, 1980.

Khan, Arshad Hassan. *Desalination Processes and Multistage Flash Distillation Practice*. Amsterdam: Elsevier, 1986.

King, Ethel M. *The Fountain of Youth and Juan Ponce de León*. Brooklyn, NY: T. Guaus' Sons, 1963.

Knight, Peter G. *Glaciers*. Cheltenham, UK: Stanley Thornes, 1999.

Konvicka, Tom. *Teacher's Weather Sourcebook*. Englewood, CO: Teacher Ideas Press, 1999.

Kuppuram, G., and K. Kumudamani, eds. *Marine Archaeology: The Global Perspective*. Delhi: Sundeep, 1996.

LaBounty, James F. *Lakes, Ponds, and Reservoirs: What They Are and How to Study Them*. Denver, CO: Bureau of Reclamation, U.S. Department of the Interior, 1985.

Lagler, Karl F., ed. *Man-made Lakes: Planning and Development*. Rome: United Nations Development Programme, Food and Agriculture Organization, 1969.

Lalli, Carol M. *Biological Oceanography: An Introduction*. London: Butterworth-Heinemann, 1997.

Landa, Edward R., and Simon Ince. *The History of Hydrology*. Washington, DC: American Geophysical Union, 1987.

Lane, Frank W. *The Elements Rage: The Extremes of Natural Violence*. Newton Abbot, UK: David & Charles, 1966.

Larson, Jennifer. *Greek Nymphs: Myth, Cult, Lore*. Oxford: Oxford University Press, 2001.

Lawliss, Chuck. *The Submarine Book*. Springfield, NJ: Burford Books, 2000.

Law of the Sea: A Select Bibliography. New York: United Nations, 2000.

Laws, Edward A. *Aquatic Pollution: An Introductory Text*. New York: John Wiley & Sons, 2000.

Leach, Marjorie. *Guide to the Gods*. Santa Barbara, CA: ABC-CLIO, 1992.

Lear, Linda J. *Rachel Carson: Witness for Nature*. New York: Henry Holt, 1998.

Lehr, Jay, et al. *Design and Construction of Water Wells: A Guide for Engineers*. New York: Van Nostrand Reinhold, 1988.

Leitner, Peter M. *Reforming the Law of the Sea Treaty: Opportunities Missed, Precedents Set, and U.S. Sovereignty Threatened*. Lanham, MD: University Press of America, 1996.

Leopold, Luna B. *Water: A Primer*. San Francisco: W. H. Freeman, 1974.

Levinton, Jeffrey S. *Marine Biology: Function, Biodiversity, Ecology*. New York: Oxford University Press, 2001.

Lewis, Charles Lee. *Matthew Fontaine Maury*. 1927. Reprint, New York: Arno Press, 1980.

Lindow, John. *Handbook of Norse Mythology*. Santa Barbara, CA: ABC-CLIO, 2001.

Lurie, Charles N. *Everyday Sayings: Their Meanings Explained, Their Origins Given*. New York: G. P. Putnam's Sons, 1928.

Lutgens, Frederick K., and Edward J. Tarbuck. *The Atmosphere: An Introduction to Meteorology*. 8th ed. Upper Saddle River, NJ: Prentice Hall, 2001.

MacLeish, William H. *The Gulf Stream*. Boston: Houghton Mifflin, 1989.

Madsen, Axel. *Cousteau: An Unauthorized Biography*. New York: Beaufort Books, 1986.

Maidment, David R., ed. *Handbook of Hydrology.* New York: McGraw-Hill, 1993.

Manning, John C. *Applied Principles of Hydrology.* 3d ed. Upper Saddle River, NJ: Prentice Hall, 1997.

Manning, Kenneth R. *Black Apollo of Science: The Life of Ernest Everett Just.* New York: Oxford University Press, 1983.

Martin, Dean, et al. *Archeology Underwater: The NAS Guide to Principles and Practice.* Denbigh, Clwyd: Nautical Archaeology Society and Archetype Publications, 1992.

Marx, Robert F., with Jenifer Marx. *The Search for Sunken Treasure: Exploring the World's Great Shipwrecks.* Toronto: Key-Porter Books, 1993.

Mason, John. *Commercial Hydroponics.* New York: Simon & Schuster, 2000.

Mayer, Ralph. *The Artist's Handbook of Materials and Techniques.* 5th ed. Revised and updated by Steven Sheehan. New York: Viking Press, 1991.

McCay, Mary A. *Rachel Carson.* New York: Twayne, 1993.

McConnell, Anita. *No Sea Too Deep: The History of Oceanographic Instruments.* Bristol, UK: Hilger, 1982.

McCullough, David G. *The Johnstown Flood.* New York: Simon & Schuster, 1968.

McCully, Patrick. *Silenced Rivers: The Ecology and Politics of Large Dams.* London: Zed Books, 1996.

McGrail, Sean. *Boats of the World: From the Stone Age to Medieval Times.* New York: Oxford University Press, 2001.

McGuigan, Dermot. *Small Scale Water Power.* Dorchester, UK: Prism Press, 1978.

McNeil, Ian. *Joseph Bramah: A Century of Invention, 1749–1851.* Newton Abbot, UK: David & Charles, 1968.

Menke, Frank G. *The Encyclopedia of Sports.* 6th ed. Revised by Suzanne Treat. South Brunswick, NJ: A.S. Barnes, 1958.

Michael, A. M. *Irrigation: Theory and Practice.* Columbia, MO: South Asia Books, 1999.

Miller, Char, ed. *Fluid Arguments: Water in the American West.* Tucson: University of Arizona Press, 2001.

Miller, John B. *Floods: People at Risk, Strategies for Prevention.* Geneva: United Nations, 1997.

Mills, Richard. *Power from Steam: A History of the Stationary Steam Engine.* Cambridge: Cambridge University Press, 1993.

Mitsch, W. J., and J. G. Gosselink. *Wetlands.* 2d ed. New York: John Wiley & Sons, 2000.

Mittal, K. L., and Promod Kumar. *Emulsions, Foams, and Thin Films.* New York: Marcel Dekker, 2000.

Monahan, Dave, ed. *World Atlas of the Oceans.* Westport, CT: Firefly Books, 2001.

Monson, Bruce A. *A Primer on Limnology.* 2d ed. St. Paul, MN: Water Resources Research Center, 1992.

Montgomery, James J. *Water Treatment Principles and Design.* New York: John Wiley & Sons, 1985.

Moore, John E., Alexander Zaporozec, and James W. Mercer. *Groundwater: A Primer.* Alexandria, VA: American Geological Institute, 1995.

Morell, James B. *The Law of the Sea: An Historical Analysis of the 1982 Treaty and Its Rejection by the United States.* Jefferson, NC: McFarland, 1992.

Morgan, Judith. *Roger: A Biography of Roger Revelle.* La Jolla, CA: Scripps Institution of Oceanography, 1996.

Morris, William and Mary. *Morris Dictionary of Word and Phrase Origins.* New York: Harper & Row Publishers, 1962.

Mörth, Hans. *Weather Modification: Prospects and Problems.* Translated by Geog Breuer. Cambridge: Cambridge University Press, 1979.

Mountain, Alan. *The Diver's Handbook.* New York: Lyons Press, 1997.

Munson, Richard. *Cousteau: The Captain and His World.* New York: Morrow, 1989.

Muson, Howard. *The Triumph of the American Spirit: Johnstown, Pennsylvania.* Lanham, MD: Johnstown Flood Museum and American Association for State and Local History Library, 1989.

Myles, Douglas. *The Great Waves.* New York: McGraw-Hill, 1985.

National Sea Grant Program. *Marine Science Careers.* Woods Hole, MA: Woods Hole Oceanographic Institution Sea Grant Program, 1996.

Newell, G. E., and R. C. Newell. *Marine Plankton: A Practical Guide.* London: Hutchinson, 1977.

Niering, W. A. *Wetlands.* New York: Alfred Knopf, 1998.

Nordstrom, Karl F., and Charles T. Rowan, eds. *Estuarine Shores: Evolution, Environments, and Human Alterations.* New York: Wiley, 1996.

Nybakken, James Willard. *Marine Biology: An Ecological Approach.* 5th ed. San Francisco: Benjamin Cummings, 2001.

O'Connor, Richard. *Johnstown: The Day the Dam Broke.* Philadelphia: Lippincott, 1957.

On-Farm Irrigation Committee of the Environmental and Water Resources Institute. *Selection of Irrigation Methods for Agriculture.* Reston, VA: American Society of Civil Engineers, 2000.

Paine, Lincoln P. *Ships of the World: An Historical Encyclopedia.* Boston: Houghton Mifflin, 1997.

Parr, Charles McKew. *Ferdinand Magellan: Circumnavigator*. 2d ed. New York: Crowell, 1964.

Parson, L. M., C. L. Walker, and D. R. Dixon, eds. *Hydrothermal Vents and Processes*. London: Geological Society, 1995.

Patrick, Ruth. *Rivers of the United States*. New York: John Wiley & Sons, 2001.

Payne, P. S. Robert. *The Canal Builders: The Story of Canal Engineers through the Ages*. New York: Macmillan, 1959.

Pearce, Fred. *The Dammed: Rivers, Dams, and the Coming World Water Crisis*. London: Bodley Head, 1992.

Penn, James R. *Rivers of the World: A Social, Geographical, and Environmental Sourcebook*. Santa Barbara, CA: ABC-CLIO, 2001.

Petrenko, Victor F., and Robert W. Whitworth. *Physics of Ice*. Oxford: Clarendon Press, 1999.

Philippi, Nancy S. *Floodplain Management: Ecologic and Economic Perspectives*. Austin, TX: RG Landes, 1997.

Picard, Barbara Leonie. *Tales of the Norse Gods*. New York: Oxford University Press, 2001.

Pillay, T. V. R. *Aquaculture and the Environment*. New York: Halsted Press, 1992.

———. *Aquaculture Development: Progress and Prospects*. New York: Halsted Press, 1994.

———. *Aquaculture: Principles and Practices*. Oxford: Fishing News Books, 1993.

Pinto, John A. *The Trevi Fountain*. New Haven: Yale University Press, 1986.

Pippenger, John J. *Fluid Power: The Hidden Giant*. Jenks, OK: Amalgam Publishing, 1992.

Polevoy, Savely. *Water Science and Engineering*. London: Blackie Academic & Professional, 1996.

Poljakoff-Mayber, Alexandra, and J. Gale, eds. *Plants in Saline Environments*. Berlin: Springer-Verlag, 1975.

Polmar, Norman. *The Naval Institute Guide to the Ships and Aircraft of the U.S. Fleet*. 17th ed. Annapolis, MD: United States Naval Institute, 2001.

Post, Austin, and Edward R. LaChapelle. *Glacier Ice*. Seattle: University of Washington Press, 2000.

Postel, Sandra. *Pillar of Sand: Can the Irrigation Miracle Last?* New York: W. W. Norton, 1999.

Prager, Ellen J. *Furious Earth: The Science and Nature of Earthquakes, Volcanoes, and Tsunamis*. New York: McGraw-Hill, 2000.

Prager, Ellen J., with Sylvia A. Earle. *The Oceans*. New York: McGraw-Hill, 2000.

The Random House Atlas of the Oceans. New York: Random House, 1991.

Raymont, John E. G. *Plankton and Productivity in the Oceans*. 2d ed. New York: Pergamon Press, 1983.

Reichert, Marimargaret, and Jack H. Young, eds. *Sterilization Technology for the Health Care Facility*. Gaithersburg, MD: Aspen Publishers, 1993.

Report of the Symposium on Water for Sustainable Inland Fisheries and Aquaculture Held in Connection with the European Inland Fisheries Advisory Commission. Rome: Food and Agriculture Organization, 1998.

Resh, Howard M. *Hydroponic Food Production: A Definitive Guidebook for Soilless Food-Growing Methods*. 5th ed. Santa Barbara, CA: Woodbridge Press Publishing, 1995.

Reynolds, Ross. *Cambridge Guide to the Weather*. Cambridge: Cambridge University Press, 2000.

Reynolds, Terry S. *Stronger than a Hundred Men: A History of the Vertical Water Wheel*. Baltimore: Johns Hopkins University Press, 1983.

Rijsberman, Frank R., ed. *World Water Scenarios: Analysing Global Water Resources and Use*. London: Earthscan Publications, 2001.

Ritter, William F., and Adel Shirmohammadi, eds. *Agricultural Nonpoint Source Pollution: Watershed Management and Hydrology*. Boca Raton, FL: Lewis Publishers, 2000.

Roberto, Keith F. *How-to Hydroponics*. Farmingdale, NY: FutureGarden, 2000.

Roberts, Fred M. *Basic Scuba: Self-Contained Underwater Breathing Apparatus*. 2d ed. Princeton, NJ: Van Nostrand, 1963.

Rowles, Raymond. *Drilling for Water: A Practical Manual*. 2d ed. Aldershot, UK: Avebury, 1995.

Ruppe, Carol, and Jan Barstad, eds. *International Handbook of Underwater Archaeology*. New York: Kluwer Academic/Plenum Publishing, 2001.

Rutland, Jonathan. *The Age of Steam*. New York: Random House, 1987.

Rys, Franz S., and Albert Gyr, eds. *Physical Processes and Chemical Reactions in Liquid Flows*. Rotterdam: A. A. Balkema, 1998.

Sanz, V. Murga. *Juan Ponce de Léon*. 2d ed. Rio Piedras, PR: University of Puerto Rico Press, 1985.

Schetz, Joseph A., and Allen E. Fuhs, eds. *Handbook of Fluid Dynamics and Fluid Machinery*. New York: Wiley, 1996.

Scheumann, Waltina. *Managing Salinization: Institutional Analysis of Public Irrigation Systems*. Berlin: Springer-Verlag, 1997.

Schlee, Susan. *The Edge of an Unfamiliar World: A History of Oceanography*. London: Hale, 1975.

Schneider, David A., and Glenn Zorpette. *Scientific American Presents: The Oceans*. New York: Scientific American, 1998.

Schreier, Carl. *Yellowstone's Geysers, Hot Springs, and Fumaroles*. Moose, WY: Homestead Publishing, 1987.

Schwarz, Meier. *Soilless Culture Management*. Berlin: Springer-Verlag, 1995.

Scott, Jeanette. *Desalination of Seawater by Reverse Osmosis*. Park Ridge, NJ: Noyes Data Corporation, 1981.

Seymour, Richard J., ed. *Ocean Energy Recovery: The State of the Art*. New York: American Society of Civil Engineers, 1992.

Sheets, Robert C. *Tropical Cyclone Modification: The Project Stormfury Hypothesis*. Boulder, CO: Environmental Research Laboratories, National Oceanic and Atmospheric Administration, 1981.

Smith, A. G. *Traditional Boats from around the World*. New York: Dover, 2001.

Smith, Keith, and Roy Ward. *Floods: Physical Processes and Human Impacts*. Chichester, NY: Wiley, 1998.

Smith, William. *Smith's Biblical Dictionary*. Nashville: Thomas Nelson Publishers, 1997.

Solley, Wayne B., Robert R. Pierce, and Howard A. Perlman. *Estimated Use of Water in the United States in 1995*. U.S. Geological Survey Circular 1200. Washington, DC: U.S. Government Printing Office, 1998.

Source Book of Alternative Technologies for Freshwater Augmentation in Latin America and the Caribbean. Osaka/Shiga: UNEP International Environmental Technology Centre, 1997.

Sourirajan, S. *Reverse Osmosis*. New York: Academic Press, 1970.

Spaeth, Frank. *Mysteries of the Deep*. St. Paul, MN: Llewellyn Publications, 1998.

Spector, Stephen, ed. *Essays in Paper Analysis*. Washington, DC: Folger Shakespeare Library, 1987.

Speidel, David H., Lon C. Ruedisili, and Allen F. Agnew. *Perspectives on Water: Uses and Abuses*. New York: Oxford University Press, 1988.

Spellman, Frank R., and Joanne Drinan. *Water Hydraulics*. Lancaster, PA: Technomic Publishing, 2001.

Spellman, Frank R., and Nancy E. Whiting. *Water Pollution Control Technology: Concepts and Applications*. Rockville, MD: Government Institutes, 1999.

Stein, Sherman K. *Archimedes: What Did He Do Besides Cry Eureka?* Washington, DC: Mathematical Association of America, 1999.

Stewart, B. A., and D. R. Nielsen, eds. *Irrigation of Agricultural Crops*. Madison, WI: American Society of Agronomy, 1990.

Stickney, Robert R. *Principles of Aquaculture*. New York: Wiley, 1994.

Stoett, Peter J. *The International Politics of Whaling*. Vancouver, BC: University of British Columbia, 1997.

Stommel, Henry. *The Gulf Stream: A Physical and Dynamical Description* 2d ed. Berkeley: University of California Press, 1976.

Stonehouse, Frederick. *Haunted Lakes: Great Lakes Ghost Stories, Superstitions, and Sea Serpents*. Duluth, MN: Lake Superior Port Cities, 1997.

Stott, Noel. *A Select Bibliography of the Impact of Large Scale Dams*. Wynberg, South Africa: Environmental Monitoring Group, 1999.

Strong, James, ed. *The New Strong's Exhaustive Concordance of the Bible*. Nashville: Thomas Nelson Publishers, 1997.

Suckling, Bob. *The Book of Sea Monsters*. New York: Overlook Press, 1998.

Sumich, James L. *An Introduction to the Biology of Marine Life*. New York: WCB/McGraw-Hill, 2001.

Symmes, Marilyn, ed. *Fountains: Splash and Spectacle. Water and Design from the Renaissance to the Present*. New York: Rizzoli International Publications, 1998.

Talsma, T., and J. R. Philip, eds. *Salinity and Water Use*. London: Macmillan, 1971.

Thain, J. F. *Principles of Osmotic Phenomena*. London: Royal Institute of Chemistry, 1967.

Thomas, William A., ed. *Legal and Scientific Uncertainties of Weather Modification*. Durham, NC: Duke University Press, 1977.

Thornton, J. A., et al., eds. *Assessment and Control of Nonpoint Source Pollution of Aquatic Ecosystems: A Practical Approach*. New York: Parthenon Publishing, 1999.

Throckmorton, Peter, ed. *The Sea Remembers: Shipwrecks and Archaeology from Homer's Greece to the Rediscovery of the Titanic*. New York: Weidenfeld & Nicolson, 1987.

Thurman, Harold V., and Alan P. Trujilo. *Essentials of Oceanography*. Upper Saddle River, NJ: Prentice Hall, 2001.

Tidal Power: Symposium Proceedings. New York: American Society of Civil Engineers, 1987.

Tunzelmann, G. N. von. *Steam Power and British Industrialization to 1860*. Oxford: Clarendon Press, 1978.

United Nations Environment Programme. *The Pollution of Lakes and Reservoirs*. Nairobi: United Nations Environment Programme, 1994.

U.S. Environmental Protection Agency. Office of Water. *The Quality of Our Nation's Waters: A Summary of the National Water Quality Inventory: 1998 Report to Congress.* Publication EPA841-S-00-001. Washington, DC, June 2000.

U.S. Geological Survey. *The Quality of Our Nation's Water: Nutrients and Pesticides.* U.S. Geological Survey Circular 1225, May 1999.

U.S. House Committee on Science, Space, and Technology. Subcommittee on Science. *Desalination Research: Hearing before the Subcommittee on Science of the Committee on Science, Space, and Technology*, 102nd Cong., 1st sess., 17 July 1991.

U.S. Senate Committee on Foreign Relations. *Current Status of the Convention on the Law of the Sea: Hearing before the Committee on Foreign Relations*, 103rd Cong. 2d sess., 11 August 1994.

U. S. Weather Bureau. *Cloud Forms according to the International System of Classification.* Washington, DC: Government Printing Office, 1938.

Upgren, Arthur R. *Weather: How It Works and Why It Matters.* Cambridge, MA: Perseus Publishing, 2000.

Van Deman, Esther Boise. *The Building of the Roman Aqueducts.* Washington, DC: Carnegie Institution of Washington, 1934.

Van der Veen, C. J. *Fundamentals of Glacier Dynamics.* Rotterdam: Balkema, 1999.

Van Melsen, Andrew G. *From Atomos to Atom.* Pittsburgh: Duquesne University Press, 1952.

Vogt, Evon Z., and Ray Hyman. *Water Witching U.S.A.* Chicago: University of Chicago Press, 1959.

Von Kotze, Astrid, and Allan Holloway. *Living with Drought: Drought Mitigation for Sustainable Livelihoods.* Bourton-on-Dunsmore, UK: Intermediate Technology, 1999.

Wadhams, Peter. *Ice in the Ocean.* Sydney, Australia: Gordon and Breach, 2000.

Wagner, Wolfgang, and Alfred Kruse. *Properties of Water and Steam.* Berlin: Springer-Verlag, 1998.

Waring, Gerald Ashley. *Thermal Springs of the United States and Other Countries: A Summary.* Revised by Reginald R. Blankenship and Ray Bentall. Washington, DC: Government Printing Office, 1965.

Warren, Bruce A., and Carl Wunsch, eds. *Evolution of Physical Oceanography: Scientific Surveys in Honor of Henry Stommel.* Cambridge, MA: MIT Press, 1981.

Waters, Thomas F. *Wildstream: A Natural History of the Free Flowing River.* St. Paul, MN: Riparian Press, 2000.

Watt, S. B. *A Manual on the Hydraulic Ram for Pumping Water.* London: Intermediate Technology Publications, 1981.

Webster, John G., ed. *Encyclopedia of Medical Devices and Instrumentation.* New York: John Wiley & Sons, 1988.

Wertenbaker, William. *The Floor of the Sea: Maurice Ewing and the Search to Understand the Earth.* Boston: Little, Brown & Company, 1974.

Wetzel, Robert G. *Limnology: Lake and River Ecosystems.* 3d ed. New York: Academic Press, 2001.

Whigham, Dennis, Dagmar Dykyjová, and Slvaomil Hejný, eds. *Wetlands of the World: Inventory, Ecology, and Management.* Dordrecht: Kluwer Academic Publishers, 1993.

Whitney, Edgar A. *Watercolor: The Hows and Whys: Complete Guide to Watercolor Painting.* New York: Watson-Guptill Publications, 1974.

Wiederhold, Pieter R. *Water Vapor Measurement.* New York: Marcel Dekker, 1997.

Wilhite, Donald A. *Drought Assessment, Management, and Planning: Theory and Case Studies.* Dordrecht: Kluwer Academic, 1993.

———, ed. *Drought: A Global Assessment.* London: Routledge, 2000.

Williams, Frances Leigh. *Matthew Fontaine Maury: Scientist of the Sea.* New Brunswick, NJ: Rutgers University Press, 1963.

Wilson, James. *Ground Water: A Non-technical Guide.* Philadelphia: Academy of Natural Sciences, 1982.

Wohl, Ellen E. *Mountain Rivers.* Washington D.C.: American Geophysical Union, 2000.

———, ed. *Inland Flood Hazards: Human, Riparian, and Aquatic Communities.* Cambridge: Cambridge University Press, 2000.

Wood, Stanley, Kate Sebastian, and Sara J. Scherr. *Pilot Analysis of Global Ecosystems: Agroecosystems.* Washington, DC: International Food Policy Research Institute and World Resources Institute, November 2000.

World Commission on Dams. *Dams and Development: A New Framework for Decision-Making.* London: Earthscan, 2000.

World Meteorological Organization. *International Cloud Atlas.* 2d ed. Geneva: Secretariat of the World Meteorological Organization, 1975.

Wright, Kenneth R., ed. *Water Rights of the Eastern United States.* Lanham, MD: American Water Works Association, 1998.

Young, G. J., James C. I. Dooge, and John C. Rodda. *Global Water Resource Issues.* Cambridge: Cambridge University Press, 1994.

Zagar, A., ed. *Hydropower: Recent Developments.* New York: Society of Civil Engineers, 1985.

Index

About the Author

David E. Newton has published extensively on chemistry and other science subjects. He is the award-winning author of numerous books, articles, and scholarly publications, including the *Encyclopedia of Fire, Recent Advances and Issues in Molecular Nanotechnology, The Chemical Elements, Science in the 1920s, The Ozone Dilemma, Encyclopedia of Cryptology, Chemistry of Carbon Compounds, Problems in Chemistry, Global Warming, Encyclopedia of the Chemical Elements*, and *Social Issues in Science and Technology*. Newton received his doctorate in science education from Harvard University.